Positional controls in plant development

EDITED BY

P. W. BARLOW AND D. J. CARR

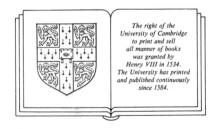

The right of the
University of Cambridge
to print and sell
all manner of books
was granted by
Henry VIII in 1534.
The University has printed
and published continuously
since 1584.

CAMBRIDGE UNIVERSITY PRESS

CAMBRIDGE

LONDON   NEW YORK   NEW ROCHELLE

MELBOURNE   SYDNEY

Published by the Press Syndicate of the University of Cambridge
The Pitt Building, Trumpington Street, Cambridge CB2 1RP
32 East 57th Street, New York, NY 10022, USA
296 Beaconsfield Parade, Middle Park, Melbourne 3206, Australia

© Cambridge University Press 1984

First published 1984

Printed in Great Britain by the University Press, Cambridge

Library of Congress catalogue card number: 83-7393

British Library Cataloguing in Publication Data
Positional controls in plant development.
1. Plant regulators
I. Barlow, P. W.   II. Carr, D. J.
581.3   QK745
ISBN 0 521 25406 X

Contents

Contributors

P. W. Barlow *Agricultural Research Council, Letcombe Laboratory, Wantage, Oxon, OX12 9JT, UK*

M. Bopp *Botanisches Institut der Universität, Im Neuenheimer Feld 360, 6900 Heidelberg, Germany*

D. J. Carr *Department of Environmental Biology, Research School of Biological Sciences, Australian National University, PO Box 475, Canberra, ACT 2601, Australia*

C. R. Goodall *Department of Biological Sciences, Stanford University, Stanford, California 94305, USA*

P. B. Green *Department of Biological Sciences, Stanford University, Stanford, California 94305, USA*

R. W. Korn *Department of Biology, Bellarmine College, Newburg Road, Louisville, Kentucky 40205, USA*

A. Lindenmayer *Theoretical Biology Group, Rijksuniversiteit Utrecht, Padualaan 8, Utrecht 3508 TB, The Netherlands*

P. M. Lintilhac *Department of Botany, University of Vermont, Burlington, Vermont 05401, USA*

C. N. McDaniel *Department of Biology, Rensselaer Polytechnic Institute, Troy, New York 12181, USA*

H. Meinhardt *Max-Planck-Institut für Virusforschung, Spemannstrasse 35, 7400 Tübingen, Germany*

D. Moore *Department of Botany, University of Manchester, Manchester, M13 9PL, UK*

R. Nozeran *Laboratoire de Botanique II, Bâtiment 360, Université Paris-Sud, 91405 Orsay Cedex, France*

T. Sachs *Department of Botany, The Hebrew University, 91904 Jerusalem, Israel*

W. W. Schwabe *Department of Horticulture, Wye College, University of London, Wye, Near Ashford, Kent, TN25 5AH, UK*

S. D. Waaland *Department of Botany, University of Washington, Seattle, Washington 98105, USA*

J. & P. M. Warren Wilson *Department of Botany, Australian National University, PO Box 4, Canberra, ACT 2600, Australia*

Preface

The idea that position within the whole guides the development of cells was explicitly stated by the zoologist Hans Driesch and the botanist Hermann Vöchting in the last century. We would like to quote their words and also a contemporary formulation of the concept of positional information by Lewis Wolpert.

'Die relative Lage einer Blastomere im Ganzen wird wohl ganz allgemein bestimmen, was aus ihr hervorgeht; liegt sie anders, so giebt sie auch Anderem den Ursprung; ... ihre prospektive Beziehung ist eine Funktion des Ortes.'

Hans Driesch, 1893 (p. 39).

('The relative position of a blastomere in the whole may well determine in general what develops from it; if it lay elsewhere, it would give rise to something different; ... the prospective value is a function of position.')

'... what actually will happen in each of the blastula cells in any special case of development experimentally determined depends on the position of that cell in the whole, if the "whole" is put into relation with any fixed system of coordinates; or more shortly, "the prospective value of any blastula cell is a function of its position in the whole".'

Hans Driesch, 1908 (p. 80).

'Daraus folgt, dass keine vegetative Zelle am Pflanzenkörper eine specifische und unveränderliche Energie besitzt ... Und zwar ist es in erster Linie der Ort an der Lebenseinheit, welcher die Function der Zelle bestimmt.'

Hermann Vöchting, 1877 (p. 170).

('From this it follows that no vegetative cell in the plant possesses a specific and unchanging determination ... the developmental function of the cell is determined primarily by its position in the living unit.')

'Geht die Entwicklung der Neubildungen in dem Gewebe jener Pflanzenteilen regellos oder nur an bestimmten Orten vor sich; und, wenn das letzte der Fall ist, durch welche Kräfte werden diese Orte bestimmt? ... Die jeweilige zu verrichtende Funktion eine Zelle wird in erster Linie durch den morphologischen Ort bestimmt den sie an den Lebenseinheit einnimmt.'

Hermann Vöchting, 1878 (p. 241).

('Do new structures form in the tissues of each plant at random or only in certain locations; and if the latter is true, what forces determine these locations? ... The developmental behaviour of a cell is determined first of all by the morphological position it occupies in the organism as a whole.')

'To put it bluntly a cell knows where it is, and this information specifies the nature of its differentiation, which will be largely determined by its genetic constitution. The main points about the concept of positional information ... are:

1. There are mechanisms whereby cells in a developing system may have their position specified with respect to one or more points in the system ...

2. Positional information largely determines, with respect to the cell genome and developmental history, the nature of the molecular differentiation that the cell will undergo. The general process whereby positional information leads to a particular cellular activity or molecular differentiation will be termed the interpretation of the positional information. The specification of positional information in general precedes and is independent of molecular differentiation.

3. Polarity may be defined in relation to the points with respect to which a cell's position is being specified: it is the direction in which positional information is specified or measured.

4. Positional information may be universal, that is, the same mechanisms that specify positional information may be operative in different fields within the same system as well as in quite different systems from different genera or even phyla.

5. The classical cases of pattern regulation, whether in development or in regeneration, ... show size invariance ... largely dependent on the ability of the cells to change their positional information in an appropriate manner and to be able to interpret this change.'

Lewis Wolpert, 1970 (pp. 201–2).

Anyone who has examined, however casually, either the external form of a plant or its internal anatomy will have been struck by the orderliness of each. The sequence of plant development is ordered in both time and space and the morphological features exhibited at any time are so particular for a given species that even a small part of a plant may suffice for its identification. Likewise, tissues and specialized cells are not aggregated haphazardly within the plant body; they have precise locations within the plant and show specific spatial inter-relationships. But this is commonplace knowledge. If we accept that cells, tissues, and organs are the result of genetic activity then we must ask, 'How is gene activity realized in time and space to cause the construction of a corresponding plant structure?'

Animal biologists have developed and employed the concept of 'positional information' to try and answer this question. The main points of this concept are (1) that within a developing system there are substances (morphogens) that directly influence cell differentiation, (2) the spatial pattern of cell differentiation corresponds to the pattern of these morphogens, (3) the pattern of morphogens is related to their site of synthesis and direction of transport, and (4) transport is often polarized. Thus, a developing system is built upon a chemical map that provides a reference system governing genetic or epigenetic activity; and the map itself also grows and develops as the cellular unit under its influence grows.

As a concept, positional information has proved a potent source of inspiration for experiments on organisms and systems as diverse as the protozoan *Stentor*, insect epidermis, and the developing wing of the chick, though in each case the chemical basis of the positional information that underlies their development remains elusive. It has also given rise to vivid images such as the 'French flag problem'. The area of the Tricolor represents a field in which differentiation occurs and the three colours represent three distinct types of cellular differentiation. The problem (or at least one aspect of it) is how any initially 'uncoloured' cells within the area of the flag would acquire the appropriate 'colour'. Information about their position along the length of the flag would enable the cells to differentiate correctly. The information is therefore a property of position. But another, and more fundamental, aspect of the problem is to identify what causes an initially uniform region to break up into separate elements situated in a definite spatial order. In terms of the French flag, this is to ask why there are three coloured areas each with a particular shape and position.

Can the concept of positional information be applied to plants, and have those who seek to understand plant differentiation and morphogenesis taken advantage of the ideas that it provides? The answer is that although positional information may, as a concept, be familiar to botanists, there has been less overt enthusiasm to adopt it. It may be premature to suggest why this should be so, but perhaps there are, in plants, few experimental systems where the concept can be applied successfully; perhaps there are few experimentalists prepared to perform the surgical interventions into tiny pieces of plant organs that validation of the concept would require; perhaps plant biologists think that 'positional information' is really not of much interest since other interpretations of development are available. This last suggestion may be nearer the mark since, in contrast to animal systems, morphogenetic substances (hormonal growth regulators) controlling plant development are well known. Thus, it might be imagined that development can be adequately explained in terms of hormone-target cell interactions. Nevertheless, the question still remains as to how a *pattern* of development arises. In this respect a cell's position within the structure as a whole would seem to be an important idea to consider, though it is scarcely an original one since Hermann Vöchting proposed such an idea over 100 years ago (see pp. vii–viii) – 'Plus ça change, plus c'est la même chose'!

It seems of interest to explore the idea of positional controls in plant development further – hence the present book. 'Positional Controls' is purposely a loose epithet. Position in an organism is, of course, relative to some point within the whole. And as for the controls of development, these operate at many different levels from the subcellular up to the domain of the whole organism; at the same time they operate within the context of, and possibly in response to, the location of the cells, or groups of cells, in which they exist. For example, significant developmental events within cells result from the interaction between macromolecular elements such as microtubules and cell walls. Cell-to-cell contact can also induce specific modes of differentiation: for instance, tannin cells may develop from parenchyma cells adjacent to already differentiated tannin cells. At another level in the developmental hierarchy, groups of cells can constitute morphogenetic fields and produce (or exhaust from their surroundings) diffusible morphogens able to influence the differentiation of other cells up to 100 cell diameters away. Erwin Bünning in his book entitled *'Entwicklungs- und Bewegungsphysiologie der Pflanzen'* (3rd edition, 1953) described many such examples; they often give rise to patterned development. In leaves many cellular patterns owe their origin to morphogenetic fields associated with tissue-complexes such as veins, or with epidermal structures such as hairs. Another example is seen in shoot meristems. The apical dome exercises an influence on the primordia it produces, determining their pathway either as leaves of different kinds, or as lateral (shoot) meristems. Finally, there may be influences that act over long distances within the plant determining the developmental pathway of root or shoot meristems. These influences may vary with time, from season to season, or over the course of many years.

Some of the morphogens involved in these developmental phenomena may be cations (e.g., Ca^{2+}) or cyclic nucleotides (e.g., cyclic AMP) which have a short range of action. Others may be natural plant growth regulators (hormones) which behave as long-range morphogens, and have long-lasting effects. In recent years technical developments have revived interest in endogenous currents generated in cells as a result of ionic fluxes, such as those of calcium. So far, such electrical fields have been studied only in connection with the polarization and growth of single cells and in tropic movements of roots and shoots. Eventually it will be possible to investigate the electric fields within structures such as meristems and leaves. These fields may well turn out to be of considerable importance in providing the gradients which cells evaluate for positional information during tissue and organ development.

The various chapters show how the unfolding of plant development is the result of the integration of all these activities at different levels of organization. A wide range of plant forms is also purposely discussed to illustrate both the unitariness of this integration and because some forms provide particularly good examples of what may be rather general principles of developmental controls.

The idea for this book originates from material presented at the XIII International Botanical Congress held in Sydney, Australia, in August 1981 in sessions on Structural and Developmental Botany. It was apparent that an underlying feature of many of the developmental phenomena discussed was a relationship with position within the plant. Unfortunately, no opportunity presented itself during the period of the Congress to explore this particular relationship in the detail it deserves and to search for consistent principles that associate development with position. Nevertheless, we felt strongly that it should be followed up, and we have been able to persuade some of those present at the Congress, as well as other authors whose research experience is relevant to the subject at hand, to contribute to this book.

The aim of the book is simply to explore the concept of position as a determinant of organized development in plants and to discuss the possible underlying mechanisms. An open-minded attitude towards the concept seems potentially the most fruitful in such an exploratory situation; however, authors were challenged with the question of whether it is a concept that is useful, or even applicable, to the study of plant development. Concepts are like hypotheses; they are artifices that do not have to be justified for their own sake; they serve to focus the intellect on a problem – in this case the problem of plant development. If this book serves to focus attention more deeply on this topic, and if, on the way, the concept of positional control proves useful in this respect, we shall be satisfied – at least until the fruits of that concept are harvested.

25 March 1983   P.W.B., D.J.C.

Driesch, Hans (1893). Entwicklungsmechanische Studien. VI. Über einige Fragen der theoretischen Morphologie. *Zeitschrift für wissenschaftliche Zoologie*, **55**, 34–62.
— (1908). *Science and Philosophy of the Organism*, Vol. 1. London: Adam & Charles Black.
Vöchting, Hermann (1877). Ueber Theilbarkeit im Pflanzenreich und die Wirkung innerer und äusserer Krafte auf Organbildung an Pflanzentheilen. *Pfluger's Archiv für die gesamte Physiologie des Menschen und der Tiere*, **15**, 153–90.
— (1878). *Ueber Organbildung im Pflanzenreich*, Vol. 1. Bonn: Max Cohen und Sohn.
Wolpert, Lewis (1970). Positional information and pattern formation. In *Towards a Theoretical Biology*, Vol. 3, *Drafts*, ed. C. H. Waddington, pp. 198–230. Edinburgh: Edinburgh University Press.

The portrait of Hans Driesch appears in *Zeitschrift für wissenschaftliche Zoologie*, Abteilung D, *Roux' Archiv für Entwicklungsmechanik der Organismen*, volume **111** (1927), and is reproduced here by permission of Springer-Verlag.
The portrait of H. Vöchting appears in *Berichte der deutschen botanischen Gesellschaft*, volume **37** (1919), and is reproduced here by permission of G. Fischer-Verlag.

1

Models of pattern formation and their application to plant development

HANS MEINHARDT

Plants are beautiful results of pattern formation events during biological development. Pattern formation – the generation of regular differences in space – occurs at very different levels of organization. Many cells together form the roots and the shoots. Each of these structures can consist of layers of different tissues. A particular group of cells at the shoot apex may receive a signal to form a leaf; the leaf will develop an upper (adaxial) and a lower (abaxial) side, then some cells of the leaf may develop into stomata, while chains of cells may form vascular strands. The individual cells may become polar. These patterns must ultimately arise from particular interactions of the constituents of the developing organism such as cells, genes or molecules.

Pattern formation is by no means a peculiarity of living systems. The formation of clouds, rivers, sand dunes, water waves or crystals are examples where the pattern formation in the inorganic world starts from rather homogeneous initial conditions. To understand what type of interactions must be involved in biological pattern formation, it is helpful to grasp that in all the given examples a minor deviation from uniformity has a severe effect on the further growth of this non-uniformity. For instance, formation of a river may have started originally from a minor depression in the landscape. The water collecting there from the (uniformly distributed) rain accelerates erosion at this location, more water runs towards this incipient valley and so on. Pattern formation must therefore have a self-amplifying, or autocatalytic, element. However, self-amplification on its own would be insufficient for pattern formation. It would lead to an explosion or to a complete conversion of one state into another, just as a burning piece of paper will be completely converted into ash (burning is an autocatalytic process). In the pattern-forming systems mentioned above, the onset of self-amplification at a particular location inhibits self-amplification in the larger environment. Water which has collected at the incipient valley no longer contributes to erosion at a distance from it.

On the basis of this principle – local autocatalysis and long range inhibition – we have proposed a theory of biological pattern formation (Gierer & Meinhardt, 1972; Gierer, 1977a, 1981; Meinhardt, 1978a, 1982). After a brief introduction to this theory, I would like to show that polar, symmetric, or periodic, pattern can be generated on this basis. Extension of the basic principle will allow the generation of stripe-like, and of net-like, patterns of differentiated cells. Conditions will be discussed which can lead to sequences of structures in space, such as are found in the sequences of structures of a flower. In an Appendix, it will be shown that the pattern-forming mechanism proposed in a pioneering paper by Turing (1952) has essentially the same basis although this is not immediately apparent from the equations he proposed.

Pattern formation by local autocatalysis and long-range self-inhibition

The principle of pattern formation by autocatalysis and lateral inhibition is easily translated into an interaction of (still hypothetical) molecules. At least two types of molecules must be involved. One mediating short-range autocatalysis we call the activator (a). The autocatalysis must be balanced by an antagonistically acting second substance, the inhibitor (h), which diffuses much more rapidly. One of the simplest interactions that leads to pattern is given by the following differential equations:

$$\frac{\partial a}{\partial t} = \frac{ca^2}{h} - \mu a + D_a \frac{\partial^2 a}{\partial x^2} + \rho_0 \qquad (1a)$$

$$\frac{\partial h}{\partial t} = ca^2 - vh + D_h \frac{\partial^2 h}{\partial x^2} \qquad (1b)$$

Such equations are a very convenient shorthand of what the assumptions really are, and they are easily read even by those not familiar with differential equations. In words, Eq.(1a) would say that we assume a concentration change of the activator per unit time ($\partial a/\partial t$) proportional to an autocatalytic production term (a^2), and that the autocatalysis is slowed down by the action of the inhibitor ($1/h$). As for any biological substance, we have to assume that some decomposition or removal of the activator takes place. The simplest assumption one can make is that the number of disappearing activator molecules is proportional to the number of activator molecules present (just as the number of people dying on average per day in a city is proportional to the number of inhabitants). This is taken into consideration by the term '$-\mu a$'. The concentration change at a particular location depends also on the concentration in the neighbourhood and how much exchange of molecules occurs between neighbouring cells. The simplest assumption would be that exchange occurs by passive diffusion. However, active transport and convection also certainly play an important role in plants. Diffusion is described by the term $D_a \partial^2 a / \partial x^2$ where D_a is the diffusion constant of the activator, and $\partial^2 a / \partial x^2$ is the second derivative of the activator concentration over the position (x). It is easy to see why the exchange by diffusion is proportional to the second derivative, that is to say proportional to the change of the concentration change in space. Let us assume three cells in a row with a concentration of 1, 2 and 3 units. The net exchange between two cells is, of course, proportional to their concentration difference. In this example, the gain of the middle cell by diffusion is zero since it gains from the neighbour with 3 units the same amount as it loses to its neighbour with only 1 unit. Therefore, for constant concentration differences (i.e., a linear concentration gradient) the change due to diffusion is zero. If, however, the concentration differences vary between neighbours a net exchange by diffusion can take place.

The last term in Eq.(1a) describes a small basic (inhibitor-independent) activator production (ρ_0) which can initiate the autocatalysis at very low activator concentrations. This term will become important if new activator maxima have to be initiated during growth.

Equation (1b), which describes the change in the inhibitor concentration, can be read in an analogous way: the inhibitor production depends in a non-linear way on the activator concentration (but not on the inhibitor itself); the inhibitor decays and diffuses.

In the Appendix, some simple calculations are given which provide an intuitive approach as to why the reaction described by Eq.(1) leads to stable patterns. These calculations show also similarities and differences in the mechanism proposed by Turing. Such equations enable us to calculate the change of the activator and inhibitor concentration in a short time-interval. This change can be added to the initial concentrations and the subsequent change can be calculated. By many such iterations, the complete time-course of reactions can be calculated. Fig. 1.1a–d, obtained in this way by computer simulation, illustrates the proposed pattern-forming mechanism. Only one cell in the centre is assumed to have an activator concentration slightly above the average. This leads to local activator increase and, via the simultaneous production of inhibitor, to a depression in the activator concentration in the vicinity of this incipient peak. Further peaks appear at a distance from this initial peak and a more

4 *H. Meinhardt*

or less regularly spaced periodic pattern emerges. The inhibitory effect can also result from depletion of a substance necessary for autocatalysis (Fig. 1.1e–h). In the following, some general properties of these mechanisms are discussed and compared with some pattern formation processes observed in plants.

Fig. 1.1. Stages in pattern formation by autocatalysis and lateral inhibition. Assumed is a two-dimensional field of cells, and an activator-inhibitor (a–d), or an activator-depleted substance system (e–h). (a) A very small elevation of activator in the centre (upper row) grows further (b) due to autocatalysis. A simultaneously-produced inhibitor (lower row) diffuses rapidly and suppresses the production of activator in the environment (c), and ultimately stabilizes the maximum itself. At a distance larger than the range of the inhibitor, a ridge of activated cells appear (c). Due to mutual inhibition, this ridge decays into individual maxima (d). Activator and inhibitor have similar distributions; however, the activator pattern has sharper maxima due to the lower diffusion rate. (e–h) The limitation of activator autocatalysis can result from the depletion of a substance (bottom row) necessary for the autocatalysis. In this mechanism, the activator may be shifted towards a higher substrate concentration (Meinhardt, 1974) In this simulation, the first-formed maximum splits into four maxima. Lacalli (1981) has used this property to simulate pattern formation in unicellular algae.

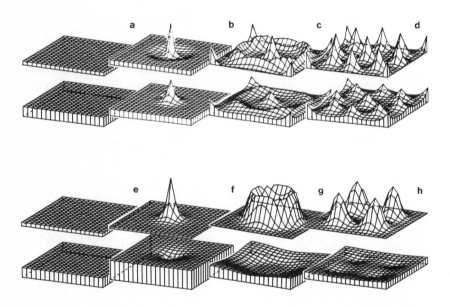

Generation of polar pattern

A condition for pattern formation is that the size of the field is at least similar to the range of the activator (i.e., the mean distance which an activator molecule is displaced between production and removal). If the field is smaller than this, any pattern is smoothed out by rapid equilibration of the activator within the field. If field size is comparable with the range of the activator, a polar pattern is preferentially formed (Fig. 1.2) since a high concentration at one boundary requires space for only one activator slope. By contrast, a symmetrical pattern would require space for two slopes which is not available at the critical field size. Since the mechanism is based on lateral inhibition, the mechanism has a tendency to set up extreme concentrations at the largest possible distance; that is, the polar pattern will be formed across the longest dimension of the field. Fig. 1.2 shows that a change of the geometry may lead to a reorientation of the pattern. Polar patterns are very frequent in plant development. especially in respect to the formation of polar cells. In most cases, the polarity of a cell is formed parallel to the axis of cell elongation. Cell division usually occurs in the plane perpendicular to the axis of polarity.

Fig. 1.2. Generation of a polar activator pattern in a two-dimensional field. (a–c) In a field the size of the activator range, random fluctuations are sufficient to generate a stable polar pattern. Polar patterns are preferred since a marginal maximum requires space for only one activator slope. (d–f) The separation into an activated and non-activated part can be connected with a major change in the geometry. Despite the strong residual pattern in both fragments (d), the pattern can reorient along the longest dimension of the field since the extreme concentrations have the tendency to appear at the greatest possible distance from each other.

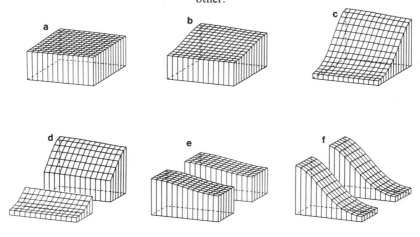

The determination of the anchor points of the spindle apparatus for a cell division is, of course, another important pattern-formation event within a cell. A high activator concentration could initiate the self-assembly of the microtubules and so determine the plane of cell division, even in an initially more or less symmetric cell. Asymmetric cell divisions, which are fairly frequent in plant cells, could be explained in this way since the new cell wall may occur closer to the activated site. Of course, a second centre of microtubule self-assembly has to be created at the opposite side of the cell. The generation of bipolar fields will be discussed below.

A well-known example of the initial generation of polarity is the developing egg of the brown alga, *Fucus* (see Jaffe, 1968). The *Fucus* egg is initially non-polar; this appears to be an unstable system, and has to generate a polarity. Any small cue from the environment can orient the outgrowth of the rhizoid. Any difference in illumination, temperature, or pH, as well as streaming of the surrounding water, or attachment to an impermeable surface, polarizes the cell. In the absence of any external stimulus the polarized outgrowth takes place at a random location on the surface of the cell, but with a substantial delay. According to the model, there is competition between each surface element of the cell and one area will eventually win. The smaller the externally-imposed asymmetry is, the longer it takes for one part of the cell to dominate.

Self-regulation and regeneration of the pattern

A very important property of the proposed pattern formation mechanism is its ability for self-regulation. For instance, an activator maximum can 'regenerate'. With the removal of the activated tissue, the site of inhibitor production is removed, the remaining inhibitor decays and a new activator maximum can be triggered in the remnant tissue (Figs. 1.3, 1.4). Assuming, for instance, that the polarity of cells is based on such a mechanism, then after a cell division each cell will restore the gradient (Fig. 1.3). In this process, the remnant polarity can orient the building-up of the new gradient and both daughter cells will have the same polarity. It could well be, however, that after a cell division the daughter cells have a different orientation to that of the mother cell. The polarity in one or both daughter cells could be reoriented during re-formation of the pattern. A 90° change of the axis of polarity would be especially probable (Fig. 1.2). The condition for polar cells would be that the inhibitor can diffuse freely within the cells but is not exchanged between cells (otherwise the gradient in a neighbouring cell could be completely suppressed) and

Fig. 1.3. Self-regulation. An activator-inhibitor distribution in a linear field is shown as function of time. (a) A polar distribution is stable indefinitely. (b) After separation into an activated and a non-activated part, the remaining inhibitor decays in the non-activated fragment, a new activator maximum is triggered and the polar pattern is restored in both fragments. (c) The same simulation, plotted in a different way. High activator concentration (high density of dots) is restored in the non-activated fragment.

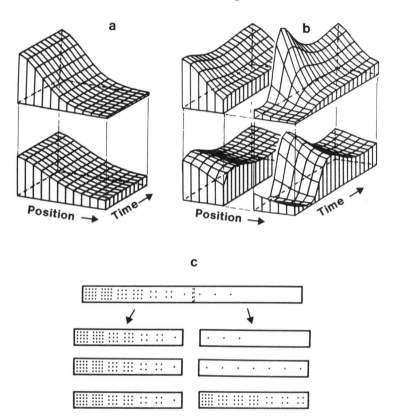

further, that the activator (or the elements of a reaction chain by which the autocatalysis is accomplished) diffuses only to a very limited extent within the cell, possibly by binding to the cell cortex.

Another example of self-regulation in plants is the formation of an apical meristem after its removal. Such an operation can cause other cells in the vicinity of the injury to become meristematic. One or several new apices may regenerate, which would be in accordance with the behaviour of an activator-inhibitor system where, after removal of an activator maximum, several maxima can appear (Fig. 1.4).

Fig. 1.4. Self-regulation can lead to the restoration of several maxima. (a) The initial activator–inhibitor distribution (activator, upper row; inhibitor, lower row). (b–d) After removal, the activator maximum can be restored (see Fig. 1.3). (e–g) After removal of a slightly larger portion of the activated area, two (or more) maxima can emerge during the re-formation of the pattern.

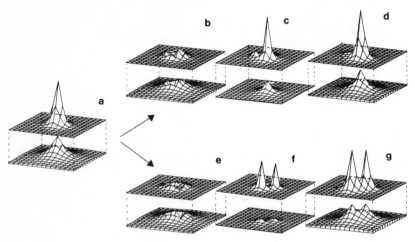

Generation of periodic pattern

The field can accommodate many maxima if its size is large compared with the range of the activator and inhibitor. The maxima will be regularly spaced if the pattern is formed during growth. An example is found in the regular spacing of leaves – phyllotaxis. Whenever the distance between the latest-formed leaf primordium and the centre of the apex becomes large enough, a new leaf primordium is inserted. The new primordium is assumed to be caused by a new activator maximum. According to the model, the inhibitor concentration decreases in the areas between existing maxima since the distance between these maxima increases due to growth. A new maximum can be initiated via autocatalysis from a low level production of activator (Fig. 1.5). After the new maximum has achieved its final height, it is surrounded by its own inhibitory field. Fig. 1.6 shows the simulation of an alternate and a distichous leaf arrangement. For a more complete treatment of phyllotaxis see Richter & Schranner (1978), Mitchison (1977) and the article by W. Schwabe (this volume).

The spacing of the maxima will be less regular if the pattern formation starts after the field has already reached a certain size. However, a certain range of maximum and minimum distances is maintained (Fig. 1.7). An example of such less-regular pattern is given in the spacing of the stomata of leaves (Fig. 1.7e). They are formed only after a leaf has essentially obtained its final number of cells (Bünning & Sagromsky, 1948). Upon further growth – caused by the enlargement of the individual cells and not by their proliferation – the distances between the stomata (assumed to be sites of activator maxima) may become so large that new stomata are initiated between the existing stomata. Sometimes, a cell which is already determined to form a stoma undergoes a further division thus leading to two adjacent stomata; this, of course, is not a violation of the lateral inhibition mechanism.

Fig. 1.5. Insertion of a new activator maximum during intercalary growth. In this process, the existing activator maxima (upper row) become separated from each other (a–c). This leads to a decrease of the inhibitor concentration (lower row) at points far from the existing maxima. In an area of low inhibitor concentration, new cells can become autocatalytic and form a new maximum which will eventually be surrounded by its own field of inhibition.

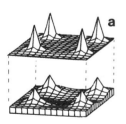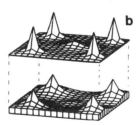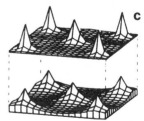

H. Meinhardt

Fig. 1.6. Examples of the formation of regularly-spaced periodic patterns during growth. Simulation of a distichous (a) and a decussate (b) leaf arrangement. Assumed is a cylindrical arrangement of cells, growing by doubling the uppermost row of cells. The first maximum (1 in a), or pair of maxima (1, 1 in b), is selected by random fluctuations. The next maxima (2,3...) appear at a distance from the previous one, either on the opposite side (a) or after each pair is rotated 90° (b). In simulation (b), an inhibitory zone at the tip of the shoot and a somewhat smaller diffusion range of the inhibitor is assumed. (c), (d) Examples of natural distichous and decussate patterns. (After Meinhardt, 1974, 1982.)

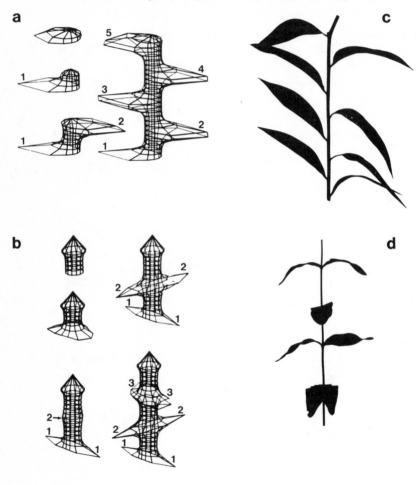

Fig. 1.7. Formation of a periodic pattern. The pattern is initiated by small random fluctuations (a–d). Assumed is a field much larger than the range of the inhibitor. The resulting pattern (d) is less regular (compare with Fig. 1.1), but a maximum and minimum distance are maintained. (e) Biological example of a somewhat irregular periodic pattern: the arrangement of stomata on the under side of a leaf (by courtesy of Dr M. Claviez).

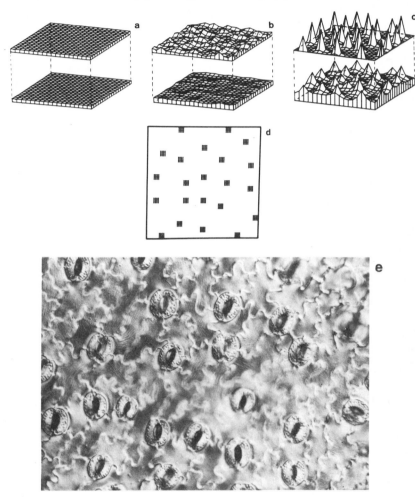

Mutual activation and stabilization of differently determined cell types

The mechanism discussed so far is appropriate to achieve a high concentration of a morphogen in a particular area and a low concentration in the surroundings. The pattern-forming mechanism to be discussed in this section should also fulfil a different task: that is, the achievement and maintenance of two (or more) different cell types in a stable community. There are many such examples in plants. The cells of the pith, cortex and epidermis of a stem form an ordered sequence of differently-determined cells. Strands of xylem and phloem cells appear in conjunction. In a leaf, the adaxial side is different from the abaxial side, and this polarization takes place very early in leaf development. We have proposed a mechanism which can generate patterns of differently determined cells in a well-balanced community. It is based on mutual long-range activation of locally exclusive cell states (Meinhardt & Gierer, 1980). We ask for an interaction which has the following properties: (i) the cells have some 'long-term memory' about their state of determination (and differentiation); (ii) the cells could be in one of several alternative determined states (1,2,3 ...), but an individual cell should make an unequivocal decision. They should be either of type 1 or 2 ... but not, for instance, half 1 and half 2; and (iii) the pattern formation should lead to predictable groups of cells: for instance, 1 borders 2, and 2 borders 3; or 1 borders 2, and both border 3.

The determination of a cell presumably consists in the activation of particular control genes and the suppression of others. This process has formal similarities with the pattern formation discussed above where a high concentration is achieved at a particular *location* and suppressed in the surroundings. Determination is, so to speak, pattern formation in the activation and inhibition of genes. To achieve a stable activation of one of the alternative states (1,2,3 ...) we will assume that each state is characterized by the activity of a particular feedback loop based on the autocatalytic substances, g_1, g_2, g_3 ... (for equations see Meinhardt & Gierer, 1980). However, each of the activated loops competes with the others via a common repressor R (in analogy to the competition via the highly diffusible inhibitor mentioned above). Genes are the most obvious substrate of these feedback loops, but other realizations are possible. Such a mechanism has the property that only one of the alternative loops can be active within one cell (Meinhardt, 1978b) and satisfies our first two requirements – long-term stability and exclusiveness. We have now to take care that the particular loops are turned on in groups of cells which have a particular spatial relation to each other. This will be the case if these states activate each other at long range. The activating substances must be distributed rapidly by diffusion, convection, or active transport. A complete reaction scheme for two feedback loops is given in Fig. 1.8, together with computer simulations in a two-dimensional field of cells. Since the stability of one cell type depends on the proximity of the other cell type, the preferred patterns are those in which long common borders are formed. For instance, in a two-dimensional array of cells, alternating stripes of high g_1 and high g_2 concentrations are especially stable. Stripes do not decay into individual maxima as is the case if the pattern results from a lateral inhibition mechanism (see Fig. 1.1).

Fig. 1.8. Mutual activation and stabilization of two different cell types. (a) Scheme of the proposed interaction. Assumed are two feedback-loops (1 and 2) resulting from the feedback of g_1 and g_2 molecules, respectively, on their own production. Both loops compete with each other via a common repressor R. This has the consequence that in a particular cell, either loop 1 or loop 2 is turned on, but never both. Both loops support each other over a long range via the rapidly diffusing molecules s_1 and s_2. A well-balanced community of both cell types with a long common border would be a stable state. (b–d) Simulation in a two-dimensional array of cells. A small increase of g_1 at the right-hand margin initiates the formation of an alternating stripe-like arrangement of cells with high g_1 and high g_2 concentrations.

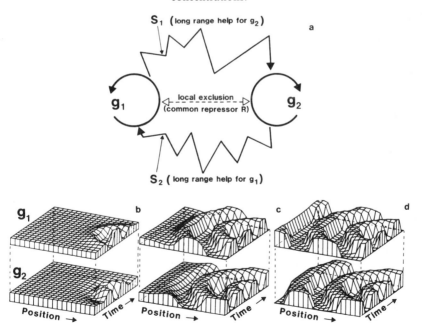

Two or more adjacent sheets of different cell types having two-dimensional surfaces as a common border could allow an efficient mutual stabilization. The concentrically arranged cell-types of the stem may arise in this way. In this radial pattern, missing structures can be regenerated and, after confrontation of different cell types by graft experiments, intervening structures can be intercalated (see the article of J. & P. M. Warren Wilson, this volume). Both regulatory features – completion of partial sequences of structures and intercalation of missing elements – are a feature of the lateral activation mechanism (Meinhardt & Gierer, 1980).

This mechanism has also a strong capability of pattern regulation. Imagine that all cells with high g_1 concentration are removed from a system of g_1 and g_2 cells. The g_1-feedback loop in the g_2 cells – normally at a very low level – would be reinforced by the active g_2 loop, while the g_2 loop itself is no longer sustained. This has the consequence that after a while, some g_2 cells are converted into g_1 cells (Fig. 1.9), thus restoring the correct g_1/g_2 ratio. For the same reason, such a system shows very good size regulation. If relatively too many g_2-cells are present, the greater support for the g_1 cells leads to an extension of the g_1 area at the expense of the g_2 area. Isolated cells can lose their differentiated character since they need the support from their differently determined neighbours. In an isolated, originally homogeneous cell population, a new pattern can be generated. Again, a certain tissue size is required for this to occur. The minimum size which allows pattern formation depends on the range of the molecules which accomplish the feedback. Below this size, the cells either oscillate between the different possible states or will exist as a mixture between the different possible states, which is presumably very close to an undifferentiated state. An example may be the formation of nodules of cambium-like cells in cultures of pure pith-tissue (Meins, Lutz & Foster, 1980). Such 'habituated' cultures arise at an elevated temperature. The nodules and pith tissue seem to support each other in such a way that no exogenous cytokinin has to be added to the medium. In agreement with the model proposed, Meins *et al.* (1980) found that the pith tissue has to have a minimum size (cubes with a length of about 3 mm, about 30 mg) if habituation (which, in our model, is the onset of a new pattern formation and generation of different cell types) is to occur.

Fig. 1.9. Size regulation and regeneration in a system of two locally exclusive cell-states which activate each other over a long range. A linear assembly of cells is shown which grows at its right-hand margin. The concentrations of g_1 and g_2 are plotted as functions of position and time. (a) The field is subdivided into an area of high g_1 and high g_2. Despite that only the number of g_2 cells increases due to growth, the number of g_1 cells increases due to the increasing help which the g_1 feedback loop receives from the g_2 loop. This causes a switch of some cells at the border from high g_2 into high g_1 and therefore to size-regulation. Even an almost complete removal of the g_1 area is regulated by a partial reprogramming of the remaining g_2 cells. (A complete g_1-removal will be regulated too, but the polarity may change.) (b) If the substance which accomplishes the local exclusion of high g_1 and high g_2 is also diffusible, the high g_1 and high g_2 concentrations appear at opposite ends of the field separated by a region in which none of the loops are active. A different high activation appears at each pole of the field (bipolar field). This system also shows good size regulation over a certain range and removal of the activated region at one or both poles (as shown) leads to their restoration.

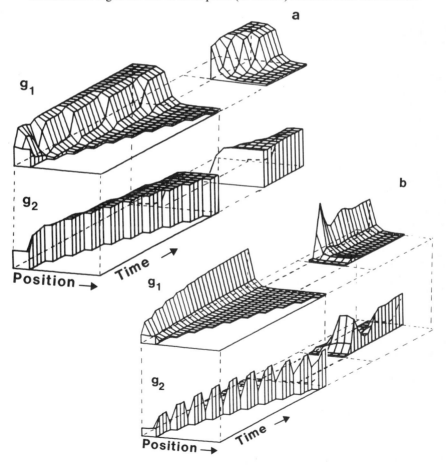

Bipolar fields

The mechanism of long-range activation and short-range exclusion discussed in the last section leads to areas of high g_1 and high g_2 concentration with a narrow zone of transition determined by the diffusion of g_1 and g_2 molecules. This mechanism leads to interesting new properties if the molecules which accomplish the mutual exclusion (R in the scheme Fig. 1.8) are also diffusible. Such an interaction corresponds to a system with two different activator molecules which compete with each other via a common inhibitor. The mutual activation of the competing loops – acting over a very long range – has the consequence that no loop can dominate the other. The result is that a g_1-maximum and a g_2-maximum appear at opposite sides of the field, separated by a (possibly large) zone in which neither the g_1 nor the g_2 concentration is high (Fig.1.9).

The shoot and root meristem systems may be generated (e.g. in the embryo) and regulated according to such a scheme. Both meristematic regions are separated by a large non-meristematic area (the stem and mature root). Due to active transport, the mutual stabilization of differentiated tissue can cross the whole plant. The shoots can stabilize the roots (possibly via auxin) and *vice versa*. Removal of one structure may lead to its regeneration; pieces of roots can regenerate shoot apices, and stems (shoots) can regenerate roots.

The polarity of a cell has been discussed above in connection with an activator gradient. However, the spindle apparatus is not assembled at only one end of a cell but at two opposite ends. In this model, the high g_1 and the high g_2 concentration could each induce the self-assembly of microtubules at the opposite poles. Nevertheless, the pattern in the cell remains asymmetric (high g_1 on one side and high g_2 on the other, but not a high g_1 at each pole). Such asymmetry is required to account for the frequently observed asymmetric cell divisions.

A case where the cell obviously has a bipolar character is *Acetabularia* with its cap at one end and its rhizoid at the other. A fragment derived from the centre of the cell will regenerate both basal and apical ends.

Periodic patterns are possible if the total size of the field is much larger than the range of the mutual activation and inhibition of both loops; g_1 maxima appear interspersed with g_2-maxima in an ordered manner (Fig. 1.10). Bünning & Sagromsky (1948) found an interspersion of stomata and hairs, indicating a system with two activators and a common inhibitor. The maintenance of the apical meristem on the one hand, and the determination of new leaves (discussed above) on the other, may, in fact, be better described as a system controlled by two different activator systems coupled in this way. New leaves can be initiated only at some distance from the centre of an apical meristem *and* at distance from other leaf primordia. However, a complete theoretical description of the interaction between meristem and leaf primordia has to account for the fact that new shoots arise only from meristematic nodes located at the apical side of the leaf base, or petiole. This is not yet included in the model.

Fig. 1.10. Alternation of two different types of activator maxima in a two-dimensional array of cells. Assumed is a system of two co-operating feedback loops (the g_1 and g_2 of Fig. 1.8a). The common repressor which accomplishes the exclusion is assumed to be diffusible also (g_1: top row of (a–d), dots in (e); g_2: bottom row of (a–d), crosses in (e)). Pattern formation is initiated by a slight elevation of g_1 in the centre of the field (a). At the location of the resulting g_1 maximum, g_2 is repressed (b), but isolated g_2 maxima appear in the surroundings (c,d). Further g_1 maxima maintain their distance from these g_2 maxima as well as from each other. This pattern is similar to the pattern of the concentric ring shown in Fig. 1.8. Due only to the diffusivity of the common repressor, the stripes are no longer stable and decay into individual maxima.

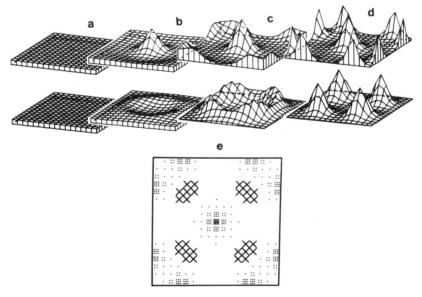

Sequences of structures

The parts of a flower – the sepals, petals, carpels and the stamen – form a
defined basal-apical sequence of structures. So far, four mechanisms have
been proposed which are able to generate such sequences.

(a) There is the mutual induction of locally exclusive states as discussed
     above.
(b) Determination is under the control of a graded morphogen concen-
     tration, i.e., by positional information and its interpretation (Wol-
     pert, 1969, 1971; Holder, 1979).
(c) There is sequential increase of a morphogen concentration at an
     outgrowing tip due to a feedback of the determined states so far
     achieved to the source (Meinhardt & Gierer, 1980).
(d) There is sequential addition, with time, of new structures at a growing
     tip involving, for instance, the counting of the number of cell divi-
     sions (Summerbell, Lewis & Wolpert, 1975).

All these mechanisms seem to occur in development of animals. Mechanism (a), for instance, is appropriate to describe the pattern formation
within abdominal segments of insects; mechanism (b) can account for the
basic body pattern of insects; and mechanism (c) or (d) for the proximo-
distal pattern of vertebrate limbs.

In plants, too few experiments are available (and even fewer have been
performed) to distinguish definitely between these possibilities. Hicks &
Sussex (1971) have made longitudinal incisions in the flower of *Nicotiana*
and observed ability to regenerate the missing parts. The regeneration of
the sepals etc. around the circumference provides only indirect evidence
for pattern formation in the basal-apical sequence. Nevertheless, the results indicate that the sepals, petals, carpels and stamen are laid down in a
sequential order. This argues against mechanism (b) – a pure positional-
information scheme – since this mechanism would be based on a reprogramming of existing cells under the influence of the morphogen
gradient, not on the accretion of new structures at a growing tip. Discrim-
ination between mechanisms (a) and (c) would require observations of
whether intercalation between normally non-adjacent cells occurs or not
after certain graft operations (Meinhardt, 1982). However, such graft
operations seem not to be feasible with plant tissue and, therefore, I will
not discuss these possibilities in detail.

Formation of net-like structures

Signals generated in the way described above can initiate a more or less irreversible determination and eventually differentiation. A determined state should be maintained independently of whether the inducing signal remains present or not. Differentiation, in turn, can influence the generation of further signals. Rather complex patterns can be generated in this way. This can be exemplified by discussing systems which allow the generation of net-like structures (Meinhardt, 1976, 1982).

The venation of leaves is a beautiful example of net-like structures in plants. Here, there appears to be a hierarchy of veins in the branching pattern. In the needles of pine trees, no branching occurs at all. Therefore, the leaf (the needle) must be narrow to allow the supply or removal by diffusion of substances via the central vascular strand. In more evolved and complex leaves, the vascular strands can bifurcate at the growing tip, but no lateral branch-veins occur from existing strands. An example is the leaf of the maidenhair tree (*Gingko biloba*). A leaf with such a dichotomous branching can enlarge by marginal growth only. Intercalary growth would lead to an enlargement of the interstices which would be insufficiently supplied with veins due to the lack of in-growing branch-veins. Only the most evolved leaves, with their ability to form lateral branch-veins as well as anastomoses, possess intercalary growth and can form an interlinked network of vessels. I will start with a description of a system which is able to generate such an evolved net and show that simplification of the system leads to the more primitive pattern in a straightforward manner. This will provide us with an intuition of which 'inventions' the plant has found to be necessary during its evolution in order to achieve the sophisticated pattern of the contemporary leaf.

How can we generate a long narrow chain of differentiated cells? We have seen how a local high concentration, an activator maximum (Fig. 1.1), can be generated. Such a maximum can be used as the signal to cause determination of the exposed cells into members of the vascular system. I will assume further that determined (or differentiated) cells repel the signal. The signal will be shifted into a neighbouring cell which will be determined also. A repetition of this process – determination, shift of the signal, determination – leads to a strand of determined cells behind a wandering activator maximum (Fig. 1.11).

We have seen that a sharp local activator maximum can be generated by the following interactions (see also Eq. 1):

$$\frac{\partial a}{\partial t} = \frac{ca^2s}{h} - \mu a + D_a \Delta a + \rho_0 y \tag{2a}$$

and
$$\frac{\partial h}{\partial t} = ca^2s - vh + D_h \Delta h \tag{2b}$$

(The terms s and $\rho_0 y$ will be explained below; $D_a\Delta a$ is the generalized diffusion term for a two-dimensional field of cells.) The local high activator concentration should initiate a more or less irreversible transition into a determined and differentiated state. This can be described by:

$$\frac{\partial y}{\partial t} = \frac{y^2}{1+fy^2} - ey + da \tag{2c}$$

The concentration of y within a cell is an indicator of whether a cell is differentiated or not. An interaction as given in equation (2c) behaves as a switch (Meinhardt, 1976). Only two stable states are possible: the concentration of y is either high or low. At low concentration of y, the negative (linear) decay term dominates and the y concentration decreases further. Above a certain threshold, y increases due to the non-linear feedback until saturation is reached. At a certain activator concentration, sufficient y is produced to turn on the autocatalysis of y. The cell switches from low to high y production and remains at the high y level even after a decrease of the activator concentration. To achieve a shift of the activator maximum away from the differentiated cells, we will assume that all differentiated cells remove a substance s which is produced by all cells of the leaf and on which the activator autocatalysis depends (Eq. 2d; see also Eq. 2a, b):

$$\frac{\partial s}{\partial t} = c_0 - \gamma s - \varepsilon sy + D_s \Delta s \tag{2d}$$

In Fig. 1.11, a computer simulation is provided for this set of differential equations showing the formation of a net-like structure. The formation of lateral branches is easily described by this process. It occurs whenever activator maxima become sufficiently remote from each other during elongation of filaments that become the veins. The inhibitor concentration can then become locally so low (see Fig. 1.5) that a new activator maximum is triggered along an existing vein due to the term $\rho_0 y$ in equation (2a). Fig. 1.11 also shows the similarity between a pattern gener-

this mechanism and a pattern of veins in a real leaf. The filaments keep their distance from each other and from the margin. This can lead to a 45° branching of the first-order branches, while higher-order branching occurs preferentially under 90°.

The model proposed shows that very little genetic information is required for the generation of very complex-appearing structures. The interaction of four substances – which can be easily encoded in the linearly arranged genetic information – is sufficient. The resulting pattern depends on the parameters involved. For instance, a larger diffusion rate of the activator leads to more curved veins. The inhibitor diffusion and the basic

Fig. 1.11. Formation of a net-like structure. Assumed is an activator-inhibitor system which generates local high-activator concentrations (■). These maxima are used to initiate the differentiation in the exposed cells (□). The differentiated cells remove a substance s (ᗱᗱ) on which activator production depends. Differentiation causes, therefore, a shift of the activator maximum into a neighbouring cell which will also become differentiated. Long and branching filaments of differentiated cells can be formed in this way. (a–e) Simulation using Eq. (2a–d) and a growing array of cells. (f) Simulation in a larger array of cells. (g) A leaf of *Fittonia verschaffeltii* in which the major veins are clearly visible. (After Meinhardt, 1978a.)

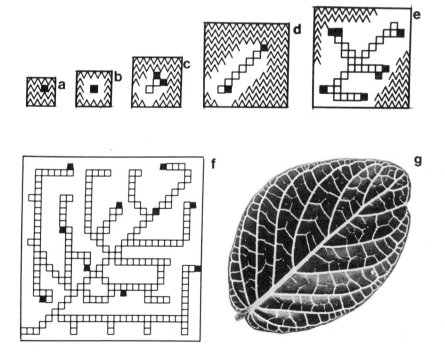

Fig. 1.12. (a) Two leaves of the same tree having a similar but not identical pattern of vascularization This occurs despite the fact that both patterns are under the control of the same genetic information. This feature – similarity but not identity – is reproduced by the model. Small differences between the individual cells can be decisive in determining whether a branching occurs to the left or right side. Such differences are introduced into the model by small random variations (up to 3%) in the constant c of Eq. 2a, b. (b–d) and (e–g) show two computer simulations at different stages with different random fluctuations but otherwise the same constants. (After Meinhardt, 1982.)

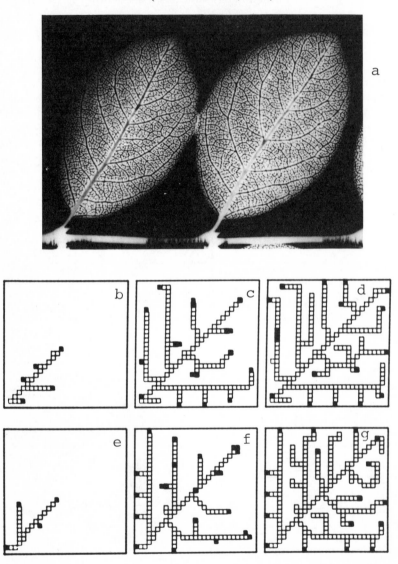

Fig. 1.13. Regeneration of a net-like structure. The model proposed has a pronounced ability of self-regulation. The computer simulation shows that a severe injury can be repaired. (a–c) Formation of a net. (d) Removal of all veins in the left half of the field. (e) Since the substrate s is no longer removed in this region, new branches grow into it until (f) the pattern is restored. Note that the new pattern is similar but not identical with the former pattern, while the remaining pattern is maintained unchanged.

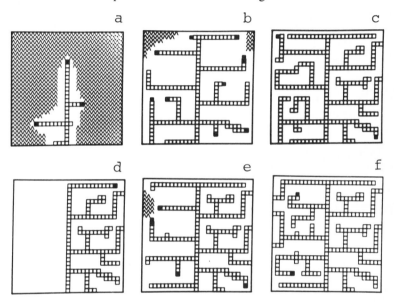

activator production ρ_0 (Eq. 2a) determines the distance between branching points, etc. The fine details of the pattern can depend on small differences between the cells. These are not necessarily genetically determined. Just as two leaves of the same tree may show a different vascularization, simulations with the same constants can lead to similar but not identical patterns (Fig. 1.12).

Another important feature of the model for the generation of net-like structures is its ability for self-regulation. Partial removal of veins leads to a sprouting of new branches into the deprived region until the pattern is restored (Fig. 1.13). (For the modelling of further details such as the formation of reconnections between veins (anastomosis) and limitation of the maximum net density, see Meinhardt, 1976, 1982.)

H. Meinhardt

A candidate for the hypothetical shift-substance, s, is the plant hormone, auxin. Application of auxin to an injury leads to the ingrowth of new vessels into the auxin-enriched area (Jost, 1942). Auxin is removed from the leaf by the vessels. Within the vessels, it is actively transported towards the roots, a feature which for simplicity is not incorporated into equation (2). Mitchison (1980) has proposed a model for vein formation according to which the necessary local autocatalysis is based on the self-reinforcement of this active auxin transport while the limitation of the autocatalysis results from auxin depletion; this is similar to the model outlined above.

In this scheme, we have employed two substances with an inhibitory action: the inhibitor of an activator-inhibitor system to generate a local signal for differentiation and, secondly, the substance s whose depletion causes the shift of the signal away from the differentiated cells. As shown in Fig. 1.1, the localization of an activator signal can also result from the depletion of a substrate. Therefore, localization *and* shift of the signal could be based on the depletion of a single substance. No separate inhibitor would be required, so equations (2a–c) can be simplified into

$$\frac{\partial a}{\partial t} = ca^2s - \mu a + D_a \Delta a \tag{3a}$$

$$\frac{\partial s}{\partial t} = c_0 - ca^2s - \gamma s - \varepsilon sy + D_s \Delta s \tag{3b}$$

This simplification has a profound influence on the resulting pattern. Due to the substance's depletion along an existing vein, it is impossible to initiate a new maximum. No branches can sprout from existing filaments. However, the activator maximum at the tip of an elongating filament can still split into two, causing a bifurcation. The result is a dichotomous branching pattern such as is observed in earlier-evolved leaves (Fig. 1.14).

Fig. 1.14. Dichotomous branching pattern. A simple system (Eq.(3a,b) and Eq.(2c)) is able to generate this evolutionarily-earlier branching pattern. (a–f) Simulation in a two-dimensional sheet of cells which grows at two margins. The activator maxima (■) move towards higher substance concentrations (∿∿), leaving behind a trail of differentiated cells (□). Splitting of the activator maximum can lead to a bifurcation of a growing filament. However, no lateral branching from an existing filament is possible. (g) A biological example: vein pattern in *Kingdonia* (after Foster & Arnott, 1960). The first fork – in the leaf as well as in the simulation – is a triple fork since, at small sizes, the relative growth by the division of marginal cells is high and two splittings of the activator maximum can follow quickly upon one another.

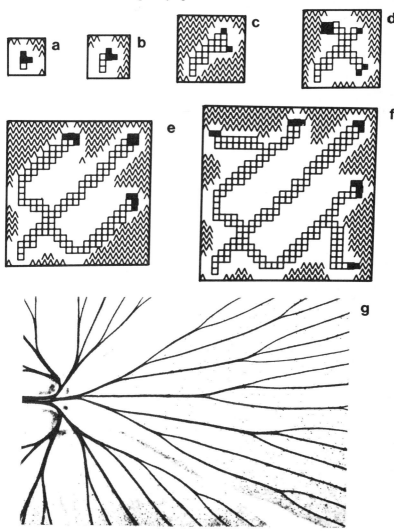

Boundaries between differently determined cells as organizing regions for substructures

A subdivision of a field into distinct patches of differentiated cells signifies that sharp borders are formed in between. If two cell types have to co-operate to produce a morphogen, this morphogen will be produced only along the common border. In a sheet of cells, three different cell types are close to each other only at a particular point (just as three countries touch each other only at one point). Such a point can become the organizing region for a substructure. The local morphogen concentration is a mea-sure of the distance to the common boundary and can supply positional information. I have shown (Meinhardt, 1980, 1982, 1983) that many ex-perimental observations concerning pattern formation in legs or wings of insects and vertebrates can be explained using this assumption. This mode of pattern formation enables the determination of a substructure at a precise position in relation to other, already determined, structures with a defined asymmetry (such as found in a left or right leg), a feature which is of obvious importance for limb determination in animals.

It is not yet clear whether a similar mechanism is used in plants. An indication of such a mechanism would be the formation of new structures at a borderline between existing structures. The formation of the root and shoot meristem may be an example. Most plants during their early embry-ogenesis pass through a heart-shaped stage. The two 'heart lobes' are the primordia for the two cotyledons. The meristem for root and shoot devel-opment is formed at the border between the two cotyledons. One possibil-ity would be that the two cotyledons are different from each other (say, type 1 and type 2) and that the condition to form the meristem is a confrontation of the two types. This hypothesis can presumably be tested by transplantation of cells from one primordial cotyledon to the other and observing whether a second site of meristematic growth is induced by this experimental juxtaposition.

Conclusion

By assuming a few specific interactions, especially local autocatalysis and long-range inhibition, we have been able to explain several basic features of pattern formation. The generation of polar and periodic pattern – both of which are very common in plant development – can be accounted for in a straightforward manner. Many such pattern-formation processes take place in the course of development. By an appropriate linkage of several such processes, systems emerge which can generate a far more complex pattern than the individual elements could do alone. The venation of leaves has served as an example. The coupling of the generation of a local maximum of an activator and an irreversible determination has enabled us to account for the generation of complex net-like structures.

Not yet included in this pattern-formation theory is the control of tissue growth and of cell proliferation. And obviously of great importance is the generation of the real form of a tissue, i.e., morphogenesis proper. Gierer (1977*b*) has developed a theory for the shaping of cell sheets. In this theory, mechanisms are proposed by which the patterned distributions of substances are transformed into bending moments within cell sheets. Mechanisms of this type can lead, for instance, to the curved shape of the petals and sepals of flowers.

We have seen that even random fluctuation can initiate pattern formation. Why, then, is development such a reproducible process? It is because the result of one pattern-formation process provides the orienting asymmetry for the next. The position of a first leaf, for instance, may be determined by random fluctuations. The position of the following one is, however, so strongly influenced by the previous one that fluctuations of the same strength have no longer an influence on leaf determination.

Further experimentation is certainly necessary to find out how these principles are realized in molecular terms, and what substances are involved. Since at least two substances are required which act in an antagonistic and non-linear way with each other to form a pattern, these systems have counter-intuitive properties. For instance, the removal of an inhibitor from a tissue can lead to an increase of the inhibitor concentration (due to the triggering of a new activator maximum). Or, in an attempt to isolate the activator, isolated activated cells would appear as non-activated since the inhibitor can then no longer escape into non-activated surroundings. The accumulating inhibitor would rapidly switch off the activator production. Without knowledge of the underlying principles on which pattern formation is based, the chance would be high that a successful experiment would be incorrectly interpreted as a failure. We hope that our models can help to design appropriate experiments.

Appendix

A precise mathematical formulation such as given in equation (1) allows considerations which provide a good intuition of why such an interaction leads to a stable pattern (Meinhardt, 1982). We can calculate at which concentration of a and h a steady state (i.e., a balance between production and decay) is reached. A steady state is characterized by a zero concentration change ($da/dt=0$; $dh/dt=0$). Let us first look at Eq.(1a). For simplicity, we assume all constants are equal to unity; then, disregarding diffusion and assuming that even the inhibitor concentration is constant, Eq.(1a) would be simplified to

$$\frac{da}{dt} = a^2 - a$$

A steady state ($da/dt=0$) would be attained when $a=1$. However, this steady state is unstable since for any concentration of a which is a bit larger than 1, $a^2 - a$ will be positive and the concentration of a will increase further. Conversely, when $a<1$, da/dt will be negative. The reason for this instability lies in the over-exponential autocatalytic production term in conjunction with a normal exponential decay.

Now let us include the action of the inhibitor. Disregarding again any constants and diffusion, Eq.(1b) would simplify to

$$\frac{dh}{dt} = a^2 - h$$

which has a steady state when $h=a^2$. If we assume that the inhibitor equilibrates relatively rapidly to a changed activator concentration, we can express the change of activator concentration as a function of the activator concentration alone:

$$\frac{da}{dt} = \frac{a^2}{a^2} - a$$

$$= 1 - a$$

Therefore, if we include the action of the inhibitor, we obtain a steady state, again at $a=1$, which is stable. If a is larger than 1, $1-a$ is negative and the concentration will return to the steady state at $a=1$.

To see why Eq.(1) can generate a pattern we have to take into consideration that the inhibitor diffuses much faster compared with the activator.

Let us assume an array of cells, all at the steady state concentrations of *a* and *h*, except one cell which should have a slightly increased activator concentration. It will also produce a bit more of the inhibitor but, since the inhibitor diffuses rapidly into the surroundings, its concentration can be regarded, in the first approximation, as constant. In this circumstance any deviation from the activator steady state will grow further, the equilibrium being unstable. However, after a substantial increase of the activator maximum, the inhibitor concentration can no longer be regarded as constant. As shown above, the action of the inhibitor leads to stabilization of the autocatalysis. A new stable-patterned steady state is reached.

Turing's mechanism

The possibility of generating a pattern by the reaction of two substances with different rates of diffusion was discovered by Rashewsky (1940) and Turing (1952). The involvement of diffusion seems to contradict our intuition since with diffusion one associates a smoothing out of any local accumulation of molecules, not the creation of differences. However, as we have seen, the different diffusivities of the substances can lead to a restriction of an increase at one location and to a decrease in the surroundings. Turing exemplified the mechanism he proposed by the following set of equations (Turing, 1952, p. 42):

$$\frac{\partial x}{\partial t} = 5x - 6y + 1 \tag{4a}$$

$$\frac{\partial y}{\partial t} = 6x - 7y + 1 \ (+\text{Diffusion}) \tag{4b}$$

Both equations, (4a) and (4b), look very similar. It is not immediately obvious why such an interaction should lead to a pattern. However, a similar consideration as undertaken above for Eq.(1) shows that the basis of the Turing pattern is also autocatalysis and lateral inhibition. From Eq.(4) we can calculate that a steady state is given at $y = 1$ and $x = 1$. If we regard y as constant when there is a small local deviation from the steady state – remember that y is the rapidly diffusing substance – we obtain

$$\frac{\partial x}{\partial t} = 5x - 5$$

which is again positive for any $x > 1$ and, therefore, the deviation will grow further. Again, the system is locally unstable due to autocatalysis.

To show how the interaction of y stabilizes the system, we have to calculate the steady state of y as function of x. By setting $dy/dt=0$, we obtain from Eq.(4b):

$$7y=6x+1.$$

Inserting this into Eq.(4a) we obtain

$$\frac{\partial x}{\partial t}=5x-\frac{36}{7}x-\frac{6}{7}+1$$

$$=\frac{1}{7}(1-x)$$

indicating a steady state at $x=1$. This state is stable since we get a negative change of x for any $x>1$. This elementary calculation shows that in Turing's mechanism the local instability results from a short-range autocatalysis. The long-range substance ensures that the total system remains stable, and that an increase in one area depends on a decrease in the surroundings. Therefore, Turing's mechanism can generate basically the same types of pattern (see Bard & Lauder, 1974; Lacalli & Harrison, 1978) as the lateral-inhibition mechanism, i.e., graded concentration profiles and isolated maxima (but not stripes or branched filaments).

The mechanism proposed by Turing has one important drawback: its molecular basis is not reasonable. According to Eq.(4a), the number of x molecules disappearing per time unit is assumed to be proportional to the number of y molecules and independent of the number of x molecules. This signifies that molecules can disappear even if no molecules are present. This can lead to negative concentrations. Turing had seen this problem and proposed to ignore negative concentrations. However, such a cut-off can cause an unlimited concentration increase. Anyway, such a cut-off is not necessary if the number of decaying molecules is proportional to the number of molecules present. This requires, as shown in Eq.(1), non-linear production terms.

Much of the work described in this article is the result of a very enjoyable collaboration with Professor Alfred Gierer. I wish to express my sincere thanks to him for the many years we have worked together. I would also like to thank Dr P. W. Barlow for stimulating suggestions and help with the manuscript.

References

Bard, J. B. & Lauder, I. (1974). How well does Turing's theory of morphogenesis work? *Journal of Theoretical Biology*, **45**, 501–31.

Bünning, E. & Sagromsky, H. (1948). Die Bildung des Spaltöffnungsmusters in der Blattepidermis. *Zeitschrift für Naturforschung*, **3b**, 203–16.

Foster, A. S. & Arnott, H. J. (1960). Morphology and dichotomous vasculature of the leaf of *Kingdonia uniflora*. *American Journal of Botany*, **47**, 684–98.

Gierer, A. (1977*a*). Biological features and physical concepts of pattern formation exemplified by *Hydra*. *Current Topics in Developmental Biology*, **11**, 17–59.

— (1977*b*). Physical aspects of tissue evagination and biological form. *Quarterly Review of Biophysics*, **10**, 529–93.

— (1981). Generation of biological patterns and form: some physical, mathematical, and logical aspects. *Progress in Biophysics and Molecular Biology*, **37**, 1–47.

Gierer, A. & Meinhardt, H. (1972). A theory of biological pattern formation. *Kybernetik*, **12**, 30–9.

Hicks, G. S. & Sussex, I. M. (1971). Organ regeneration in sterile culture after median bisection of the flower primordia of *Nicotiana tabacum*. *Botanical Gazette*, **132**, 350–63.

Holder, N. (1979). Positional information and pattern formation in plant morphogenesis and a mechanism for the involvement of plant hormones. *Journal of Theoretical Biology*, **77**, 195–212.

Jaffe, L. F. (1968). Localization in the developing *Fucus* egg and the general role of localizing currents. *Advances in Morphogenesis*, **7**, 295–328.

Jost, L. (1942). Über Gefäßrücken. *Zeitschrift für Botanik*, **38**, 161–215.

Lacalli, T. C. (1981). Dissipative structures and morphogenetic pattern in unicellular algae. *Philosophical Transactions of the Royal Society of London*, B, **294**, 547–88.

Lacalli, T. C. & Harrison, L. G. (1978). The regulatory capacity of Turing's model for morphogenesis, with application to slime moulds. *Journal of Theoretical Biology*, **70**, 273–95.

Meinhardt, H. (1974). The formation of morphogenetic gradients and fields. *Berichte der deutschen botanischen Gesellschaft*, **87**, 101–8.

— (1976). Morphogenesis of lines and nets. *Differentiation*, **6**, 117–23.

— (1978*a*). Models for the ontogenetic development of higher organisms. *Reviews of Physiology, Biochemistry and Pharmacology*, **80**, 48–104.

— (1978*b*). Space-dependent cell determination under the control of a morphogen gradient. *Journal of Theoretical Biology*, **74**, 307–21.

— (1980). Cooperation of compartments for the generation of positional information. *Zeitschrift für Naturforschung*, **35c**, 1086–91.

— (1982). *Models for Biological Pattern Formation*. London & New York: Academic Press.

— (1983). Cell determination boundaries as organizing regions for secondary embryonic fields. *Developmental Biology*, **96**, 375–85.

Meinhardt, H. & Gierer, A. (1980). Generation and regeneration of sequences of structures during morphogenesis. *Journal of Theoretical Biology*, **85**, 429–50.

Meins, F., Jr., Lutz, J. & Foster, R. (1980). Factors influencing the incidence of habituation for cytokinin of tobacco pith tissue in culture. *Planta*, **150**, 264–8.
Mitchison, G. J. (1977). Phyllotaxis and the Fibonacci series. *Science*, **196**, 270–5.
— (1980). A model for vein formation in higher plants. *Proceedings of the Royal Society of London*, B, **207**, 79–109.
Rashewsky, N. (1940). An approach to the mathematical biophysics of biological self-regulation and cell polarity. *Bulletin of Mathematical Biophysics*, **2**, 15–121.
Richter, P. H. & Schranner, R. (1978). Leaf arrangement. Geometry, morphogenesis, and classification. *Die Naturwissenschaften*, **65**, 319–27.
Summerbell, D., Lewis, J. H. & Wolpert, L. (1975). Positional information in chick limb morphogenesis. *Nature*, **244**, 492–6.
Turing, A. (1952). The chemical basis of morphogenesis. *Philosophical Transactions of the Royal Society of London*, B, **237**, 37–72.
Wolpert, L. (1969). Positional information and the spatial pattern of cellular differentiation. *Journal of Theoretical Biology*, **25**, 1–47.
— (1971). Positional information and pattern formation. *Current Topics in Developmental Biology*, **6**, 183–224.

2

Cell shapes and tissue geometries

ROBERT W. KORN

Historical perspective

The first account of cells by Robert Hooke in 1665 was not only from plants but also included drawings from two views: a transverse section that indicated the isodiametric shape of cork cells and a longitudinal cut which illustrated their elongate appearance. The early plant microscopists likened the appearance of cells in tissue to froth, and Kieser in 1815 (whose description was quoted by D'Arcy Thompson, 1942, pp. 545–6) was the first to suggest that they had a specific shape, that of a dodecahedron with twelve lozenge-shaped quadrilateral facets (Fig. 2.1A). Lord Kelvin (1894) modelled the shape of soap bubbles and found the ideal packing polyhedron to be the orthoid tetrakaidecahedron, a figure noted two thousand years ago as one of the thirteen semi-regular Archimedean solids (Fig. 2.1B). The tetrakaidecahedron has fourteen sides, eight regular hexagons and six squares, all bounded by vertices with 120° angles. This is the ideal packing figure because it can be repeated without leaving any small spaces and it requires a minimum of material for partitioning of space.

Lewis (1923) first counted the number of sides of packed plant cells and found an average close to the Kelvin ideal of 14.0. Later, Matzke and many of his associates (see Dormer, 1980, for details of the survey) inspected the tissues of many plants and consistently found averages approximating 14.0. These workers also stressed that although most cells (~98%) were deviant of the Kelvin figure they tended toward the ideal value and this tendency was caused by intercellular compressive forces as they found for soap bubbles, compressed lead shot and peas swelling in a bottle. The primary causal explanation was a physical force and consequently was reported in detail by D'Arcy Thompson (1942) to support his general idea of physical forces as the primary determinant of biological form.

R. W. Korn

Recently, the author has proposed an alternative hypothesis for the explanation of cell shapes because of two problems with the Matzke-Thompson scheme (Korn, 1974, 1980; Korn and Spalding, 1973). First, the soap bubble appearance of cells is true only of parenchymatous cells, and the twenty or so other types of plant cells are not considered. For physical forces to explain these other shapes, different force vectors would be operative in closely-nested regions of the plant. Second, soap bubbles, compressed shot, and swollen peas undergo slippage to assume states with minimum free surface energy; for the most part, plant cells are fixed together with no freedom to slide. My alternative explanation is that cell shape is determined by a few simple biological rules of growth and division. Not only does the tetrakaidecahedron come out as the most frequent type but the same probability distribution is also generated for frequencies of cells with different numbers of facets as has been scored for actual cells by Matzke (1946). These explicit rules can be precisely modified and new shapes will be generated that have their counterpart in other types of plant cells, i.e., elongate cells, stellate cells, etc.

This approach of postulating a set of simple rules, that is, a developmental algorithm, was first explored by Lindenmayer (1968, 1978) to generate the branching pattern of algae and this method appears to be a powerful tool for describing complex form by means of the rules which generate that form. Traditionally, form of the adult state has been given a geometric description, but through algorithmic procedures a parallel and equally legitimate approach is to generate the pattern such that the algorithm itself becomes a kind of description. More recently, the author has suggested how a number of related algorithms can generate a variety of plant cell types through 'positional specificity' of facets, edges and vertices (Korn, 1982). Hence, a cellular façade is both structural and informational, the latter specifying how the former changes during growth, division and differentiation.

Fig. 2.1. A, dodecahedron; B, tetrakaidecahedron.

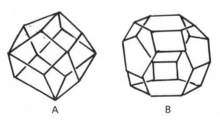

A   B

Adaptive features of cell form

A planar array of epidermal cells averages 6.0 sides and their vertices average 120°. Geometrically, a planar surface can be constructed of regular repeating triangles, quadrilaterals or hexagons (Loeb, 1976). With compressive force applied to one side of the array, the hexagon becomes the most flexible of the three. The triangle is a rigid body with no degrees of freedom to shift; it is a strut, and hence will collapse under pressure while the quadrilateral can shift in either one of two directions and the hexagon can flatten because of two simultaneous directions of adjustment (Fig. 2.2, D–F). Also, during the compressive state, the lines of stress or tension will follow the edges of the figure, passing from edge to edge when they are joined by angles less than 90° (normal to the direction of force) (Fig. 2.2, A–C). Forces of stress can be dissipated within a triangular or hexagonal array but not over quadrilateral units because the edges are normal to one another. The hexagonal array is adaptive both in ability to change shape and direct forces dissipatively and, therefore, it is the most

Fig. 2.2. Effects of compression on planar arrays. Lines of force (arrows) are well distributed over a triangular array (A), are poorly spread through an array of quadrilaterals (B), and are well dissipated through a hexagonal array (C). Freedom to adjust to new shapes under force of compression is non-existent in a triangular array leading to bends and breaks (D), is minimal in a quadrilateral array leading to a one-sided shift (E), and is great in a hexagonal array resulting in a flattening of the sheet.

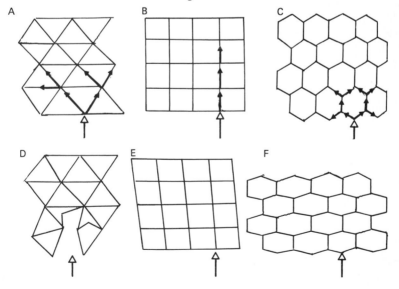

stable unit. This consideration is justifiably applicable to plant cells as long as vertices have the capacity to rotate, a feature easily demonstrated.

A second feature of stability of cell arrays is the size of the individual cells. A daughter cell is approximately half the size (volume) of its parental cell and it will double in size before division. Hence, there is a limited range of cell sizes within a homogeneous tissue. An old mason's rule is that in building a wall or floor never use a stone that is more than twice the size of the smallest stone. Since there is the geometrical necessity for a two-dimensional array of polygons with three-rayed vertices to average 6.0 sides, a larger-than-average cell will have a tendency to have more than six sides and so there must be a compensating small cell, or pair of cells, having fewer than six sides. Small cells with three sides behave as struts and limit tissue flexibility, and one with four sides will flatten in only one direction. The most advantageous population of cells will be those with from five to seven sides, and most cell populations have about 0.892 of their cells within this range. Since cell wall number is correlated with size, it is also significant that a population of cells has 0.65 of the cells within one standard deviation of the average cell size (Korn, 1980).

A third feature of geometric stability of cells pertains to the internal restriction of their growth. An elongating cell undergoes extensive growth because the end walls are programmed not to expand and the lateral walls enlarge in one direction, parallel to organ growth. A parenchymatous cell expands in three directions because each facet can expand in two directions. Similarly, collenchyma cells do not change shape because of vertex thickening. The rate and direction of local extensibility collectively determine the pattern of cell growth, and apparently this restriction is immediately due to the orientation of cellulosic microfibrils (see Green, this volume).

Proliferative cells

Degree of order

Meristematic cells are constantly undergoing changes in size and shape. For example, a quadrilateral epidermal cell will increase continuously in size by logarithmic growth and will increase in number of side walls because of cell division of adjacent cells. Four simple rules of behaviour can be attributed to a cell to account for this change in geometry:

(1) Walls expand logarithmically;
(2) Cell division halves the size of a cell by a cell plate positioned perpendicular to the longitudinal axis of the cell;

(3) Cell cycle is of determinate duration;

(4) Cell plates do not grow for one generation.

In order to inspect the consequences of each rule on cell geometry a method of measurement of cell order is required. That of Clark & Evans (1954) can be applied to indicate how uniform cells are deployed by the non-scalar value R. In the equation, $R = l\,(d)^{\frac{1}{2}}$, l is the nearest neighbour distance measured in absolute distance and d is the distance measured by the number of cells per unit area. An R value of 2.19 is perfect ordering of sites according to a hexagonal array (Fig. 2.3C), an R value of 1.0 indicates random ordering (Fig. 2.3A), while an R value of 0 suggests a clustering effect (Fig. 2.3B). A typical field of cells gives an R value of around 1.6 to 1.7, meaning cells that are more ordered than random.

When rule (1), which states that wall growth is exponential but cell cycle time and wall placement occur at random (selecting random numbers for time in hours and orientation of cell plate in radians), is applied to a paper model of cells, an R value of 1.2 is obtained (Fig. 2.3E). What

Fig. 2.3. Order in various arrays of sites as measured by the method of Clark & Evans. (A), a random array with an R value of 1.0; (B), a clustered array with R equalling zero; and (C), a perfect array having an R value of 2.19. (D), cells of the epidermis of the cotyledon of *Lupinus albus* giving an R value of 1.64. Modelled cell arrays: (E) with only exponential growth having an R value of 1.2; (F), the rule above and equal cell division give an R value of 1.45; (G), determinate cycle time and rules above generate a pattern with R equalling 1.55; and (H), all rules above plus no growth of new cell plates for one generation give an R value of 1.62.

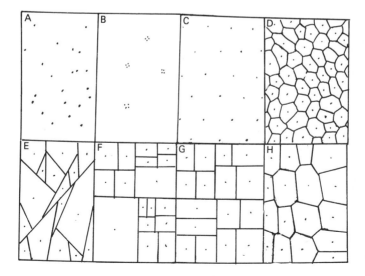

little order emerges (> 1.0) comes from a Gaussian distribution of daughter cell size, or radii, rather than a negative exponential distribution. When rule (2) of halving cell size perpendicular to the longitudinal axis of the cell is added to the model along with rule (1), an R value of 1.45 is obtained. Addition of rule (3) of determinate cell cycle duration increases the R value to 1.55. An initial daughter cell of size S_i is ideally half the final size it will attain, S_f, so that S_f/S_i is 2.0. Realistically, a daughter cell, arising from an approximate halving upon division of a mother cell, is of size S_d and will increase in size by a factor of S_f/S_d. This factor will be generated exponentially over I intervals at a factor of increase of a per interval. Hence $a^I = S_f/S_d$, or $I \cdot \log a = \log(S_f/S_d)$, or $I = \log(S_f/S_d/\log a$ in which I, again, is the number of intervals of the cell cycle. As duration of the cell cycle is derived from the initial size of the cell which, in turn, is determined by the division pattern, cell geometry strongly dictates cell cycle time.

Rule (2) can be further partitioned into three rules and the corresponding R values determined. Thus, to summarize:

(1)   Exponential growth. ($R = 1.2$)
(2a) Halving of cell size. ($R = 1.33$)
(2b) New plate perpendicular to cell axis. ($R = 1.40$)
(2c) Plates avoid existing, old vertices. ($R = 1.45$)
(3)   Determinate cycle duration. ($R = 1.55$)
(4)   No plate growth for one cycle. ($R = 1.62$)

This fourth rule means that the 90° angles of vertices between an old and a new plate will shift to 120° because only the former enlarges. It is of particular interest that the Clark & Evans method finds that the regular hexagonal array generating the highest value of R is also the pattern described earlier as the most stable in responding to external forces.

Stability of form

A proliferating tissue maintains dynamic stability of form in two respects, size and shape. An unusually small cell early in the development of a tissue will not lead to a patch of small cells, and an unusually large cell does not generate a sector of larger-than-average cells (Korn, 1980). The range of cell sizes remains constant during cell proliferation. Disorder in cell size comes from growth, while the compensating process of increasing order is from a determinate cell cycle time, established by the initial cell size. Cell division is also a disordering feature as most divisions are clearly unequal (not to be confused with asymmetrical divisions to be discussed later).

Similarly, the shape of a growing cell changes during the cell cycle. A new daughter cell is younger than its adjacent cells (except for its sister cell), and these adjacent cells will divide before this reference cell divides. Each division of an adjacent cell possibly adds another facet to the reference cell through bisecting one of its facets (Fig. 2.4). Growing cells of a tissue periodically increase in number of sides from adjacent cell divisions and this change can be taken as a disordering feature. The compensating, ordering feature is cell division when the number of sides per daughter cell is less than the number for the parental cell. Specifically, the relationship is $N_d = ((N_{pe}/\,2)+2) + ((N_{pe}/2)+2)$, or $((N_{pu}/2)+2) + ((N_{pu}/\,2)+1)$ where N_{pe} and N_{pu} are an even number of walls and an uneven number of walls in the parental cell, respectively, and N_d is the number of walls in the daughter cells. With $6\rightarrow5+5$ or $7\rightarrow5+6$, the number of walls always decreases except in the rare cases of parental cells with four side walls. All

Fig. 2.4. Effects of adjacent cell division on a cell. Initial quadrilateral cell assumes four, five, or seven sides according to number of adjacent cells dividing. The results are different sizes of the cell at the time of dividing.

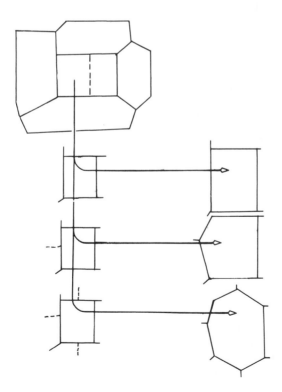

possible cell shapes, combinations of adjacent cell divisions, and cell cycle durations have been calculated in tabular form and it is found that a stable distribution of cell shapes is realized (Korn, 1980).

A third mechanism of stabilizing cell form has been found and, while subtle in expression, it is of importance in maintaining tissue integrity. This feature states that edges are the primary element of cell growth, while facets and volumes have only an accompanying role in growth. The motivation can be demonstrated by a simple example of a hypothetical square cell (Fig. 2.4). This cell will enlarge at a constant rate until it divides, but if during its cycle it experiences an adjacent cell division it will become a five-sided cell, more circular in outline and, therefore, of greater size. The length of the margin remains unchanged, only the inscribed area increases. If the cell has three adjacent cell divisions it becomes seven-sided and of even greater size. As adjacent cell division is an event independent of the growth of the cell, new shapes of the cell and increases in size are both probabilistic and polytonic. The feature that remains constant regardless of adjacent cell divisions is a monotonic increase in length of the margin.

This edge growth can be understood as an inherent extensibility status of the edge, capable of extending due to internal turgor force. As the edges around a facet extend so will the area of the facet increase; and as facets of a cell expand so will the volume of the cell. By taking edges as the primary, constantly increasing element of growth, a cell will enlarge in a particular manner, a tissue of cells will grow in a co-operative fashion (since edges of adjacent cells are fixed together), and adjacent but unlike tissues can develop as an apparently singular activity. The integrating feature of co-operatively growing, adjacent tissues is clearly based on rates of edge growth (Korn, 1980). On an average, edges will extend at a rate of 0.52138 for old edges, and zero for new edges, per cell generation. This pair of rates averages out to 0.414 increase per cell generation or by a factor of 1.414 which is the cube root of the doubling of cell size per cycle ($(2)^{\frac{1}{3}} = 1.414$) for parenchyma cells; for epidermal cells it averages out to 1.732 which is the square root of doubling ($(2)^{\frac{1}{2}} = 1.732$). Hence, edge growth rates of 0.521 and zero lead to equal rates of expansion for facets, including those facing each other at the epidermal–parenchymatous interphase. A particular primary rate of growth of edges results in harmonious development of several types of tissues.

Presently, unpublished data indicate another subtle feature of proliferating cells. A single edge of a cell will undergo different rates of extension during the cell cycle and all the edges of a cell undergo extension at varying rates independently of each other. Apparently this juggling of rates and, therefore, cell shape, is due to what few degrees of freedom a cell has for expanding when embedded within a tissue. Surface cells of an organ have the highest number of degrees of freedom to change shape and express growth and, through their juggling, provide freedom for subdermal cells progressively deeper within the tissue.

A cell is neither a block, a cube, nor a cage, and is not solely restricted by internal and external forces of compression and tension. It is best seen as an ensemble of rods (edges) and sheets (facets) assembled into distinct forms. These rods and sheets have specific properties for behaviour of growth rates, growth directions, and capacity to separate from one another. The deployment of these structural elements and their combination of states becomes the architecture of the cell as well as a co-ordinated array for generating specific types of growth. This combination of structure and states has been termed positional specificity: each vertex, edge and facet has both structure (occupancy of space) and one of several different states for growth or separation. This positional specificity is the basis for persistence of cell type through proliferation, transformation from one type of meristematic cell to another, and the emergence of new co-ordinated structure during differentiation.

Form in differentiated cells

General cell types

Three basic types of proliferating cell can be recognized in plant tissues (Fig. 2.5A–D). (1) Elongating cells grow in only one dimension because the end facets expand little, if at all, and the side walls expand anisotropically only in the direction of organ growth (Fig. 2.5D). (2) Epidermal cells enlarge mostly in two dimensions as the periclinal (top and bottom) facets expand isotropically while the anticlinal (side) walls grow in only one direction (Fig. 2.5B). (3) Finally, parenchymatous cells enlarge in three dimensions because all walls expand isotropically (Fig. 2.5C).

Each type of cell can be described by the type of elements of which it is composed. Facet elements may be assigned different states (A, B, C, etc.); edge elements are assigned different states (a, b, c, etc.); and vertices are also of different states (α, β, γ, etc.). Growth can be described by the formation of filaments along these elements. For example, filaments are initiated at vertices type α and grow along edges type a ($\alpha \rightarrow a$); branches from these filaments are formed at edges a and extend over facets type A

($a \rightarrow A$). Other rules can be assigned for other vertices, edges and facets (Fig. 2.5B–D).

These rules include not only restrictions for growth but also rules of persistence of cell type during cell division, transformations from one type to another and for changes during cell differentiation. Through the combination of the ontogenetic, or cell lineage, deployment of these states and their associated nuclear states (I, II, III, etc.) the system is parallel to the L-systems of Lindenmayer (1968, 1978). While it is difficult to demonstrate the existence of these states, cells can be clearly identified as to types (Fig. 2.6A, C, D), and the separate existences of certain elements are realized (Fig. 2.6B).

The simplest types of differentiation can be seen in parenchymatous cells (Fig. 2.9A, p. 46). Usually, chlorenchyma is composed of cells which have partially separated from one another by the specific separation of vertices, edges and some facets to form small intercellular air spaces. If the nuclear state changes ($III_p \rightarrow III_c$) then the wall, vertex and edge states

Fig. 2.5. Basic types of cells as found within a shoot apex (A). Epidermal cell (B), parenchymatous cell (C) and elongating cell (D) with elements specified and rules of growth designated.

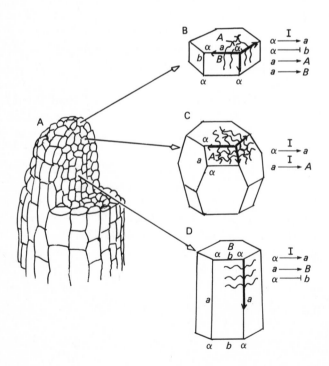

Fig. 2.6. Cell types. A, Epidermal cells of *Crassula argentea*, × 120. B, Epidermis of *C. argentea* peeled away leaving edges of epidermis remaining on subepidermal chlorenchyma, × 120. C, Flat, elongating files of cells in *Typha latifolia* root, × 60. D, Egg-beater glands of *Salvinia rotundifolia* with long, elongating cells of the stalk, × 40.

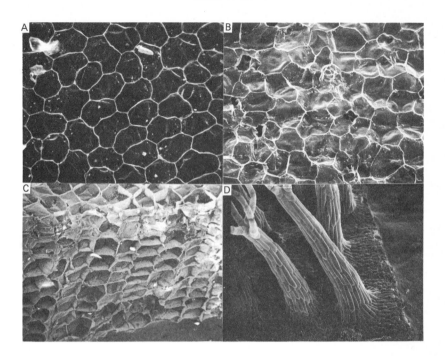

change as a result (*A→C, a→c, α→γ*). Despite the variety of shapes that these cell can assume, the change in states of their elements produces a co-operative change in cell pattern.

Stellate cells

An interesting and complex cell-form is that of stellate cells (Figs. 2.7, 2.9B). These cells are derived from flat, elongating cells arranged in files (Fig. 2.6C). They have come about by rapid division of elongating cells and, according to the assumptions with respect to positional specificity, the top and bottom walls would be of one type (*A*) and the lateral walls of another (*B*). Transformation of elongating cells into stellate cells can be explained by the elongating cell walls serving as a pre-pattern for the change (Fig. 2.7).

Specific rules governing this transformation would be as follows. Facets of type *A*, along with edges *a* and *b*, as well as vertices α, separate from similar elements in adjacent cells leaving cells held together only by facets of type *B*. Next, facets of type *A* expand radially while intact facets of type *B* do not grow; consequently the cell is directed into the typical star-shape configuration. Each cell averages twelve arms since there was an average of twelve side walls of type *B* which remain intact during the differentiation process.

Fig. 2.7. Development of stellate cells from stacked, flat elongating cells (A) by separation of end facets and all edges and vertices (B), followed by enlargement through growth of lateral walls (C).

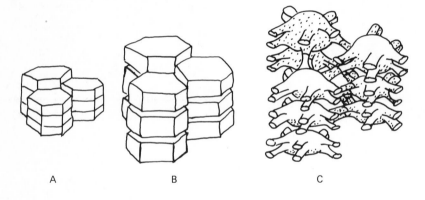

A   B   C

Apical cells

Special meristematic cells identified with organ development in mosses and ferns are apical cells. These entities are isolated sites of active growth and cell division in which the latter is characterized by a distinctive type of asymmetric division. One common type of apical cell is a four-sided pyramidal, or tetrahedral, form with three cutting faces and a superficial facet as in moss stems, fern stems, and several fern roots. In most fern root apices all four facets are cutting faces. Two features of asymmetric division in apical cells are a directionality to the sequence of cuts, counterclockwise or clockwise, and the division itself is gnomonic, that is, one daughter cell is of the same shape as the mother cell. This new pyramidal daughter cell is the new apical cell in the lineage during organ development and the other daughter cell is a five-sided, sector-shaped cell that passes through a series of divisions to form many vegetative cells, or a merophyte (illustrated in Fig. 2 of Chapter 10).

The rules governing the pattern of cell division, which also explain cell shape, can be explicitly stated (Fig. 2.8). In cases with three cutting walls, each wall is of a different age as measured in cell generations and these three walls are specified according to their age. During division the old wall (O) is removed into the merophyte initial, the medium-aged wall (M) becomes an old wall (O) and the new wall or recent cell plate (N) becomes the medium-aged wall. With old walls serving as the cutting face, the directionality of divisions is established. As the walls are specified according to age, so the vertices can be uniquely recognized. The vertex bordering the N and M walls as well as the superficial facet could produce a set of three filaments that extend along the three edges. The ends of these

Fig. 2.8. Pyramidal apical cell (A) with three cutting faces. A, vertices (●) and edges (—) of cell indicated along with old (O), medium-aged (M) and new (N) facets designated. B, hypothetical method whereby the vertex between N and M facets sends out filaments (arrows) along its three contiguous edges. At their termini a pre-prophase band (---) encircles cell. C, cell arrangement after four division cycles indicating consistent counterclockwise partitioning pattern. Numbers refer to the sequence in which the cells were produced.

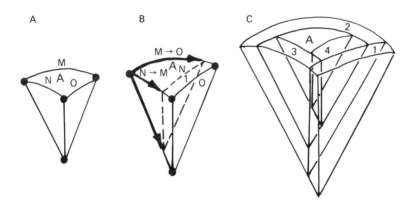

filaments serve as the initiation sites for the pre-prophase band formation and, hence, the placement of the cell plate (Korn, 1982). In the cases of apical cells with four cutting facets, the mechanism is modified as to where walls of four ages, N, M1, M2 and O, are programmed uniquely by the cell.

Complex tissues

Many tissue patterns are established by a series of changes involving two or more pathways of differentiation. In the most complex reorganization of meristematic cells, perhaps as in the formation of vascular bundles, many cell fates exist, but there are also cases where as few as two or three fates are possible indicating that there is a wide range in the degree of complexity of tissue specialization.

A relatively simple tissue pattern is specified by heterogenetic induction (Bünning & Sagromsky, 1948) which unfortunately describes a hypotheti-

Fig. 2.9. Cell types. A, partially separated chlorenchyma of *Crassula argentae*, × 50. B, stellate chlorenchyma of the mature leaves of *Dryopteris thelypteris*, × 270. C, aerenchyma of the petiole of an *Aponogeton* sp., × 60. D, distribution of hairs on the adaxial epidermis of the petal of *Salvia splendens*, × 120.

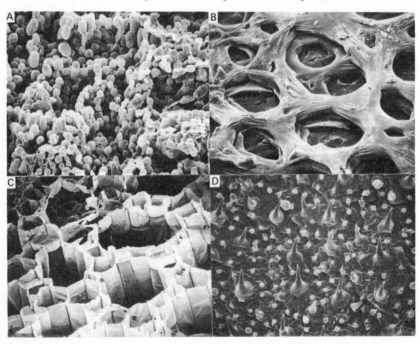

cal mechanism rather than the arrangement of cell types. It is characterized by a checkerboard pattern of two types of cells: general epidermal cells and either stomates, hairs, glands or fern archegonia (Korn, 1981). Hence, it appears to be of universal exploitation in plants.

The concept of heterogenetic induction has been modified by the author (Korn, 1972, 1981) as the random induction of a cell to become specialized followed by inhibition of the adjacent cells from becoming specialized. Computer models of this scheme adequately parallel observed patterns of stomates and hairs.

It is difficult to give a description of this pattern (Figs. 2.9D, 2.10) but it can be described algorithmically either by computer printouts of cell arrays or by recursive algebraic equations. In the method of recursive formulation, each general, or free, cell (F) becomes a specialized cell (S) at a certain rate (frequency per unit time, or iteration), a. Hence the equation is

$$\Delta S_i = aF_{i-1}$$

In words, the increase in frequency of new specialized cells at one iteration is the induction rate, a, times the frequency of free cells at the previous

Fig. 2.10. Schematized drawing of the arrangement of cotton fibres (●) in a field of epidermal cells on the ovule. $R = 1.67$ when density is measured in fibres per 100 cells and $R = 1.55$ when density is measured in fibres per 10 000 μm^2.

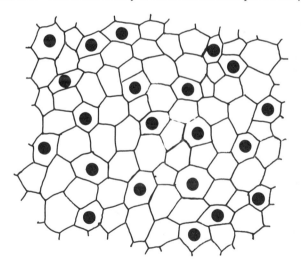

48   *R. W. Korn*

interval. The frequency of newly inhibited cells (I) adjacent to the new specialized cell becomes

$$\Delta I_i = 6aF_{i-1} \times aF_{i-1} = 6a^2_{i-1}F^2_{i-1}$$

using 6.0 as the coefficient because there are, on an average, 6.0 adjacent cells to any cell. The new frequency of free cells then becomes

$$F_i = F_{i-1} - \Delta S - \Delta I.$$

With the value of a taken from 0.0001 to 0.05 the frequency of specialized cells ranges from 0.37 to 0.05. The complexity of the description is not in following the appearance of the two types of cells but seeing it in multiplicity with considerable overlap of the basic seven-celled pattern. This complex pattern when measured by the Clark & Evans method gives an R value around 1.65, similar to that for cells.

Glands

A slightly more elaborate pattern of development is that for gland formation, particularly as found in leaves of most mints (Fig. 2.11). While the arrangement of glands follows a heterogenetic pattern, the ontogeny of separate glands is of interest, too. A general epidermal cell begins to have enlargement of the superficial wall followed by an asymmetric division in which the cell plate attaches to the superficial wall to form a large basal cell and a small terminal cell. This basal cell again divides asymmetrically with the plate attaching only to the remains of the superficial wall to form a new

Fig. 2.11. Schematized drawing of the development of glands in various *Mentha* species. Atypical divisions in C and D, but typical divisions in E, characterize the ontogeny.

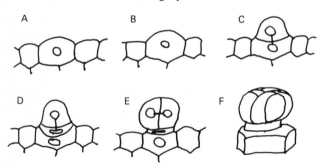

basal cell and a stalk cell. This division is most peculiar as the cell plate is strongly convex in order to attach to the region of the superficial wall. The terminal cell enlarges greatly and passes through three successive cell cycles to form eight cells, each approximately an octant of a sphere.

The algorithmic rules according to positional specificity can be stated simply as (1) the first two divisions result in cell plates that have freedom to shift and can only attach to the superficial wall, call it of type *C*. The final three divisions follow the four rules specified earlier.

Aerenchyma

Among the most visually complex tissues are the variety of aerenchymas in plants. One common pattern is that found in petioles of *Aponogeton* spp. (Figs. 2.9C, 2.12). Large air spaces are laterally bordered by files of elongating cells and terminated at the ends by diaphragms composed of small cells with intercellular pores.

Fig. 2.12. Development of aerenchyma of the petiole of *Aponogeton elongatus* in a centripetal (left to right) direction. Stage A is the stacked, flat elongating cell; B is the period of all vertices opening; stage C occurs when transverse division takes place; D is characterized by radial divisions and divisions to form diaphragm initials; and E completes the ontogenic sequence with cell elongation and division in diaphragm cells.

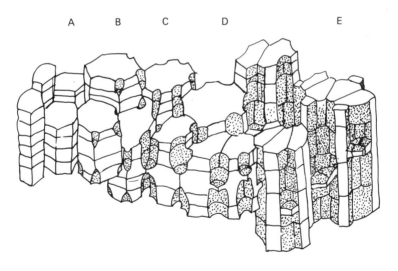

Ontogenetically this complex pattern is derived from files of flat elongating cells as were stellate cells. First, all vertices open up into small air spaces bordered by six cells. The walls of cells facing these air spaces expand at a faster rate than did the original walls, resulting in enlarging air spaces. Cell divisions are atypical in that cell plates attach only to walls facing air spaces, and these divisions account for the large hexagonal cells bordered by three smaller quadrilateral cells. The original lateral walls remain intact but grow out as diaphragms, and highly asymmetrical divisions produce very large columnar cells and very small triangular diaphragm initial cells.

By interpretation of this developmental sequence according to positional specificity of the cell elements, the pattern can be explicity explained. Vertices change their type to one that separates from adjacent vertices ($\alpha \rightarrow \gamma$). These vertices become new types of walls ($\gamma \rightarrow G$) that undergo rapid expansion and are specific for cell plate attachment. Lateral edges, of type a, of the original elongating cell, become a new type, g, which attracts mitotic spindles, resulting in the highly asymmetrical divisions for diaphragm initiation.

While it may be possible to explain this sequence of spatial changes by tension forces originating from the expanding epidermis, the peculiar manner of cell plate orientations and spindle alignments appears to require more directional control than that which physical forces might create.

Hierarchy of form

The advantage of developmental algorithms for the description of cell, tissue and organ geometries is in the vast variety of forms that can be addressed. Traditional approaches by geometry and simple non-recursive equations treat only standard forms, i.e., hexagons, tetrakaidecahedra, arcs or parabolas. This is perhaps why ideal forms became central to understanding botanical geometry rather than accepting the observed range of variations.

Another advantage of developmental algorithms for describing shapes is that it is inherently an hierarchical approach while the traditional methods are single-levelled. The lowest level of form is the macromolecular which leads into cell elements of facets, edges and vertices, then into cells as polyhedra, then into tissues, up to organs and finally the individual plant. This ascendency is a step-wise continuum – step-wise in that levels are discrete and non-overlapping, and a continuum in that features at one level can be traced explicitly into the next higher level.

The hierarchy of plant form must be seen from two perspectives. One view is the vertical, or truly hierarchical, in that components build structures which become super-structures. A paradigm of molecular biology is that assembly is clearly directional; order emerges in an upwards fashion in a definite analytical framework. This position also seems to hold for plant structure and development, but one must be cautious in drawing conclusions about multicellular systems that seem to hold for macromolecular complexes.

The second view is a lateral one. Those components that assemble spontaneously together in a growing or dividing cell may have different origins; different activities occur at various sites within the cell. When distances an order of magnitude greater than cell size are considered, i.e., simultaneous events in adjacent tissues or organs, co-ordinated assembly is no longer possible by chance interaction because of membrane barriers and a low probability of interaction of molecules in low concentrations. At these distances carriers of information become necessary and the roles of plant-regulating substances become critical (see Sachs, this volume).

Were one to simulate the entire structural development of a higher plant by computer graphics, positional specificity would generate much, but not all, of the desired form. Both short- and long-distance controls and interactions become important as two different kinds of phenomena. Algorithmically, this may become the ultimate challenge to the plant developmentalist.

References

Bünning, E. & Sagromsky, H. (1948). Die Bildung des Spaltöffnungsmusters in der Blattepidermis. *Zeitschrift für Naturforschung,* **3b,** 203–16.

Clark, P. J. & Evans, F. C. (1954). Distance to nearest neighbor as a measure of spatial relationships in populations. *Ecology,* **35,** 445–53.

Dormer, K. J. (1980). *Fundamental Tissue Geometry for Biologists.* Cambridge: Cambridge University Press.

Kelvin, Lord (1894). On homogeneous division of space. *Proceedings of the Royal Society of London,* **55,** 1–16.

Korn, R. (1972). Arrangement of stomata on the leaves of *Pelargonium zonale* and *Sedum stahli. Annals of Botany,* **36,** 325–33.

– (1974). The three-dimensional shape of plant cells and its relationship to tissue growth. *The New Phytologist,* **73,** 927–35.

– (1980). The changing shape of plant cells: transformations during cell proliferation. *Annals of Botany*, **46**, 649–66.

– (1981). A neighboring-inhibition model for stomate patterning. *Developmental Biology*, **88**, 115–20.

– (1982). Positional specificity within plant cells. *Journal of Theoretical Biology*, **95**, 543–68.

Korn, R. & Spalding, R. (1973). The geometry of plant epidermal cells. *The New Phytologist*, **72**, 1357–65.

Lewis, F. T. (1923). The typical shape of polyhedral cells in vegetable parenchyma and the restoration of that shape following cell division. *Proceedings of the American Academy of Arts and Science*, **58**, 537–52.

Lindenmayer, A. (1968). Mathematical models for cellular interactions in development. *Journal of Theoretical Biology*, **18**, 280–99, 300–15.

– (1978). Algorithms for plant morphogenesis. *Acta Biotheoretica*, **27**, *supplement* **7**, 37–81.

Loeb, A. (1976). *Space Structures: Their Harmony and Counterpoint*. Reading, Mass.: Addison-Wesley.

Matzke, E. (1946). The three-dimensional shape of soap bubbles in foam – an analysis of the role of surface forces in three-dimensional shape determination. *American Journal of Botany*, **35**, 323–32.

Thompson, D'A. W. (1942). *On Growth and Form*, 2nd edn. Cambridge: Cambridge University Press.

3

Analysis of axis extension

PAUL B. GREEN

The issue of positional controls in plant development must be examined within some concept of the nature of plant growth. The best known frame of reference, that of gene expression, deals with a chain of causation leading from DNA sequence to protein and to self-assembly. The nature of the pertinent levels of molecular structure, and the mechanism for transition between them, are relatively well known. In extending this frame of reference to development at large, two complications arise. First, there are obviously more levels beyond protein and assembled protein structure, and second, developmental activity is heterogeneous over long distances in space. Specific geometrical progression on a large scale is evident in the rigorously-inherited phyllotactic patterns, tree branch pattern, etc. Even the cylindrical corn root displays a consistent complex pattern of cell activity and histology along its extensive growth zone. What frame of reference is suitable to connect such activities to the familiar processes of transcription and translation?

We will connect morphogenesis to the gene by an analysis which starts at the phenotypic end of the causal chain. The usual developmental phenotype is a time-course of chemical and geometrical change. The progression is complex yet reproducible. This time-course is regarded, literally, as an integration in space and time. Thus a reasonable way to start a series of links from such a progression 'back' to conventional gene activity is first to differentiate the time-course into the local activities that generate it. For plant morphogenesis this means to characterize, and to measure, the local behaviour in organs (e.g., rapid directional extension in a root meristem). The second step is to analyse the regional behaviour in terms of the appropriate cell physics. The cell physics in turn will have a basis in cytology and metabolism. These latter two processes finally bring us within range of conventional mechanisms of gene expression. Thus one can proceed from the phenomenon itself through ever more distant causes until a chain is built to the genome. When read from the phenotypic end, this chain is the recommended sequence for analysis. When read from the genome outward, it is a causal chain. It is only within an extended frame of reference of this sort that controls on large-scale development can be comprehended.

This chapter will present several such hierarchical analyses. In such treatments one must settle for successive levels of explanation, each explanation being complete within its own frame of reference but leading to questions that can be settled only at the next level. First we will consider the gross basis of extension, then cell patterns in elongating organs, and finally morphogenesis.

Axis extension

Elongating cylindrical plant structures (roots, stem internodes, hyphae) constitute a prominent developmental phenotype. The first step in analysis is to differentiate the activity in space and time; that is, one must characterize the local behaviour of the growing structure. Even crude marking experiments reveal two basic modes of plant elongation. In one, activity is restricted to a rounded terminal region; this is tip growth. In the other, growth activity takes place along a cylindrical growth zone; this is diffuse extension growth. These two modes will be treated in turn. Several steps on the path from morphogenesis to the genome are evident in tip growth. More complete analysis is available for diffuse extension.

Tip growth

In tip growth, cylindrical shape is extended by localized activity at the structure's dome-like tip; the cylindrical part is inactive in growth. Because of the radial symmetry, extension of volume may be simplified to the issue of the extension of surface. If a dome-shaped region shows a gradient of surface expansion that falls in a certain way from a maximum at the dome's tip to zero at the doine's base, the net result of its activity will be (a) continuous production of new cylindrical structure at the base of the dome, and (b) preservation of the size and shape of the generative dome.

As shown in Fig. 3.1a, one can visualize this gradient in local behaviour by considering the performance of a small circle of surface, initially put just off the tip of an idealized cell. The circle will get relatively bigger very rapidly at first, but then its relative rate of increase will fall conspicuously. The rate reaches zero as the patch approaches the equator, the base of the dome. The patch undergoes a relative migration from the dome onto the cylinder because of the growth between the patch and the tip of the dome. During the change in relative position the circular patch remains circular as it increases in size. Growth is viewed as isotropic. Details of the specific gradient in rate can be found in Green & King (1966) and Erickson (1976).

Trinci & Saunders (1977) point out that the tip of a hypha is not

Fig. 3.1. Models of the two major forms of extension. (a) In idealized tip growth, surface expands only between the top of the dome and the equator (EQ.). Successive stages of development are shown for a small circular patch, initially just below the tip. The vertical or meridional (v) and horizontal (h) relative rates are always equal, so the circle stays a circle. The relative rate of growth in each direction falls from a high value at the tip to zero at the equator. The rate at several positions is shown by the crosses and by the graph below. The cross is the differential, describing current activity. The circles show the integral, or progressive effect of previous growth. (b) In simple diffuse growth of a cylinder all regions behave alike and retain their relative positions on the axis. At all times and at all locations the vertical relative rate of growth is a fixed multiple of the horizontal rate. This is shown by the crosses and the graph. The crosses reveal what is happening at the moment. They show the principle axes of the strain-rate ellipse (Fig. 3.2). The open ellipses show the cumulative effect of growth (integral). The constantly biased local growth converts a circle to an ever-more elongate ellipse, and elongates the cylinder.

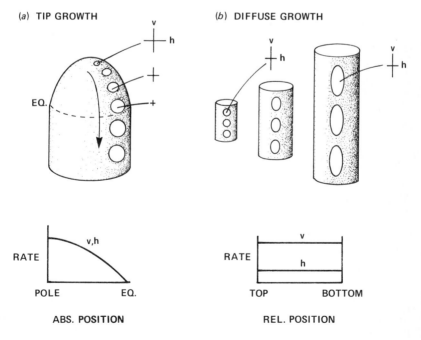

hemispherical. They give gradients for a more appropriate outline – that of half an ellipse. The assumption of growth as isotropic is also a simplification. Stretch is somewhat transverse at the base of the dome, but this does not alter significantly the idea that tip growth is a function of local growth rates.

To summarize the first level of explanation, tip-growing cells such as root hairs, hyphae, and pollen tubes extend their cylindrical form by maintaining a specific gradient in the rate at which area expands within the growth zone; this rate falls, generally as a cosine or cotangent function, to zero at the base of the zone.

The biophysical basis for the gradient and the cessation of growth at the base of the dome is not obvious. The stresses tending to enlarge the wall are actually greater there than at the tip. The cell wall cross-section shows no increase in thickness or density in the electron microscope, which could explain the halt of growth. One suggestion is that the gradient and the ultimate cessation of growth result from two contrasting properties of the carbohydrate-containing vesicles which fuse with the wall to enlarge it (Green, 1973). The vesicles would fuse preferentially at rapidly-expanding regions. Upon arrival the carbohydrate would 'loosen' the wall, promoting extension and hence the fusion of still more vesicles. This positive feedback activity enhancing growth would be checked generally by an upper limit on the rate of vesicle production and locally by the strain (deformation) hardening properties of the carbohydrate once it is in the wall. The cumulative stiffening effect would increase with distance from the tip to bring on stoppage at the base of the dome.

The paradox that expansion is rapid in regions of low stress (low apparent driving force) while it is slow in regions of high stress could be explained by the above causal loop involving strain rate. Strain would affect both the rate of incorporation of new materials and the subsequent stiffening of the wall as it is stretched.

A different suggestion is that of Hejnowicz, Heinemann & Sievers (1977) who consider the inverse correlation between local expansion and stress to be explained by a loop involving stress directly. There would be high synthesis rate (and growth) at regions of low stress, cessation of synthesis at regions under high stress. The main difficulty with this idea is that local stress level, while readily calculated, is in practice 'sensed' or transduced only via the strains it causes. Strain, on the other hand, is readily transduced by the thinning or spreading of structure as in conventional strain gauges.

A related suggestion by Koch (1982) couples high synthetic activity to apparent low surface tension at the tip, while low synthesis is found in regions of high tension. How tension is to be sensed independently of the ongoing strain is again a problem. In brief, the inverse correlation between stress and growth rate is an attractive basis for mechanism. It is hard, however, to envisage the details of how cytoplasmic activity can be coupled to stress *per se* (divorced from strain).

The cytology of tip growth, including the geotropic response, is well covered by Sievers & Schnepf (1981).

The mechanism by which intracellular growth gradients for tip growth are first set up is relatively obscure. The most incisive work is that of Jaffe (1982) on the initiation of rhizoid growth in the zygote of the seaweed *Fucus*. His idea is that the tip is a site of ion-leaks; these allow calcium to enter the tip. The calcium is immobilized there and its positive charge is thought to draw vesicles, by self-electrophoresis, to the tip. These vesicles would contain not only the wall substance to generate new wall volume but also new ion-pores. This view of tip growth as a self-sustaining loop process, which would bring on the appropriate rate gradient, is a major conceptual advance; it is the third, or cytoplasmic, level of explanation. Positional controls, relating to the initiation or variation in tip growth, would have to operate on components of such loops. A striking control by light is found in branch initiation in the alga *Vaucheria*. Here localized blue light initiates new current loops and ultimately a new lateral axis (Weisenseel & Kicherer, 1981).

At the level of organs, there are few well studied examples of tip growth. The trunk of a palm tree appears as a massive cylindrical assemblage of leaf-bases. Assuming these leaf-bases have changed their shape little since being produced at the stem apex, the trunk is in fact a collection of such uniformly enlarged structures. The parallel with the root hair should be evident. The hair is a collection of enlarged circular patches (conceptual only, not evident to the eye). The palm trunk is a collection of bases of enlarged leaf primordia. In both cases the effective extension of cylindrical form – the first level of explanation – is by a gradient in the rate of expansion of initially small structures, the rate reaching zero near the tip of the growing cylinder.

Diffuse extension growth

In this mode of growth the elongation of the cylindrical structure is explained, at the first level, by the fact that the surface (and volume) of the structure shows directed extension all along the cylindrical axis (see Fig. 3.1*b*). This growth is characteristic of typical plant cells in tissues

(Wilson, 1964) and of the internode cells of *Nitella*. It stands in contrast to the tip growth generally seen in superficial or hair cells. The successive levels of explanation of diffuse extension are shown in Fig. 3.2.

The first step in analysis, characterizing the local behaviour, is easy in

Fig. 3.2. A series of levels linking elongation to the genome. (a) is a progression of form generated by integration from a constant strain-rate ellipse, (b). The asymmetry of the ellipse can be accounted for, algebraically, by cell physics involving circumferential hoop reinforcement, (c). The directional cellulose reinforcement is tied to the alignment of microtubules, (d). Microtubules are polymers of tubulin, (e), which arises in the usual way from DNA, (f, g). Read from top to bottom, this series is an analysis of a developmental progression. Read from bottom to top, following the arrows, it gives the essentials of a causal chain. (Reprinted with permission of Alan R. Liss, Inc.)

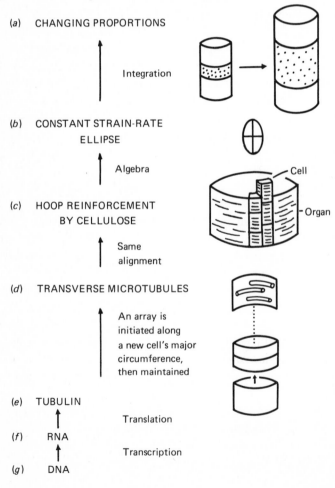

(a)   CHANGING PROPORTIONS

                    Integration

(b)   CONSTANT STRAIN-RATE
              ELLIPSE

                    Algebra

(c)   HOOP REINFORCEMENT
           BY CELLULOSE

                    Same
                    alignment

(d)   TRANSVERSE MICROTUBULES

                    An array is
                    initiated along
                    a new cell's major
                    circumference,
                    then maintained

(e)   TUBULIN

                    Translation
(f)   RNA
                    Transcription
(g)   DNA

Nitella. Here, marks can be applied to the exposed cylindrical surface; their separation shows that length increases at four to five times the rate of increase in girth. This is true for all regions of the cell and is constant with time. The ratio of the two rates can be made precise when each is characterized as a compound interest rate (relative rate), e.g., length increases at 5% per day, girth at 1% per day. The gross course of cell elongation is clearly the integral of this biased, or directional, expansion.

The most reasonable explanation of the oriented growth is that the cell wall is highly directionally reinforced by cellulose microfibrils (see Frey-Wyssling, 1959). A cylindrical cell with high internal pressure would be expected to enlarge into a sphere unless its wall were specifically reinforced to resist swelling. The *Nitella* wall, and most cortical cell walls in higher plants, have a predominantly transverse arrangement of cellulose microfibrils which almost certainly plays the necessary reinforcing role. Thus directional extension is explained by transverse reinforcement in the wall.

The third level of mechanism, accounting for the oriented reinforcement, involves the cytoplasm. Oriented deposition of cellulose is associated with microtubules of similar orientation (see Gunning & Hardham, 1982). The microtubules probably act on cellulose synthesis through an action on membrane rosettes that actually extrude the cellulose (Staehelin & Giddings, 1982). At any event, reagents which disrupt microtubules generally disrupt cellulose orientation. It follows that the initiation and maintenance of a transversely aligned microtubular array is essential to directional extension. The array, consisting of overlapping short microtubules, arises from cell edges after mitosis and, in growing cells, generally keeps its orientation during interphase. During later differentation, as of stomata, abrupt changes can occur without mitosis (Palevitz, 1980).

The next deeper level of explanation, the basis of the oriented microtubular array, is important but relatively obscure. Clues come from organogenesis. During organ formation the direction of the arrays in a tissue can change abruptly. When this happens the new direction is frequently seen in pairs of new daughter cells after a mitosis of unusual orientation (Green & Lang, 1981). In these cases the long axis of the daughter cell coincides with the new microtubular alignment. In such a configuration new microtubule arrays would suffer the least sharp bending (i.e., would have minimal strain energy) because they follow the longest circumference that can accommodate the whole array. In the established meristem where the cells are short and broad the microtubules also follow the long line of girth of the short cells. In other cases, however, particularly in the root of

Azolla, microtubules are transverse regardless of the proportions of the cells or the orientation of cell division (Gunning, 1981, 1982). This behaviour could reflect a tendency for new arrays (a) always to be parallel with the mother cell's array, or (b) to be aligned with arrays of neighbouring cells. In any case, the factors influencing the orientation of the microtubular array appear to be the critical cytological ones to explain diffuse extension growth. This brings the chain of analysis to the organelle.

The still more remote levels of explanation are, fortunately, much less elusive. The best-known controls in this sequence act via microtubules which are made of tubulin (see Gunning & Hardham, 1982), a well-known protein produced *via* RNA and DNA. The alkaloid colchicine causes disappearance of microtubules. The ramifications of this loss, leading to random cellulose synthesis, bring on swelling (Taiz, Métraux & Richmond, 1981). The natural growth regulator, ethylene, causes a shift to longitudinal orientation of microtubules (Lang, Eisinger & Green, 1982). Swelling results, but by somewhat different physical mechanisms.

It is thus possible to deal with diffuse extension growth through a causal chain of seven explicit levels running from the observed phenomenon, elongation, to DNA. Only the most pertinent features have been pointed out. Countless additional activities, such as ion uptake and carbohydrate metabolism, are necessary. The chain shown in Fig. 3.2 concentrates on the morphogenetic aspects.

To sum up, the elongation of most plant organs is the result of many hoop-reinforced cells carrying out diffuse extension growth. Such cells constitute the organ's growth zone. The cytoplasmic basis of the longitudinal growth is found in the transverse microtubule array which governs cellulose alignment. It is on the microtubules that the best known controls, colchicine and ethylene, have their effect.

Examples of organ extension

The shoot of the alga *Nitella* and the root of the corn plant offer useful examples of organ extension based on diffuse extension growth. The appropriate way to describe and measure local behaviour in the two cases will be presented first. A valid description is of course pre-requisite to proper explanation and assessment of the role of controls.

Several major features are shared by the two organs. In both cases cell division is restricted to a region near the tip; in both cases growth zone activity ceases when cells reach a large, generally fixed, size.

The Nitella *shoot*

This plant (Fig. 3.3) is generated with great parsimony in cell division; it consists mainly of the giant internode cells discussed above. Of special interest is the way that the alternating pattern of leaf-whorls and inter-nodes is generated at the shoot tip. The organ terminates in a single apical cell. The whorls of laterals (leaves) originate from a single cell called a node cell; the internodes are always single cells. It would seem to be a foregone conclusion that the shoot is formed by the apical cell, (a), simply producing first a node cell (n) and then an internode cell (i). The flow scheme, or algorithm, which would suffice is:

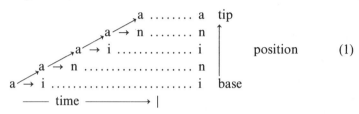

This yields the necessary string of alternating nodes and internodes. Surprisingly, this is not the actual developmental sequence. The reason may be that the above alternating scheme requires that the apical cell must in some way remember which of the two cell types it had produced at its last division. This requirement, apparently a formidable one for an alga, is by-passed in the actual development. A fourth cell type is in-volved. The apical cell *always* enlarges and divides to produce another apical cell plus a transitory segment cell (s). The segment cell divides to make an upper node cell and a lower internode cell, the segment cell disappearing in the process. Thus each cell type always does the same thing. The actual flow scheme is:

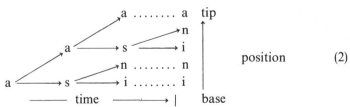

which also yields alternating nodes and internodes (Fritsch, 1948).

Most development mechanisms achieve results by the proper coupling of relatively simple steps. In this case the second scheme has more steps to it, but these steps are less demanding. There is an appropriate analogy to modern computers which achieve their ends by the judicious coupling of the simplest of all possible decisions (on *vs* off, or 1 *vs* 0). Hence it is appropriate to consider the encoding of developmental schemes as equivalent to reduction of a sequence to machine language.

Because the full elongation of the internode cell takes up to ten days, while the apical cell divides about every three days, there will be a gradient of ever-longer internodes below the apical cell. The gradient ends where

Fig. 3.3. A *Nitella* shoot. The main axis is a series of giant internode cells. Nodes consist of six 'leaves' and one or two branches. This is clearly seen at the bottom node. The alternating nodes and internodes are produced from single apical cells at the tips of the shoot and branches. Scale bar is 3 cm. (From Taiz *et al.*, 1981, with permission.)

these cells reach their final length. This is the typical internode pattern for a higher plant shoot. It requires no special explanation, or controls, other than that the development of the stem segment be long relative to the cycling time for the apical cell. In higher-plant shoots this means that internode development must take several plastochrons (a plastochron is the time between successive leaf initiations).

It is a striking feature of *Nitella* that the leaf axes also arise from single apical cells. These cells follow sequence (2) with two changes: (a) the leaf apical cell is smaller, and (b) it divides only a few times. Perhaps these differences are connected. The small apical cell may lose developmental potential, in a way that big apical cells do not, at each division. Chemical suppression is contra-indicated by the fact that leaves do not readily revert to shoots upon fragmentation of the plant. The programmed termination of leaf outgrowth is essential to the generation of normal form. If all six leaf apical cells at each node were to operate indefinitely in the style of a shoot apical cell, there would be an excessive proliferation of organs. This would yield a dense ball-like tangle of cells rather than the typical lacey configuration.

Finally, the production of true branches (laterals of indefinite growth) involves a secondary swelling of certain cells in the basal node of the leaf. This is analogous to axillary bud formation in higher plants.

In brief, it is seen that the basic form of the *Nitella* shoot can be explained by simple recursive behaviour with regard to cell division and cell extension. Much of this behaviour appears to be intrinsic to the cells in question, as against being based on exogenous controls. This puts the burden of explanation on cell behaviour itself and on the steps that generate specific cell types.

The corn root

Roots are generally devoid of the useful natural landmarks that allow one to visualize specific material regions of a stem or *Nitella* shoot. In the angiosperm root the only convenient geometrical frame of reference for development is distance from the tip, or from the junction of the root proper with its cap. The pattern of activity down this axis, fortunately, is constant for long periods of time. What are the most general activities? How can they be measured, so that controls can be studied? Here we will discuss two major processes, extension and division, along with the visible product of their interaction: mean cell length. This last can be thought of as longitudinal histological pattern, characteristic for each tissue in a root.

At a given time each cell along the root growth zone has a characteristic extension rate. For a given cell (or tissue element), the rate will change with time. For a given distance from the tip, however, the rate will remain constant. How can one describe the varying pattern in extension activity along the root axis in such a way that the activity of a given cell, through time, can also be measured? The problem of dealing with dynamic systems where the pattern of behaviour is constant while the units of behaviour pass through it, has been dealt with by other branches of science. The hydraulic equivalent is a fountain. Here the outline is constant, as is the velocity of water at particular heights in the fountain. The experience of a given drop of water is of course quite variable with time. The botanical solution to this general problem was initiated by R. O. Erickson decades ago (see Erickson & Goddard, 1951). Growth processes were treated as instantaneous rates occurring in infinitesimal regions. Later, Silk & Erickson (1979) and Gandar (1980) applied continuity principles to the problem. A simplified account will be given here.

The rate of extension at a given distance must be described as a compound-interest rate. Two points, close together and both on the axis, will separate somewhat over a short period of time. The relative rate of growth between these two points is found by taking the natural logarithm of their separation both before and after the time period. One then takes the difference of the logarithms. This difference between the logarithms is divided by the time interval to give the relative rate. Thus if the separation went from 10 to 15 μm in 10 h, the rate would be (ln 15 − ln 10)/10, 0.405/10 h, or 4.05% per h. Unfortunately, this rate would apply to the cell or region as it moved down a sizeable section of the root during the 10 h. The shorter the region and the briefer the period, the more useful is the measurement. By analysing many adjacent small regions, one can characterize the local extension activity along a growth zone (Figs. 3.4a, 3.6). When measured ideally, as Erickson & Sax (1956) did by another method, the rates are indeed exact for single points.

To get a valid rate for a given point at a given time, one needs to know the point's distance (from the organ tip) at three closely spaced times. A parabola is fitted to this three-point curve of distance *vs* time. The slope at the central point gives the exact velocity for the central point. This taking of slopes is repeated at many adjacent points to yield a plot of velocity *vs* position. The slopes on this second plot give the local relative (elemental) rate for each point.

The meaning of a relative rate is precise, but slightly abstract. The value 4.05% per h means that the length of a cell would be changed during one

hour by a factor equal to the antilogarithm of 0.045, namely 1.046. There
is no other terminology suitable for describing the activity at a point in a
growth gradient.

These relative rates can be converted to doubling times. When the rate
is divided into ln 2 (0.693), the result is the doubling time. Thus a rate of
0.045 h^{-1} becomes a doubling time of 15.4 h. Recall that this rate gave a
50% increase in 10 h; due to compounding, doubling comes in less than

Fig. 3.4. Graphs which allow complete characterization of cell extension in a root
growth zone with extension activity constant in time but varying down the root axis.
For example, a cell is located at d_1 with relative rate X. How long will it take for the
cell's length to double? Where will it be when its length has doubled? What will its
relative rate be then? In (a) the relative elemental (local) rate is given. The cell at d_1
has a relative rate X at this moment, but such rates, as a function of distance, do not
allow characterization of cell extension. To predict doubling time one first uses
graph (b), a distance *vs* time relation, to get t_1 an initial time. This time is then found
on plot (c), a graph of local relative rate vs time. A line is extended from the time axis
to the curve. With time this line sweeps out an area which is related to the fractional
change in the cell's length. The anti-logarithm of the area is the fractional change. In
this case the area is ln 2, so the cell has doubled at t_2. The doubling time is $(t_2 - t_1)$.
Using the same time-distance relation (now at graph (d)) one finds the new position
of the cell at t_2 to be d_2. Returning to (a), one finds that the cell is now elongating at
the higher relative rate, Y. In this way both immediate and long-term performance
can be assessed throughout the growth zone.

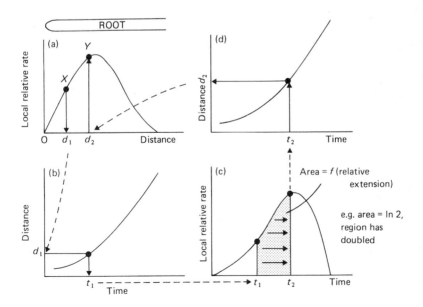

20 h. When the rate gradient down the root is gentle, such characterization is adequate. The rate would not change much during a brief measurement period. When the gradient in rate is steep, however, the cell's performance will vary with time. Under these conditions the rate value at a point tells only what a cell would do, were its performance to remain constant. This would be the case for cells at d_1 and d_2 in Fig. 3.4a. Their relative rates are X and Y.

To find out what a cell would do as it traverses a known steep gradient (described as relative elemental rates) one must take further steps. We will follow the growth of a cell at d, with initial relative growth rate of X (see Fig. 3.4a). First, the relation between position of a mark and the passage of time must be known (Fig. 3.4b). On this plot one picks the location of the cell in question, and then finds the time characteristic for that location. This plot is also used to regraph local relative rate against time (Fig. 3.4c). On this new plot of relative rate *vs* time one draws a vertical line between the initial time value on the x axis and the curve above. As time passes, this vertical line sweeps out an area which is a pure number (has no units). The meaning of this area is precise. It is the natural logarithm of the relative change in length of the cell. Thus, if one wants to find out how long it takes the cell to double its length, one sweeps out an area of 0.693 (ln 2). The vertical line is now at t_2. The interval $(t_2 - t_1)$ is the time required. By using the time and distance plot, Fig. 3.4d, t_2 is associated with d_2, the position of the cell after doubling. The cell now has a relative rate equal to Y. Thus the growth performance of any cell (or region), over any period of time, can be calculated from the appropriate set of graphs (see Green, 1976).

In analogous fashion one can deal with the rate of new cell production. Each region along the root has a characteristic ability to produce new cells. If a region has the value 0.693/10 h, this means that the region doubles its number of cells in 10 h, hence the cell cycle is 10 h.

With both the extension process and the division process expressed as relative rates, one can grasp the exact way by which these processes determine the variation in mean cell length for a given small region of tissue as it moves down the root (see Fig. 3.5). Whenever the mean cell length is constant in the region, the relative rate of extension must equal the relative rate of cell production. Clearly if a region doubles its length while it also doubles in cell number, mean cell length must be constant. Any fall in mean cell length must be the result of the division rate exceeding the extension rate. Any rise in mean cell length must be the result of the division rate being below the extension rate. The standing pattern of mean cell length down the root is also a function of the extension and partitioning processes. At each point the slope on the mean cell length curve is proportional to the difference of the two processes. The slope is also proportional to the quotient: 1/(mean cell length × velocity).

With the two major processes, extension and division, viewed as being combinable in all proportions, one can account for any kind of change in cell length dimension. Unfortunately this frame of reference, where the

Fig. 3.5. Characterization of a root growth zone for division, extension, and cell length. Extension occurs throughout the growth zone. Division accompanies extension in the meristem. In (a) the local relative rates of extension (solid line) and division (dashes) vary with distance from the tip. In (b) the changes in mean cell length are shown. This histological feature is related to the relative rates of extension and division. When the division rate exceeds the extension rate (slashed area), mean cell length must fall. When the rates are equal (constant or co-varying), cell length is constant. This is seen in the central region between dashed lines. When extension dominates (cross-hatched region), cell length must rise. Thus cell length pattern is controlled in a known way by two distinct, measurable processes. Technically the slope at a point in curve (b) is found by taking the difference in values in (a) and multiplying by mean cell length over the velocity of the tissue of that point. (See also the Appendix.)

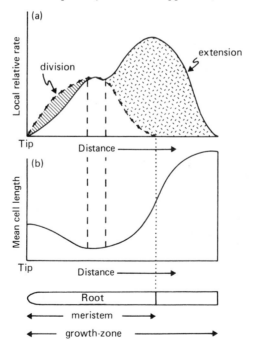

two processes can be combined in any ratio, is recent. Traditional botanical terminology refers to 'growth by cell division' and contrasts it with 'growth by cell elongation'. Such statements imply that the fully compatible processes are mutually exclusive. The confusion is generated by the fact that growth 'by cell division' *includes* the extension that occurs between divisions. Ambiguity is avoided if one simply converts the traditional duality to 'growth with cell division' and 'growth without cell division'; this keeps growth and division as processes which are distinct and superimposable. (The rates may vary independently.)

The two major histogenetic processes, division and extension, each have their own characteristic distribution down the root (Fig. 3.5). One may now ask, what controls the character of these functions? Very little is known, but since the functions represent valid quantitative pictures of what is going on, it is reasonable to speculate. The division process always involves protein and nucleic acid synthesis, while elongation has an emphasis on water uptake and carbohydrate synthesis. It could be that the maturing transport capacities in the vascular system deliver different substances with varying efficiency. Perhaps the pertinent carbohydrate or

Fig. 3.6. (a) The local relative rate of extension varies as shown along the axis of a growing wheat root. The area under the curve is equal to the overall elongation rate, e.g., 2 mm h^{-1}. (b) When auxin is added the elongation rate falls greatly (reduced area) but, at certain apical regions, the local rate of extension has been increased. This format reveals both the inhibitory and stimulatory effects of the control of elongation by auxin. (Redrawn from Hejnowicz, 1961.)

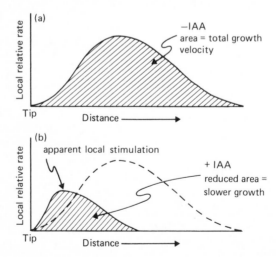

water is released effectively farther from the tip (extension rate peak) than are the nitrogen-containing compounds (division rate peak).

Growth rate itself influences cell differentiation pattern (Peterson, 1967). Rapidly growing roots (moist air) first make metaxylem cells at 47 μm from the tip. Slower roots (submerged) do so at 18 μm from the tip. The growth rates were 17.6 and 5.5 mm d^{-1} respectively. The relative advance in the site of differentiation was not the same for all tissues. The first phloem normally appears 278 μm from the tip. In the slowed root it was at 160 μm. Thus growth velocity (the area under the relative rate curve) acts on histogenesis. It would be interesting to know if this new positioning of cell types acts back on the relative rate curves for extension and division. The peaks in these curves may advance in accord with the terminus of specific cell types – as intimated in the previous paragraph.

A striking early study of hormonal control carried out in the modern format is that of Hejnowicz (1961). He studied the course of inhibition of wheat root growth by applied auxin. For any root, the graphical representation of relative growth rate *vs* distance from the tip embodies the overall growth rate of the whole root in the form of the area under the curve. This area naturally has the units of velocity, e.g., 2 mm h^{-1}. When auxin is added, the curve for elongation activity should show reduced area above the *x*-axis. This was indeed observed as in Fig. 3.6. A surprising finding that could only be made by using the current analysis was that certain regions of the root, defined by distance from the tip, actually showed stimulation by the hormone. Thus the inhibition, or control, by the hormone was brought about by a highly non-uniform action on the relative elongation rate pattern. (Possible changes in diameter were not studied.)

Analysis of persisting features in growing organs

Dynamics of persisting gradients

Virtually all higher plant axes contain longitudinal gradients of various sorts (protein or nucleic acid content) along the axis. The gradient persists, in fixed position relative to the tip, during long periods of growth. Such a fact is readily taken for granted until one recognizes that remarkable activity must be carried out to maintain such a dynamic gradient.

Silk & Erickson (1979) have pointed out that two processes are forever tending to obliterate, or smooth out, such gradients. The first is readily recognized as extension itself. Any substance in a growth zone is always being thinned out by the expansion of the area or volume containing it. Thus, one cost of maintaining a gradient is the synthesis to compensate for expansion. This cost is measured as a net synthesis rate. The synthesis rate is calculated by multiplying the region's relative rate of extension by the amount of substance. Thus, if a short region of a root has protein at a level of 10 mg/100 μm, and that region doubles its length every hour (rate is ln 2/time in h), then synthesis must proceed at 6.93 mg/100 μm h^{-1} solely to maintain this level of protein in the region.

There is a second, sometimes even more significant, factor influencing the maintenance of gradients. This is convection. All points in the growth zone are continually moving away from the tip. If there is an ascending gradient of anything, say carbohydrate content, with distance from the tip, then the growth process will simply tend to move the gradient out of the growth zone. For the gradient to be maintained, synthesis must compensate for this re-locating factor also.

The cost to combat convection can readily be appreciated from an analogy with a conveyor belt carrying sand. A single worker loads it continuously at one end. As long as the sand layer is to be level on the belt, the belt could be of any length with just one worker. Obviously if the belt speeds up and the same level is to be maintained, extra workers are needed at the start of the belt in exact proportion to the increase in speed. Back at the original slow speed, if it is desired always to have an ascending gradient all along the belt, clearly one must also have additional workers *all along* the belt. Furthermore, the steeper the gradient, the more workers are required per unit length of the belt. Thus, the cost of combating convection (movement of the belt) is the product of the velocity of the belt times the steepness of the gradient. (See Appendix for further discussion.)

Since it is a universal property of plant growth zones that (a) two marks on the axis of the growth zone always separate from each other (expansion), and (b) the pair of marks moves ever farther from the tip of the growth zone (convection), these two dissipating processes are always present. All persisting gradients must be maintained by synthetic activity to compensate for dissipation.

A beneficial by-product of this kinetically complex situation is that the rates of (net) synthesis required can be calculated on the basis of relative rate and velocity data, plus simple measurements of the gradient itself. Since the time component for the rate comes from growth analysis, a *rate* of synthesis of a substance is obtained from measurements of *amount* of the substance. This method can complement, or even supplant, difficult conventional methods of direct synthesis-rate estimation which use radioisotope incorporation and are susceptible to effects of pool size of the labelled precursor.

Dynamics of persisting curvatures

The careful application of the concepts of kinematics (above) to the bending of organs can lead to a grasp of 'what's going on' which often differs from one's intuition. For example, the highly curved hook of a bean seedling is another system where the configuration (the hook) is essentially constant while successive populations of cells pass through it. Since the radius of the hook is larger on the outside than the inside it seems reasonable to assume that the hook is maintained simply by a greater growth rate in the outside tissue.

Someone interested in control would therefore look for relative growth stimulation (e.g., a hormone) in peripheral cells. This mechanism seems inevitable because the periphery of a wheel moves faster, in absolute velocity, than the centre. The same is true of cars in a multi-lane roundabout or traffic circle. This concept of the basis of hook existence is quite misleading. Velocity is not the key issue, nor is the hook part of a circle.

Fig. 3.7. The bean hook retains its form as tissue moves through it. In (a) the overall progression and growth of a given disc of tissue is depicted. This shows the integral of the disc's growth. In (b) the progression is broken down into local behaviour differentials. The left side (1) of the hook shows predominant extension at the periphery of the organ (locally the tissue 'bends'). On the right side (2) a similar disc expands more on the inner side (locally the tissue 'unbends'). In (c) is shown the approximate distribution (shaded area) of relative growth stimulation (presence of stimulator, lack of inhibitor, or both).

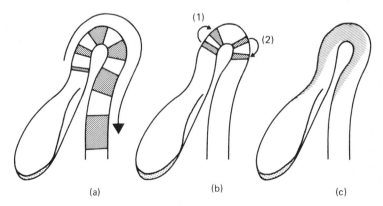

again the local behaviour, expressed in relative rates, is what is pertinent (see Fig. 3.7).

Regions of the axis entering the hook must increase their curvature. Here the relative stimulation must indeed be toward the periphery (outside) of the forming hook. But, as the same tissue passes the tip of the hook, it must obviously *un*bend, and show curvature change in the opposite sense. In this half of the hook the relative stimulation must be on the *inside* of the hook. Thus the most rational expectation for the stimulator distribution requires a shift in its maximum, from outside to inside, the change being at the tip of the hook. The greater over-all velocity of the outside of the hook is only a very small feature of what is happening. It is accounted for if, over all, there is slightly more stimulator toward the outside. For a more thorough treatment see Silk & Erickson (1978).

Morphogenesis

The main theme of this chapter is that extension of plant axes should be investigated with an initial analysis which breaks down the overall process into local behaviour, otherwise subsequent efforts towards mechanism may be misdirected. The utility of first knowing the pattern of local behaviour is equally valuable in morphogenesis, where form is not constant.

One might think that shape change alone might suffice to reveal the basic, local features of the growth pattern. That this is not so is evident in Fig. 3.8. Here a cylinder grows into a trumpet shape. The natural assumption, in terms of local behaviour, is that the top grew faster than the base. To test this, one needs to characterize expansion behaviour on the surface. In this model idealized circles are drawn on the cylinder and their subsequent deformation, to bigger circles or to ellipses, is shown.

It is seen that the intuitive pattern where non-directional growth increases from one end to the other will indeed bring on the observed change in form. The lesson, however, is that a very different pattern, where the vertical and horizontal components are in opposing gradients (not similar), will also account for the same morphogenesis. One pattern, (a), has a gradient in local rate of growth, all of it of non-directional. Here circles stay circles, the rates of increase differ. The other pattern, (b), has a gradient in the directionality of growth, all of it at the same rate. Here circles become ellipses of similar area. It follows that shape change must be broken down into regional behaviour by direct measurement if deeper causes are to be sought effectively.

Organogenesis

The principle of characterizing local growth activity has been applied to organ formation in plants. An early technique was to apply small round drops of viscous ink to the formative surface and record how these circles deformed into ellipses (Green & Brooks, 1978). A later method uses epidermal cell outlines as the basis for calculation of the major axes of the ellipse that characterizes the growth of a given cell (Green & Poethig, 1982). The cell outlines are visualized, each day, by applying a negative stain. A water-soluble dye, with detergent, is applied to the epidermis. As the water evaporates, the dye concentrates in the indentations between cells, to accentuate the cobblestone surface of the organ.

In the formation of a single new lateral organ from a parent organ, the arrangement of strain ellipses, or crosses, must be converted from a parallel pattern to one with radial symmetry, the crosses or ellipses radiating from a pole. This is indeed the case (Green & Brooks, 1978). The interesting question is whether this definitive pattern of growth follows or precedes major changes in other parameters, such as cell division direction,

Fig. 3.8. Different local behaviour can bring on the same shape change. In (a) a small cylinder, marked with circles, transforms into a trumpet shape because of an increasing gradient in local relative rate of surface expansion, from left (L) to right (R). The gradient is graphed at right. The two components, *v* (vertical) and *h* (horizontal) are in the same gradient and always equal. The local relative rate of increase in area, the sum of *v* and *h*, is a steeper linear gradient. In (b), the same shape change is brought on by a gradient in directionality. The horizontal component dominates at left, the vertical dominates at right. There is no gradient in local relative rate of area expansion. The mechanism of morphogenesis relates to the local behaviour pattern, not to shape change *per se*.

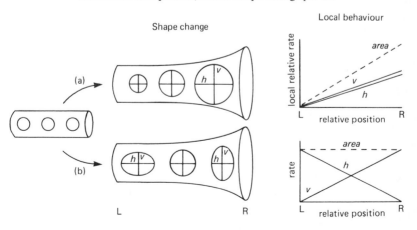

the direction of cellulose reinforcement in the cells, and the bulging of the organ. It is found that the final growth pattern is seen only after the new symmetry has been achieved in all these other features.

The local character of the transition from parallel to radial patterns is revealing. Initially all ellipses are parallel. The new pattern is started by an abrupt shift, by 90°, on two sides of the pattern. Thus four 'arms' of the ultimate radiating pattern are generated. Two opposite arms are a simple continuation of the initial pattern, two other arms come from a 90° shift (see Fig. 3.9). This rough orthogonal pattern is then smoothed out into the radial symmetry of the new organ.

Fig. 3.9. Local behaviour in organ formation. The progression deals with the *de novo* formation of the mid-line leaf in (a). The initiating area (b) has parallel cell files and transverse reinforcement. An abrupt shift in polarity (c) forms a band with reoriented polarity. This region becomes a growth centre which bulges (d) to produce a surface with transverse reinforcement across three domains. After further growth the axis is more evident (e) and full radial symmetry is attained (f). (From Green & Poethig, 1982, with permission.)

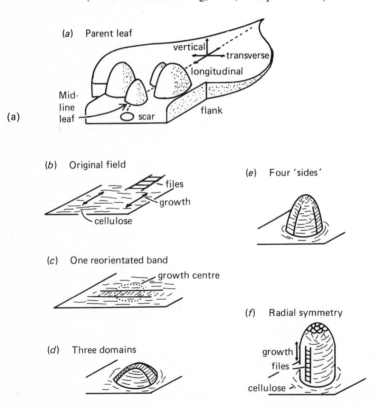

The picture for organogenesis, elaborated in Green & Poethig (1982), is that an early event is an abrupt 90° shift in cellulose reinforcement in a band of cells. As studied so far, the shift occurs in paired daughter cells right after the mitosis that formed them. This discontinuity, a re-oriented band of cells in the midst of the original alignment, becomes a centre for growth and a bulge appears. At the same time the lines where cells of the different cellulose orientations 'confront' each other are quickly smoothed away by cell divisions where the alignments in the daughter cells take on an intermediate angle. By this time the bulge has arisen and the organ, especially its epidermis, has gained over-all hoop reinforcement. This accounts for the elongation which will characterize the organ (a leaf or stem) indefinitely.

The above picture indicates that a three-step sequence, (a) abrupt shift in polarity, (b) smoothing, and (c) bulge, characterizes organogenesis. The subject of phyllotaxis can thus be re-examined. A reasonable mechanism would have the early development of one leaf bring on the polarity discontinuity required for the production of the next leaf. Hence the long-recognized influence of nearby leaves on the apical dome would be one of constructive involvement in the mechanism of leaf initiation, as against the generally assumed role of inhibition. In traditional views of phyllotaxis, leaf formation is viewed as a natural activity of the apical tissue which it carries out unless inhibited. That is a rather non-specific mechanism. Other views (Meinhardt, 1978) generate a chemical pre-pattern for the leaves, having the leaf form at a peak of concentration of a postulated morphogen. That also leaves unaddressed the issue of how a leaf is actually constructed. The concept of a general positional information grid (Holder, 1979) specifying individual cell activity also skirts this issue.

The message of this chapter is that growth and morphogenesis, even up to the issue of phyllotaxis, must be expected to have many levels of explanation. The fundamental levels of gene expression are obviously essential and are comparatively well understood. It is the additional levels, which deal explicitly with gradients in spatial behaviour, that are the current challenge.

The features of gross development, that is the stages remote from the gene, are such that algebraic concepts are often insufficient. Viewing the growing system as an integration in space and time, and applying the appropriate calculus concepts to it, has proved fruitful for description and mechanism. Preliminary analytical chains linking morphogenesis to DNA have been presented. The issue of controls, from that of ethylene causing swelling to the influence of one leaf primordium on the location of the next, can be couched in this hierarchical frame of reference.

Appendix

Factors influencing mean cell length

Colin R. Goodall and Paul B. Green

Change in mean cell length can occur in two senses. In the first, Lagrangian, sense one asks, 'How, and by what local activities, does mean cell length vary in a tissue region as it moves away from the apex of a plant organ?'. In the second, Eulerian, sense one asks, 'What local activities explain the gradient in mean cell length, seen for a given cell type, in a longitudinal section of tissue?'. We answer both questions for the most straightforward case, that of an idealized, steadily growing, plant organ, varying in one dimension only, for example a root.

This treatment considers cell length to be analogous to price per share of an issue of stock. Two things can happen to the issue. Its value may go up (extension), or stock splits may be declared (division or partitioning). The balance, or lack of balance, between these two processes determines the price per share.

With the Lagrangian question, we seek to characterize changes occurring in a material region as it moves. To understand changes in mean cell length, we need plots of the local relative rates of extension and division, as shown in Fig. 3.5. Each such rate is a compound-interest rate. The difference between the two rates is also a compound-interest rate; it is the compound-interest rate, i.e., the local relative rate, for change in mean cell length (see Green, 1976). For steady one-dimensional growth, we give the algebraic expression in its Lagrangian form below (equation 15), after introducing the necessary notation.

When the difference of the local relative rates of extension and division is plotted against time, rather than distance, the resulting graph provides answers to the Lagrangian question posed above. For example, cell length increase occurs only when the extension rate exceeds the division rate. Then the time required for mean cell length to double is simply the time required for an area of ln 2 to be generated under the differences plot (see Fig. 3.4). In general, the time required varies with the initial location, and therefore the developmental stage, of the cells. On the other hand, when the division rate is greater than the extension rate, cell length falls. The time required for the generation of an area of ln 2 produces a halving of mean cell length. In this way we answer questions dealing with the activities of a group of cells as they, and their progeny, move down a growing axis. The frame of reference moves with the cells of interest.

The other question, the Eulerian one, deals with changes that occur with position. A natural objective is to explain a persistent, characteristic gradient in an axis. Here we examine mean cell length along a root. We depart from a simple difference of rates, as above, because the frame of reference no longer moves with the cells. Rather, the cells move past the observer. This introduces a role for convection, essentially in place of the local relative rate of mean cell length increase.

The pertinent mathematics can be followed in terms of the conveyor belt analogy used previously in the main text (p. 70). We first deal with the problem in terms of *any* commodity whose amount varies down the growing root axis, or conveyor belt. To begin with, we assume this commodity is the height of sand on a moving, and elastic, conveyor belt. The question becomes, 'What factors govern a constant pattern of height of sand on the moving belt?'. In plant growth zones all pairs of points on the axis both move away from a reference point and separate from each other, so the analogy (displacement and stretch) is valid.

We need a source of sand. This consists of a series of little pipes that drop sand onto the belt; the rate of deposition of sand may vary along the series of pipes. We assume the sand does not spill spontaneously from the site of deposition. Under conditions where the pattern of height (H) of sand on the belt does not change (despite belt motion and stretch), the rate of deposition at each point must exactly compensate the twin thinning effects of convection and dilation.

With x the distance from the apex, the gradient in height (the slope) of sand on the belt is dH/dx, and the velocity of the belt at the point is dx/dt. Then the convection term is

$$\frac{dH}{dx} \cdot \frac{dx}{dt} \tag{1}$$

The local relative rate of stretch (the one-dimensional divergence) is $(d/dx)(dx/dt)$. Then the dilation term is

$$\frac{d}{dx}\left(\frac{dx}{dt}\right)H \tag{2}$$

With constant height of sand at a point, the flow, dS/dt, to compensate the two dissipating terms is

$$\frac{dS}{dt} = \frac{dH}{dx} \cdot \frac{dx}{dt} + \frac{d}{dx}\left(\frac{dx}{dt}\right)H \qquad (3)$$

This is a form of the continuity equation, familiar from fluid mechanics (see Silk & Erickson, 1979).

The slope of the sand, dH/dx, embodies changes in spatial pattern along the belt. Knowing this slope at all points, we can integrate to form the entire profile of sand on the belt. So the question is, 'What processes or factors determine the slope?'. To find out, we solve Eq. (3) for dH/dx. Let $dx/dt = V$. Then

$$\frac{dH}{dx} = \frac{dS}{dt} \cdot \frac{1}{V} - \frac{d}{dx}\left(\frac{dx}{dt}\right)\frac{H}{V} \qquad (4)$$

When the middle term is multiplied and divided by H we get

$$\frac{dH}{dx} = \left(\frac{1}{H} \cdot \frac{dS}{dt}\right)\frac{H}{V} - \frac{d}{dx}\left(\frac{dx}{dt}\right)\frac{H}{V} \qquad (5)$$

Now $(1/H)(dS/dt)$ is the local relative rate of accumulation of sand and $(d/dx)(dx/dt)$ is the local relative rate of the stretch of the belt. It thus turns out that the slope of the sand is, in part, governed by a familiar difference of two relative rates (compare the Lagrangian solution), i.e.,

$$\frac{dH}{dx} = \frac{H}{V}\left[\frac{1}{H} \cdot \frac{dS}{dt} - \frac{d}{dx}\left(\frac{dx}{dt}\right)\right] \qquad (6)$$

Note that the difference term (in brackets) determines the sign of the slope, because H and V are always positive. The quotient H/V influences only the magnitude of the slope.

To convert from the sand analogy to mean cell length, we substitute spatial frequency of cross-walls, F, for the height of the sand. F is the reciprocal of mean cell length. In the sand analogy, the input of sand comes from pipes above the belt. With cells, however, changes in F come from cell division, an internal activity. A conceptual difficulty is that frequency is intangible, and cannot be physically reproducing. However, the total number of cells in the axis to a distance x, written $c = c(x)$, *is* tangible. We have

$$F = dc/dx \qquad \text{and} \qquad c(x) = \int_0^x F(z)\, dz$$

Setting $S = F$, some calculus manipulations* show that

$$\frac{1}{F} \cdot \frac{dS}{dt} = \frac{d}{dc}\left(\frac{dc}{dt}\right),\qquad (7)$$

which is the local relative rate of cell division. The use of this logarithmic double derivative of c is consistent with the notation of Erickson & Sax (1956).

The change in spatial frequency of cell walls with position is

$$\frac{dF}{dx} = \frac{F}{V}\left[\frac{d}{dc}\left(\frac{dc}{dt}\right) - \frac{d}{dx}\left(\frac{dx}{dt}\right)\right]\qquad (8)$$

Since our interest is in mean cell length, M, which is the reciprocal of spatial frequency of cross-walls, we change variables; i.e.,

$$\frac{dF}{dx} = \frac{d(M^{-1})}{dx} = -\frac{1}{M^2} \cdot \frac{dM}{dx}\qquad (9)$$

$$-\frac{1}{M^2} \cdot \frac{dM}{dx} = \frac{1}{MV}\left[\frac{d}{dc}\left(\frac{dc}{dt}\right) - \frac{d}{dx}\left(\frac{dx}{dt}\right)\right]\qquad (10)$$

$$\frac{dM}{dx} = -\frac{M}{V}\left[\frac{d}{dc}\left(\frac{dc}{dt}\right) - \frac{d}{dx}\left(\frac{dx}{dt}\right)\right]\qquad (11)$$

$$\frac{dM}{dx} = \frac{M}{V}\left[\frac{d}{dx}\left(\frac{dx}{dt}\right) - \frac{d}{dc}\left(\frac{dc}{dt}\right)\right]\qquad (12)$$

Thus the change in mean cell length in space is a function of the difference between the local relative rates of extension and cell division. Notice the change of sign: cells lengthen with higher extension rates. The magnitude of the slope is also a function of the quotient M/V. Equation (12) can be written as

$$\frac{dM}{dx} = \frac{M}{V}\left(\frac{\ln 2}{t_{d(l)}} - \frac{\ln 2}{t_{d(c)}}\right)\qquad (13)$$

where t_d is the local doubling time for the region's length (l) or cell number (c).

---

*In outline these manipulations are

$$\frac{1}{F} \cdot \frac{dS}{dt} = \frac{1}{F} \cdot \frac{dF}{dt} = \frac{dx}{dc} \cdot \frac{d}{dt}\left(\frac{dc}{dx}\right)$$

Interchanging the order of differentiation, and applying the chain rule

$$\frac{1}{F} \cdot \frac{dS}{dt} = \frac{dx}{dc} \cdot \frac{d}{dx}\left(\frac{dc}{dt}\right) = \frac{d}{dc}\left(\frac{dc}{dt}\right)$$

The second differentiation, d/dc, implicitly involves an increment in distance along the axis.

The difference term embodies the current local biological activities at the point in question. These are the local relative rates of cell production and tissue extension. In contrast, the quotient term deals with activity throughout the portion of the root towards the apex. The numerator of the quotient term, M, the local mean cell length, is a function of history. The denominator, V, the local rate of movement away from the apex, is a function of extension activity distally. It is as if the difference term, the activity at a point, cannot alter cell length effectively (dM/dx is small) if the cells are passing through the point rapidly. This is true when V, the velocity, is high, or when M, the mean cell length, is small. It is an important consequence of continuity principles (Silk & Erickson, 1979; Gandar, 1980) that spatial gradients are not simply a function of local activity.

Equations (12) and (13) are Eulerian, for an observer fixed in space. Note, however, that in this, the steady growth case, the difference of local relative rates is the same difference as appears for the Lagrangian form, discussed at the start of the Appendix. We may easily manipulate the Eulerian equation to give the Lagrangian equation (and the reverse). Multiplying Eq. (12) on both sides by V/M, and writing dx/dt for V, gives

$$\frac{1}{M} \cdot \frac{dM}{dx} \cdot \frac{dx}{dt} = \frac{d}{dx}\left(\frac{dx}{dt}\right) - \frac{d}{dc}\left(\frac{dc}{dt}\right) \tag{14}$$

The steady growth assumption allows us to 'cancel' the dx's on the left side of Eq. (14). (For the same reason we avoid the complications of partial derivatives throughout this discussion.) Thus

$$\frac{1}{M} \cdot \frac{dM}{dt} = \frac{d}{dx}\left(\frac{dx}{dt}\right) - \frac{d}{dc}\left(\frac{dc}{dt}\right) \tag{15}$$

Since differentiation of M in Eq. (15) is with respect to time, the natural interpretation is Lagrangian. In the steady growth, Eulerian, setting we have $dM/dt = 0$. The term on the left side of Eq. (14) was originally, in Eq. (3), a convection term. In Eq. (15) it becomes the local relative rate of increase in cell length. In moving the velocity term across the equation, our natural frame of reference changes to one moving down the root, that is, from Eulerian to Lagrangian. A gradient in mean cell length becomes dilation or contraction of material cells.

Equation (15) is the Lagrangian form of the continuity equation. It is derived without the use of algebra, by analogy with share prices, at the start of the Appendix. It is also derived directly by Erickson & Sax (1956) and Green (1976).

To sum up, the time course of change in mean cell length in a small group of cells moving down the axis (the Lagrangian view) is determined by the difference between two rates. These are the local relative rate of extension and the local relative rate of cell production. Both rates are inversely proportional to the respective doubling time. Changes in mean cell length along the axis (the Eulerian view) are also influenced by this same difference in rates. Knowledge of this difference (knowing which rate is the greater) determines whether the cells lengthen or shorten at the point of interest. The gradient of mean cell length is a function of the difference, but also of the mean cell length at the point, and the velocity at which the cells move past the point.

References

Erickson, R. O. (1976). Modeling of plant growth. *Annual Review of Plant Physiology*, **27**, 407–34.

Erickson, R. O. & Goddard, D. R. (1951). An analysis of root growth in cellular and biochemical terms. *Growth*, **10** (supplement), 89–116.

Erickson, R. O. & Sax, K. B. (1956). Elemental growth rate of the primary root of *Zea mays*. *Proceedings of the American Philosophical Society*, **100**, 487–98.

Frey-Wyssling, A. (1959). *Die Pflanzliche Zellwand*. Berlin: Springer-Verlag.

Fritsch, F. E. (1948). *The Structure and Reproduction of the Algae*, vol. 1. Cambridge: The University Press.

Gandar, P. W. (1980). The analysis of growth and cell production in root apices. *Botanical Gazette*, **141**, 131–8.

Green, P. B. (1973). Morphogenesis of the cell and organ axis – biophysical models. In *Basic Mechanisms in Plant Morphogenesis. Brookhaven Symposia in Biology*, **25**, 166–90.

– (1976). Growth and cell pattern formation on an axis: critique of concepts, terminology, and modes of study. *Botanical Gazette*, **137**, 187–202.

Green, P. B. & Brooks, K. E. (1978). Stem formation from a succulent leaf: its bearing on theories of axiation. *American Journal of Botany*, **65**, 13–26.

Green, P. B. & King, A. (1966). A mechanism for the origin of specifically oriented textures in development, with special reference to *Nitella* wall texture. *Australian Journal of Biological Sciences*, **19**, 421–37.

Green, P. B. & Lang, J. M. (1981). Toward a biophysical theory of organogenesis: birefringence observations on regenerating leaves in the succulent, *Graptopetalum paraguayense* E. Walther. *Planta*, **151**, 413–26.

Green, P. B. & Poethig, R. S. (1982). Biophysics of the extension and initiation and plant organs. In *Developmental Order: Its Origin and Regulation*, ed. S. Subtelny & P. B. Green, pp. 485–509. New York: Alan R. Liss.

Gunning, B. E. S. (1981). Microtubules and cytomorphogenesis in a developing organ: the root primordium of *Azolla pinnata*. In *Cytomorphogenesis in Plants*, ed. O. Kiermeyer, pp. 301–25. Vienna & New York: Springer-Verlag.

– (1982). The root of the water fern *Azolla*: cellular basis of development and multiple

roles for microtubules. In *Developmental Order: Its Origin and Regulation*,
    ed. S. Subtelny & P. B. Green, pp. 379–421. New York: Alan R. Liss.
Gunning, B. E. S. & Hardham, A. R. (1982). Microtubules. *Annual Review of Plant
    Physiology*, **33**, 651–98.
Hejnowicz, Z. (1961). The response of the different parts of the cell elongation zone in root
    to external β-indoleacetic acid. *Acta Societatis Botanicorum Poloniae*, **30**, 26–42.
Hejnowicz, Z., Heinemann, B. & Sievers, A. (1977). Tip growth: patterns of growth rate
    and stress in the *Chara* rhizoid. *Zeitschrift für Pflanzenphysiologie*, **81**, 409–24.
Holder, N. (1979). Positional information and pattern formation in plant morphogenesis
    and a mechanism for the involvement of plant hormones. *Journal of Theoretical Biology*,
    77, 195–212.
Jaffe, L. F. (1982). Developmental currents, voltages and gradients. In *Developmental
    Order: Its Origin and Regulation*, ed. S. Subtelny & P. B. Green, pp. 183–215. New
    York: Alan R. Liss.
Koch, D. L. (1982). The shape of the hyphal tips of fungi. *Journal of General
    Microbiology*, **128**, 942–51.
Lang, J. M., Eisinger, W. R. & Green, P. B. (1982). Effects of ethylene on the orientation
    of microtubules and cellulose microfibrils of pea epicotyls with polylamellate walls.
    Protoplasma, **110**, 5–14.
Meinhardt, H. (1978). Models for the ontogenetic development of higher organisms.
    Review of Physiological and Biochemical Pharmacology, **80**, 48–104.
Palevitz, B. A. (1980). The structure and development of stomatal cells. In *Stomatal
    Physiology*, ed. P. G. Jarvis & T. A. Mansfield, pp. 1–23. Cambridge: Cambridge
    University Press.
Peterson, R. L. (1967). Differentiation and maturation of primary tissues in white mustard
    root tips. *Canadian Journal of Botany*, **45**, 319–31.
Sievers, A. & Schnepf, E. (1981). Morphogenesis and polarity of tubular cells with tip
    growth. In *Cytomorphogenesis in Plants*, ed. O. Kiermeyer, pp. 265–99. Vienna & New
    York: Springer-Verlag.
Silk, W. K. & Erickson, R. O. (1978). Kinematics of hypocotyl curvature. *American
    Journal of Botany*, **65**, 310–19.
– (1979). Kinematics of plant growth. *Journal of Theoretical Biology*, **76**, 481–501.
Staehelin, L. A. & Giddings, T. H. (1982). Membrane mediated control of cell wall
    microfibrillar order. In *Developmental Order: Its Origin and Regulation*, ed. S. Subtelny
    & P. B. Green, pp. 133–47. New York: Alan R. Liss.
Taiz, L., Metraux, J-P. & Richmond, P. A. (1981). Control of cell expansion in the *Nitella*
    internode. In *Cytomorphogenesis in Plants*, ed. O. Kiermeyer, pp. 231–64. Vienna & New
    York: Springer-Verlag.
Trinci, A. P. J. & Saunders, P. T. (1977). Tip growth of fungal hyphae. *Journal of General
    Microbiology*, **103**, 243–8.
Weisenseel, M. H. & Kicherer, R. M. (1981). Ionic currents as control mechanism in
    cytomorphogenesis. In *Cytomorphogenesis in Plants*, ed. O. Kiermeyer, pp. 380–99.
    Vienna & New York: Springer-Verlag.
Wilson, K. (1964). The growth of plant cell walls. *International Review of Cytology*, **17**,
    1–49.

4

Positional controls in meristem development: a caveat and an alternative

PHILIP M. LINTILHAC

In this essay I would like to approach the question of the appropriateness and applicability of positional thinking to developmental problems in plants. In particular I will address the problems associated with the application of the theory of positional information as originally conceived by Wolpert (1971). I realize that the scope and intent of this book is more loosely defined, allowing for a broader usage of words such as 'position', 'pattern', and 'movement' than that defined by Wolpert. I also realize that more recent contributions to the literature on positional information have diverged considerably from the original concepts put forward by Wolpert, and now encompass a number of different models including the polar coordinate model proposed by Bryant, French & Bryant (1981), which I will not discuss here.

My concern is that the original concepts put forward by Wolpert, having served so well in directing thinking about development in animal science, will be applied uncritically to plant systems, which I consider to be organized in a fundamentally different way.

The question is not whether these models can be applied to plant development, because they can. I doubt that they could be proved wrong, in fact. The question which I will address is one of the appropriateness; particularly the appropriateness of the basic and original idea that position itself, as measured by some arbitrary scalar or morphogen, is the principal effector which, along with the genome, regulates developmental patterning and plant form.

I include in my criticism a discussion of the concept of cell movement in order to throw into relief the organizational differences between plant and animal systems, even though cell movement, in the sense of change-of-neighbours, may no longer play a central role in the theory of positional information as applied to animals.

There are two ways to read the term 'positional controls' used in the title of this book. First, it can be taken to mean those controls which refer

to the position of a cell as the primary determinant of its developmental
behaviour; in this sense 'positional controls' must be taken to mean very
nearly the same thing as Wolpert's term 'positional information'. A second meaning can be drawn which refers to any control mechanism which
affects the position of a structure, be it a leaf or a cell wall. Clearly this
second meaning is beyond criticism, since there must be controls of some
sort.

Caveat

I do not subscribe to the theory of positional information as applied to
plant development. I feel that the language of positional information has
provided us with a semantic framework within which certain developmental processes can be studied, but that it lacks the predictive power that
other more empirically-derived perspectives can give. I regard the theory
of positional information as a metaphor which has been stretched to
cover many developmental problems, but which will not in the end contribute to our ability to predict and manipulate the real variables of plant
growth and development. In the case of apical meristems in particular, I
think that the basic mechanisms for the generation of form and pattern
will be more directly understood as a process of structural epigenesis,
which can be directly modified by experimental intervention in the stress-
mechanical environment of growth.

Superficially, the idea of a control system based on positional information can be applied directly to plant systems where developmental events
occur at regular intervals along the growing axis. My difficulties with such
a direct application of pre-existing theory are two-fold. First, the semantics of positional information theory becomes awkward when applied to
plant materials. The words 'position', 'movement', and 'patterning', perhaps three of the most central elements of the original theory, have somewhat equivocal meanings when applied to plant tissues. Second, the application of this model to plant systems may serve only to obscure their
extraordinary sensitivity to structural and mechanical stimuli; this sensitivity is qualitatively quite different from the positional sensitivities of
animal cells, and depends on the mechanical coupling provided by shared
rigid cell walls. I feel that any real understanding of plant morphogenesis
will have to be phrased in terms of sophisticated architectural responses to
structural stimuli, and that the application of models which are inappro-
priate will serve only to confuse the issue.

My plan for this chapter, then, is first to attempt a limited critique of the concept of positional information in the ontogeny of plant apical meristems, and then to offer in its stead a concept of structural epigenesis which I believe will prove to be more directly applicable and semantically consistent, and which will serve to focus our attention on questions relating directly to the special developmental tools which are perhaps uniquely available to plant cells.

Positional information reviewed

The concept of developmental control by means of positional information was elaborated by Lewis Wolpert (1971) and was first applied to the events associated with amphibian limb regeneration. It was presented as a model which permitted an understandable expression of some of the questions and problems associated with cell behaviour following partial limb amputation.

According to Wolpert's original conception, the control of development by positional information requires two basic elements. The first of these is a reliable method of assigning a 'positional value' to each cell in the developmental field. This is assumed to involve the establishment of diffusional gradients of some morphogen scaled so that each cell in the matrix can be considered to reside at a unique intersection. The second element essential to this system is a means by which each cell can interpret its assigned positional value in terms of an appropriate developmental protocol. The cell is assumed to possess some form of 'reference library' in which it can turn to the appropriate page, specified by its positional value, and read what it must do to complete its part of the overall developmental process.

There is one major advantage to a developmental system acting in this way. As Wolpert himself put it, there need be '... no unique relationship between the form of the gradient in positional value and the pattern which results from the cells' interpretation of it'. This means, in principle, that no complicated arrangement of sources and sinks is necessary to shape the morphogen concentrations into a 'prepattern' which directly predisposes the different regions to develop along certain paths; precisely defined sources and sinks may, however, be required for the unequivocal assignment of positional values.

Positionally controlled development, then, relies on a complex reference library which is available to each cell in the developmental field. This library provides the developmental instructions necessary to define the final distribution of cell types in the organism. Furthermore, it is clear that a prime candidate for such a developmental library is the information encoded in the genome of each cell.

Consequences

One consequence of this line of thinking is that the cell must be regarded as being information-rich (the reference library). Further, if one regards the initial interpretive aspect as being referred to this genomic library, then the first cellular events associated with any developmental change must involve gene transcription and translation (i.e. the synthesis of proteins) in which case nuclear activity must always precede cytoplasmic activity and patterning. The impression one gets then, is that developmental cues are never dealt with directly. They are always first referred to the interpretive rule-book and from this follows the manufacture of new developmental machinery.

Specific criticisms

The body of Wolpert's theory was based on a view of development according to which '... the genetic information in the egg does not contain a description of the adult but a set of instructions on how to make it' (Wolpert, 1971). The mature organism could then be treated in terms of the generation of form and pattern which are distinct to the extent that 'form involves cell movement and changes in shape (meaning that form results from shape changes of the multicellular mass, which in turn result from cell movement) ... whereas pattern does not involve changes in shape or cell movement, but rather the specification of spatial differences'. In plants, however, movement must be carefully defined; relative movement of neighbouring cells is generally impossible. The only cellular movement which occurs in a growing apex is due to the continuous flow of material away from the apical pole; neighbouring cells do not slip past each other to any great extent, and hence nearest-neighbour relationships are conserved through time. The generation of form is thus largely due to shape change at the cellular level (directional enlargement) rather than to cell movement and change in position, or to the slippage of layers of cells past each other. The word 'movement' is thus less easily applied to plant cells than to animal cells.

Similarly, the word 'position' is ambiguous since neighbouring cells retain a constant relationship to each other while continually changing their location relative to the apical pole. Does the word 'position' refer to the relationship of a cell to its neighbours, or its relationship to the boundaries of the structure as a whole? If it is taken to be the former then a cell's position is invariant through time and the concept of positional information degenerates, since position is unchanging. If, on the other hand, position is referred to the structure as a whole, then it is constantly changing and has only an instantaneous meaning. In this case the temporal as well as the physical limits of a given position must be defined and carefully related to the time-scale of the developmental event associated with that position. In most cases, a critical event would inevitably be smeared over a number of positions if looked at in fine enough detail, in which case the word 'position' requires redefinition in terms that belie the significance of position itself as the developmental effector.

To put it another way, if the 'life-line' of an apically-derived cell consists of a continuum of different positions, then what singles out one position as the trigger which initiates the developmental event in question? Only when we have this effector in hand will we truly understand developmental controls in the apex. My thesis is that it is the structural and stress-mechanical relationships within the apex which serve this function, and that they constitute the real developmental controls. Therefore, the question 'Why is it that plant cells in certain positions do certain things?' should be rephrased to ask 'Why are certain developmental activities associated with certain locations on the growing apex?'.

My most serious reservation concerns the concepts underlying the mechanism of pattern formation. Wolpert defines pattern formation as 'the specification of spatial differences'. Here the key word is 'specification' since it implies that such differences are genetically determined. In plants however, and in apical meristems in particular, I believe that a good argument can be made that most pattern is epigenetic in origin, and that genetic specification provides only the basic biochemical tools common to all cells in the apex. In this respect I differ in a most basic sense from Wolpert's view that the genome includes a 'set of instructions' for the construction of the adult. (The concept of an 'adult' is itself somewhat suspect since plants have 'juvenile' and 'mature' forms, but they do not have an 'adult' life-stage in the sense intended by Wolpert; they grow or they die.)

Not only may there be aspects of pattern which need no specification in plants since they result directly from the mechanics of the division process, but the entire vegetative life-history of the meristem, that is, the initiation of successive leaves and branches, may be epigenetically controlled, even though the characters of constituent cells are genetically specified.

Positional information theory as applied to plants

The most direct application of positional information theory to plant development was put forward by Holder in 1979. Using the concepts and terminology advanced by Wolpert, he considers that 'The central problem of the study of morphogenesis is how the genetic potential of undifferentiated cells is expressed in a controlled manner to produce the specific spatial array of cell types characteristic for the species'.

I feel that such statements, although strictly speaking quite valid, can be misleading to the extent that they can be taken to mean that an understanding of developmental events will come only by considering them as results of differential gene activity. Placing the entire responsibility for development with the genome undercuts a whole level of events which do not refer directly to gene activity but which are controlled epigenetically through direct interaction with the cellular environment. I believe that it is a mistake to assign patterning in plants solely to a genetically-derived protocol. Many aspects of spatial regularity, and even some instances of structural differentiation, may arise as a direct and highly sophisticated architectural response to specific mechanical stimuli.

The alternative: structural epigenesis

Theoretical considerations

Plants can be considered to be architectonic structures. That is to say, their plane of cell division must be controlled so that new division walls help to support the weight of the structure and resist internally-generated stresses, just as the supporting walls in a building must be oriented to support the weight of the structure. The question then arises: How do plant cells sense and respond to mechanical forces? The significance of this question goes much deeper, however, because the development of form itself, and the control of certain aspects of morphogenesis, can also be traced to the orientation of divisions in the growing regions (Sinnott, 1943).

The orientations of the accumulated divisions determine not only the shape of localized reservoirs of cells which through subsequent distortion and differentiation become the mature plant body, but they also determine the polarity of the distortions themselves. This is because the axis of elongation of the individual cells is determined by the nature of the microfibrillar reinforcement of their walls (Green & Brooks, 1978) and this, in turn, is generally related to the division plane in a consistent way (Sinnott, 1960; Green & Lang, 1981). It is the highly-ordered transverse microfibrillar reinforcement of the individual cells which, in the aggregate, confers upon many plants the ability to rapidly extend an axis solely through the action of cell enlargement.

To take the argument one step further it should be recalled that the cell-plate-phragmoplast complex, although apparently free-floating within the cell, is pulled into an alignment which is structurally adaptive; this result must be brought about by some mechanism that directs mechanical stimuli to the interior of the cell.

At a purely phenomenological level, the most acceptable hypothesis is that the partition walls are oriented perpendicular to one of the principal stress trajectories which describe the local stresses within the tissue. This would bring the cell wall orientations into conformity with commonly understood principles of structure in that they would then be subject only to pure tension or compression within the plane of the wall. Translating this into a rule for the behaviour of the single cell, we can say that the cell-plate-phragmoplast (or its precursor) searches for a final orientation which is free from shear in the plane of the partition. Such a rule still does not tell us anything about the actual macromolecular machinery that is capable of (a) sensing the structural requirements of the cell, and (b) orienting the cell plate.

What, then, constitutes an appropriate model in which to frame questions of cell patterning and organogenesis in plants? In particular, is it possible to arrive at a model which can connect events at the phenomenological level to realistic cellular mechanisms?

Four models of cellular patterning

Historically, the most acceptable model of cell patterning is that it is a response to hormonal diffusion gradients (Sachs, 1978). From this one must conclude that the cellular activities involved in cell plate orientation somehow interpret chemical gradients through the tissue, and presumably within the cell also. Appropriate structural orientations of partition walls would therefore be specified through an hormonal intermediate. This would seem to be very difficult to apply since the 'structurally-adapted' orientation would first have to be determined by some unknown mechanism, then encoded into a hormonal messenger (presumably resulting in a diffusion-maintained 'pre-pattern'), and finally decoded within the cell and used to specify the orientation of the cell plate.

In a second hypothesis, put forward by Holder (1979), cell wall patterns are 'positionally controlled', each cell responding to its unique positional value by means of specific genomic instructions. This, too, says nothing of the actual orienting mechanism. A more sophisticated scheme relating developmental activities to position, is a 'phase-shift' model (Cohen, 1971).

These models prove difficult to test because they rely on mechanisms which are difficult to visualize: gradients of some chemical species would need somehow to control cell plate orientation, or assign 'unique' positional values to cells which would then interpret this information spatially. Both these mechanisms involve the interpretation of concentration changes in a population of intracellular molecules. According to the first proposal, hormone concentrations are directly assessed upon arrival in the cell. According to the second, populations of gene products are released into the cell in such a way that they co-ordinate the assembly of the spindle apparatus and the new cell plate. In either case I find it difficult to see how adding a new population of molecules to the intracellular environment can lead to the erection of a precisely-oriented structure within the cell, without reference to external structural cues.

A third and more direct mechanism might be proposed whereby a cell reacts directly to environmental cues by determining the axis along which it is being subjected to the greatest force. This is quite reasonable if one considers that cytoskeletal elements are presumably capable of generating forces within the cell, and that such internally-generated forces could be used as restoring forces to equilibrate with, and measure, externally-applied forces in different directions.

The fourth possible mechanism, and the one I favour, is that the individual cells simply measure slight intracellular dimensional changes in different directions. I feel that this is the simplest alternative of the four. As I will show, it is also the one that is most easily interpreted in realistic cytoskeletal terms.

Transcellular dimension as a structural measure

Any change in the balance between internal turgor forces and externally applied directional forces will result in cellular deformation and dimensional change. A dimension-sensing mechanism within each cell would therefore be the most direct method of determining the directions of the principal stresses and strains, effectively transforming each individual cell into a three-dimensional, multi-axial strain-gauge. (P. B. Green has always maintained that the intracellular sensor must be measuring strain rather than stress (Green 1980).) An intracellular mechanism capable of accurately measuring changes in the distance across the cell in different directions, will be able to determine the axis of maximum dimensional change, and a new cell plate organized in a plane perpendicular to this axis will necessarily be in the shear-free plane.

I propose that the most likely candidate for such an intracellular measure of dimensional change would be a system of relatively stiff, non-contractile elements embedded in the cytoplasm (Fig. 4.1). First, let us consider that these semi-rigid unbranched elements (microtubules perhaps) are randomly oriented throughout the cytoplasm, spanning the cell wherever possible but not penetrating vacuoles, and that they are capable of attachment to the plasmalemma at both ends. Then, when subjected to axial forces (compressive *or* tensile) which cannot be accommodated by their normal growth or degrowth processes (subunit turnover), they tend to shatter into fragments which no longer span the cell and which subsequently recombine into new elements, although not necessarily in the same orientation.

Now, for a cell whose shape is changing, this provides a means of discriminating the direction of maximum dimensional change since those elements running along that axis will be the most likely to shatter; after this new linkages will be formed randomly and the structural continuity of the elements will be re-established. Thus, a period of extended dimensional change will result in the gradual depletion of structural elements along the axis of maximum dimensional change and the accumulation of elements in the orientation which is the most protected from shattering forces. This protected orientation will be transverse with respect to the direction of maximum dimensional change.

The accumulation of these transverse elements in the equatorial region of the cell poses another problem. This aggregation into a 'pre-prophase band' could also be accomplished by passive mechanisms which rely on the reduced rotational freedom of a linear element near the end walls, so that a free-floating segment (a piece of microtubule) would be more likely to link up with another segment after it had moved away from the end walls. Here again, the proposed mechanism could be modulated by controlling the 'stickiness' of the free ends, the approach angle which would permit re-attachment, etc. A slight concentration of transverse elements

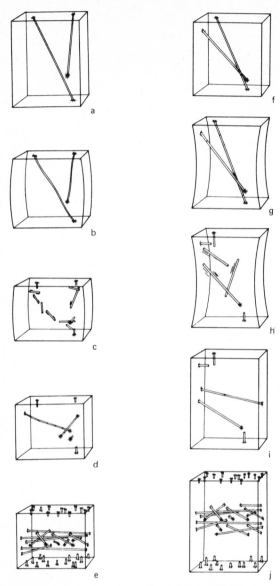

Fig. 4.1. Selective shattering and reformation of transcellular structural elements aligned near the axis of maximum dimensional change. (a–c), two elements show compressive shattering during cell shortening. (f–h), two elements show tensile shattering during cell lengthening. (d), (i), elements show reformation in a more transverse orientation. (e), (j), elements show aggregation in the plane (transverse) which is most protected from dimensional change. In tightly-packed, multicellular plant tissues cell-dimensional change in the equatorial plane (the plane of the cell plate) would be minimal.

near the equator could also be amplified over the long term by 'sieving' effects, whereby recently-shattered segments tend to become trapped in the web of transverse elements near the middle of the cell. One might even invoke some kind of active 'combing' mechanism relying on contractile elements within the cytoplasm to bring about the equatorial accumulation of transverse elements, which I am equating with the pre-prophase band.

The mechanism I have proposed to account for cell plate orientation is thus a statistical one which says that the time it takes the structural elements to break apart and re-aggregate is so short relative to the duration of the cell cycle that they can effectively accumulate in an orientation which protects them from further shattering. The only requirements for this proposal are the well known propensity of plant cells to undergo polarized and continuous cell enlargement, and the presence of transcellular structural elements whose physical properties cause them to break apart and re-aggregate in an appropriate manner when subjected to axial forces.

The microtubule, with its stiffness (relative to other cell organelles), its propensity to disassociate into sub-units (Inoue & Sato, 1967), and its demonstrated susceptibility to mechanical pressure (Salmon, 1975), could be precisely structured to make such a mechanism workable. A mechanism based on these unique properties, such as the one I propose here, would effectively link cell plate orientation to the structural demands of a growing, walled, cellular tissue.

With this hypothetical, but hopefully testable, model to support our observation that plants in general and meristems in particular are architectural structures, we can proceed to speculate about the limits of their sensitivities to mechanical forces, and define a theory of development which takes these sensitivities directly into account. (Such a theory could be tested by manipulation of the various physical parameters which affect plant structure and pattern.)

The limits of structural epigenesis

Plant cells are, in part, hydrostatic structures in which an internal medium, under high pressure, is constrained by a cellulose wall with variable mechanical properties. Meristematic tissue is a tightly packed mass of these turgid units. Pressures built up as a result of cell packing and mutual proximity tend to raise the internal pressure of the cells, and such pressures, especially in cells bounding an external surface, increase the tensile stresses in the unsupported external wall. Thus, even the simplest model of such a tissue must take into account the fact that all plant tissues harbour opposing forces which must be brought into equilibrium with one another. Meristematic activity under normal growth conditions thus involves considerable structural adaptation, a task for which plant tissues are particularly adapted.

Primary plant tissues achieve structural adaptation by three principal means. The first is the selection of appropriate orientations for new division walls, the second is the fine control of the physical properties of the wall itself, and the third is the local adjustment of cell turgor.

The mechanical forces which act through a meristem are shaped by several factors. The first of these is the distribution of growth in the tissue, which in turn results from turgor pressure changes and selective wall weakening. Localized regions of cell growth can thus be highly directional, depending on both the anisotropy of wall reinforcement of the individual cells, and on the number of cells in the field which share a similar orientation. This results in local sources of directional stress which are distributed through the tissue. The second factor that shapes forces within the meristem is the shape of the meristem itself (Lintilhac & Vesecky, 1980). Thus, surface morphology relates directly to development, since the stresses which influence partitioning and wall orientation are a function of overall organ morphology; that is, the directions taken by the principal stresses acting through a structure are a function of the shape of that structure.

We thus have the possibility of establishing an epigenetic control loop where the shape of the apex, which is a result of distributed divisions and directional enlargement, plays a role in controlling partition wall orientation in the subsequent round of divisions and enlargement. A modification to the shape of the apex follows; this causes a redistribution of growth stresses, and the process is repeated – a cyclic chain of causality not to be confused with circular logic.

From this point of view, the apex is a structural engine capable of repeatedly rehearsing stress-mechanical events to its own advantage.

Consequences

Once partition wall orientation is linked directly to mechanical stress, then certain kinds of pattern (cell wall patterns and primary meristematic form) can also be traced directly to stress-mechanical antecedents. Genetic control comes into the picture only insofar as it provides the cells with the appropriate biochemical machinery to sense and respond to mechanical forces. Individual structural events at the cellular level need not be coded for directly. Routine morphogenetic behaviour can thus be dealt with directly by the cytoplasmic machinery of growth; no new gene transcription or translation is necessary except when needed to maintain the cellular titre of biochemical equipment and to maintain critical tolerances and rates.

Secondly, the question of organogenesis in relation to position can be resolved without need for positional messengers since the stress-mechanical activity at a given location on the apex is necessarily location-specific. Furthermore, the geometrical precision with which stress-mechanical stimuli can be focused far exceeds that possible with molecular diffusates, making it much easier to account for structural adaptation at the cellular level.

Thirdly, the problem of deriving specific cell wall orientations from changes in intracellular molecular populations disappears, since these orientational events can be considered to be controlled directly, by physical manipulation from without, rather than from within by statistical interpretation of changes in intracellular molecular populations.

The thrust of developmental enquiry therefore shifts from trying to identify an appropriate system of morphogens which could bestow upon each cell a unique positional value. The question now becomes how the growing tissue as a whole maintains its own stress-mechanical environment. The cells are all the same and have no need to know their locations on the apex, or their positions relative to each other, in order to determine what to do, because the stress-mechanical environment varies in predictable ways from location to location within the apex.

By asserting that the patterning abilities of plant tissues are directly related to their unique structural sensibilities we at once remove much of the mystery from ontogenetic sequence.

This approach to the understanding of plant development does not preclude the action of traditional hormonal mechanisms, nor does it mean that plant growth and development proceed without genetic input. It is clear that large numbers of genes must be turned on and off during embryogenesis and development, each contributing its own nuance to the basic developmental sequence. It does mean, however, that a genetic mechanism need no longer be held responsible for direct control of basic structural adaptations at the cellular level.

The repetitive aspects of the generation of form can thus be considered to evolve naturally from the sensitivity of plant cells to the particular pattern of mechanical and structural stimuli that are present at every stage in ontogeny. Thus, if organ morphology is a controlling factor in defining subsequent organ morphologies (i.e., if shape begets shape), then the entire sequence of vegetative forms, from the zygote through embryogenesis, to germination, the formation of the cotyledons and the first apical meristem, and then continuing in the mature vegetative meristems – all this can be considered to be consecutive expressions of the same structural-epigenetic control cycle.

Structural epigenesis in ontogeny

Following fertilization, the initial divisions of the young embryo are highly predictable in their sequence and pattern, a fact noted by many and discussed specifically by Wardlaw under the heading 'Segmentation Patterns' in his book on embryogenesis (Wardlaw, 1955). These patterns tend to be conserved across taxonomic boundaries and fall into a handful of basic and quite similar segmentation schemes. The reason for this is that embryos share very simple and very similar shapes, and the forces generated within and around them are therefore analogous from case to case.

In the higher plants, the development of cotyledons, which can also show great similarities from taxon to taxon, represents the next stage in the ontogeny of mature form. The developmentally significant homologies lie not in the shape of the cotyledons themselves, however, but in the cotyledonary axil. This confers upon the embryo the important morphological element of a notch or concavity which shapes the forces acting through the embryo in a predictable way (Fig. 4.2), orienting and (perhaps) stimulating division activity in the axil (Lintilhac & Vesecky, 1980). This controlled division activity is what organizes the first apical meristem in the plumule.

When considered in this light, the apical meristem itself, once fully organized, serves as a kind of morphogenetic engine, capable of producing an infinite series of leaves, axils and new meristems, all arising from the necessary interaction of shape, stress, and partition wall orientation. The cellular patterns that we see in sectioned meristems thus represent instantaneous views of a continuing structural adaptation to the stress-mechanical conditions constantly sweeping through the apex.

I consider that the concept of structural epigenesis extends beyond vegetative development, however, and that it constitutes an integral part of all plant growth and patterning, culminating in a basis from which to approach aspects of reproductive anatomy such as sporangiation which, from my point of view, shows striking structural homology throughout the multicellular plants.

Sporangiation and the cell plate dilemma

The cell plate dilemma is this. If mechanical forces acting through a cellular tissue serve to orient new partition walls, or otherwise assist in the organization of the mitotic apparatus, then a cell which finds itself lacking clear directional cues for the orientation of a new cell plate may be inhibited in its attempt to organize the division process. Conversely, it follows that mechanical forces may also serve to stimulate partitioning under certain circumstances.

How does the cell decide where to place its new division wall when it senses no prominent directional stress, or when stresses are symmetrically focused around the cell? A mitotic apparatus which is dependent on directional stress or strain for its organization will be confounded by such a situation. This means that certain carefully constructed morphological situations which focus growth stresses radially around a single point (an isotropic point, stress-mechanically speaking) could be used to trigger a divergent mode of development. A single cell, or a group of cells, could be inhibited or delayed in its attempt to organize the mitotic apparatus and install a new partition wall because its location and the geometry of the surrounding tissues are such that it is subjected to forces whose vector sum is zero or nearly so. This could result in isodiametric enlargement, perhaps followed by membrane stretching, permeability change, and the tipping of the cell towards a new fate.

In a previous paper (Lintilhac, 1974), under the heading 'Idioblastic

Fig. 4.2. Photoelastically-derived stresses in a model embryo, showing stress directions in the region of the cotyledonary axil induced by imbibition of water around the perimeter of the model. (From Lintilhac & Vesecky, 1980.)

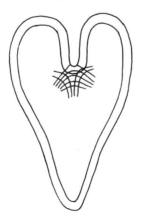

98   *P. M. Lintilhac*

differentiation', I raised the possibility of a mechanical role for the sporangium in the induction of sporogenesis. Such a role avoids some significant theoretical contradictions that arise from attempting to attribute differentiation events to immediate genetic antecedents. What I hope to add here is a reasonable hypothesis which can link an ultrastructural mechanism of cell plate control with the developmental behaviour of cells in certain physical locations.

Is the multicellular sporangium of the tracheophytes a stress-mechanical device for the induction of meiosis? Is gametogenesis in all the embryophytes linked by this mechanism? The clue to the mechanical role of the sporangium as a whole lies in the fact that the meiocytes are always internal (Fig. 4.3), which is to say, they are always covered by a tapetal or sporangial layer at least one cell layer thick. There does not appear to be a single example of an embryophyte meiocyte which is truly superficial in origin, that is, having an external-facing cell wall.

The geometry of the tapetal layer (variously referred to as tapetum, nucellus, or sporangial wall) makes it a prime candidate as a mechanical structure capable of generating and focusing mechanical forces on a group of internal cells, or even on a single cell as in *Gossypium* (Lintilhac & Jensen, 1974). All the cells of the sporangium are initially identical; but division in the centrally-located ones is slowed or stopped because of the isotropic nature of the forces there.

Fig. 4.3. The sporangium as a stress-mechanical device. (a) shows a hypothetical sporangium, seen in section, with two tiers of anticlinally dividing cells surrounding a single central cell. (b) shows the results of a completed cycle of division and enlargement. Dashed lines indicate the axes of maximum dimensional change. In the central cell such an axis cannot be defined.

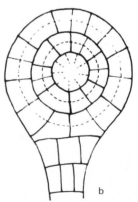

This could be considered to be as genuine an example of true positional control as one could define, but it differs from Wolpert's or Holder's conception in that it does not require the specification of a unique positional value for each cell in the structure, nor does it entail the action of diffusible morphogens or require immediate reference to genetic information for instructions. It acts through the inherent logic of a universal ability of plant cells to discriminate directional forces passing through a tissue.

Goebel (1905), and others of his era, considered the ubiquitous presence of sporangial cells overlying the meiocytes as being required for their nutrition and protection. But it is the fact that the meiocytes-to-be share common walls with the cells of the sporangial wall or tapetum which makes the sporangium an effective mechanical device, since the common wall provides an essential mechanical coupling for the transmission of forces into the interior of the structure.

Evidence for structural epigenesis

I believe that there is very little evidence that positional information, in Wolpert's original sense, serves as a major developmental control in plants. In a structurally-organized system such as a plant tissue, many genuine effectors may appear to correlate with position, so that it becomes possible to create a set of rules which predicts events on the basis of position; but that does not mean that position is the effector. Nobody has firmly identified the positional morphogen, or demonstrated how positional values are assigned in a specific case. It is insufficient simply to point to auxin as a probable morphogen. Auxin and other growth regulators are without doubt powerful effectors in development, but their ability to specify where something will happen is limited by the difficulty of focusing a diffusion-mediated process. Their ability to contribute directly to the understanding of location-specific events in development is also limited by the apparently dispersed nature of their sources and sinks, which is difficult to reconcile with the requirement for precisely controlled, yet easily resolvable, gradients in an effective positional morphogen.

Direct experimental evidence for structural-mechanical controls is also incomplete at present. However, it is becoming increasingly clear that division wall orientation in multicellular plant tissues can be affected, and even controlled in a limited sense, by the application of external mechanical forces (Lintilhac & Vesecky, 1981). The possibility exists, at least in principle, for demonstrating and dissecting these controls by structural-mechanical intervention in the development process.

The investigation of these problems has been severely hindered by the lack of appropriate experimental tools. In particular, the lack of a non-destructive method for mapping the stresses in living meristematic tissue is a constraint which leaves us with no direct way of correlating specific orientational and developmental events with their stress-mechanical causes. Ideally, one would want to be able to construct a stress-mechanical hypothesis, such as our central idea that partition wall orientations tend to fall perpendicular to the principal stress lines (Lintilhac, 1974), and test it by direct measurement of the relative forces generated in different directions at localized points in the growing meristem, thus essentially mapping the stress trajectories in the living meristem and presenting them for comparison with the same material seen in section. Photo-elastic methods, for instance, would provide such a non-destructive method if they could be applied to living meristematic tissues. However, the strong intrinsic birefringence of cell walls makes this impossible at present; and even the smallest implantable force transducers are larger than the entire apical meristem of most plant species.

An experiment to test the coupling between stress and the orientation of cell division

In my laboratory we have taken what seems to be the only other viable approach, which is to apply a force and look for oriented division walls. We can then correlate this known pattern of externally-applied stresses with the patterns resulting from the cell division behaviour of the tissue under conditions where we can be reasonably sure of the directions and intensities of the propagated forces. Although confirmation of our general hypothesis is less direct by this method, it does have an applied aspect which may evolve a methodology for actively controlling morphogenetic events in developing plant tissues.

The main thrust of our effort has been the use of an instrument capable of presenting mechanical forces to a tissue with the delicacy and control necessary for the isolation of stress-mechanical variables without causing too much damage to the tissue. This instrument (Fig. 4.4) has been developed in our laboratory for the specific purpose of applying known forces to delicate tissues growing in sterile culture. It is a feedback-controlled device which operates by taking continuous measurements of the force which is applied to the tissue, and comparing these to the desired force level, the set-point. The applied force is then continuously adjusted to minimize any discrepancy between the applied and the desired force, thus taking into account the growth of the tissue. Also, by programming continuous changes in the value of the set-point, the applied force can be varied to match any smooth curve.

Figures 4.5 and 4.6 show the effect of continuous compressive force applied to a sterile plug of *Nicotiana* pith, freshly explanted into a standard growth medium containing indoleacetic acid (IAA) and kinetin, and grown for five days at 28°C. The stressed region (Fig. 4.6) shows highly co-planar divisions at the site of force application when compared to the control (Fig. 4.5) which is an actively proliferating but unstressed region of the same explanted pith. These co-planar divisions have accumulated to a depth of approximately 10 cells during a five-day growing period, giving the experimental region an almost cambium-like appearance when seen in section. The repetitive serial divisions (periclinal) which are appar-

Fig. 4.4. Sterilizable pneumatic forcing-frame assembly for the application of delicate mechanical forces to plant tissues grown in culture. Movable teflon tissue-interfaces hang from a bellows assembly, which in turn is suspended from the lid of the culture vessel. A bridge balancing circuit and a pre-amplifier are attached to the right-hand side of the culture vessel.

4.4

ent in Fig. 4.6 seem to be characteristic of a narrow range of applied forces. The externally-applied forces must be of sufficient magnitude to offer some restraint to the radial growth of the explant in the experimental region, resulting in a visible depression in the otherwise unrestrained surface of the explant after completion of the experiment, but must not be so great as to suppress all radial growth. These conditions seem to be fulfilled in our system with applied compressive stress intensities between 500 and 800 mg/mm^2.

It is the observation that ordered divisions are optimized between certain limits of dimensional growth and active restraint that leads us to the hypothesis that the intracellular sensor is responding to *dimensional change*. Too much pressure maintains the highly directional nature of the stress trajectories propagating through the tissue but reduces polarized dimensional growth and eliminates serial divisions, resulting in cell hypertrophy. Too little pressure permits rapid cell expansion but provides no directionality; enlargement tends to be isodiametric, which confounds the intracellular sensor and again results in hypertrophy or in more random divisions.

These findings are significant in that they constitute the first concrete

Figs. 4.5 and 4.6. Experimental control of division plane *in vitro* in a pith core from a young *Nicotiana* internode explanted into a growth medium supplemented with IAA and kinetin. Transverse sections. Fig. 4.5. An unstressed region of the explanted core after one week in culture. Note the lack of ordered divisions. Fig. 4.6. A stressed region of the same explant subjected to continuous compressive stress during the growth period. Note that repeated coplanar divisions have occurred in the region underlying the point of force application (arrows) giving the stressed region a cambium-like appearance. Both figures, × 50.

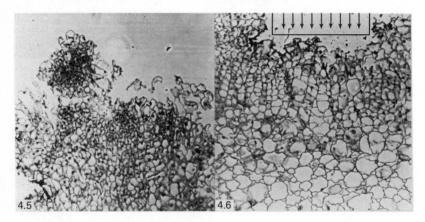

example of the active mechanical control of division activity in explanted plant tissue maintained in sterile culture.

Clearly, this evidence is just a preliminary demonstration of a relationship which we hope to be able to describe more definitively. It does provide another example, however, of the ability of mechanical forces to bring partition walls into alignment in an actively dividing tissue.

We hope to be able to probe the cell partitioning process in greater detail once we can isolate cellular dimensional change and applied force as components of cell plate orientation. When this is possible we shall be able to test directly the notion of an intracellular dimension sensor as being the principal determinant of pre-prophase band orientation which in turn predicts the final orientation of the cell plate; we shall also be able to correlate consistency of partition wall orientation with dimensional growth of the tissue (tissue strain rate).

The concept of the 'cell plate dilemma' may be more difficult to test directly (e.g., setting up an experiment which simulates real life focusing forces on a single cell and thereby stimulates meiotic behaviour); but it should be possible to demonstrate that highly directional forces accelerate mitotic activity, from which one could infer that the lack of directional forces inhibits mitotic activity by confounding the intracellular sensor. Such a demonstration would constitute a confirmation of the most essential ingredient of the cell plate dilemma: that the rapid and effective organization of the mitotic apparatus depends upon the presence of unequivocal directional cues.

Conclusion

My aim in this essay has not been to deny the logic of the theory of positional information as originally proposed by Wolpert, but rather to question its applicability and its potential for leading us to a direct understanding of the development of form and pattern in plants.

Specifically, my contention is that although the syntax and logic of positional information can be brought to bear on the problems of organogenesis and development in plants, it will prove to be mis-applied and may only delay the emergence of a developmental theory that evolves, as it must, from an understanding of the unique structural and mechanical sensitivities of plant tissues and the epigenetic possibilities which they offer. Indeed, from my point of view, the principal difficulty with the theory of positional information as originally applied is that it commits one to understanding development in terms of diffusible morphogens and

immediate genetic determinism. The alternative which I offer here has the
genes producing similar but versatile decision-making machinery in each
cell but playing no direct part in morphogenesis. The coupling of predictable responses to mechanical input is the principal element in the generation of both form and pattern.

I have tried to show that, especially for the growing shoot apex, the
questions which must be asked of an effective control system based on
positional information are too severe. What is the morphogen, and how is
it focused and sensed? Where are the sources and sinks? How can
individual cell plate orientations be specified from genetically-encoded
information? Philosophically too, the information-heavy role which is
required of the genome in supplying a more or less complete set of instructions for the elaboration of mature form in plants, seems unnecessary.

Lastly, I have tried to provide an alternative that relies on the more
natural logic inherent in the essential structural relationship of a cell to
the tissue in which it is an element. I have outlined a control cycle whereby
form, which itself is an agent actively shaping the flow of internally-
generated growth stresses, conspires to control the pattern of cell division
and the subsequent form; thus shape begets shape.

References

Bryant, S. V., French, V. & Bryant, P. J. (1981). Distal regeneration and symmetry.
    Science, **212**, 993–1002.
Cohen, M. H. (1971). Models for the control of development. *Symposia of the Society for
    Experimental Biology*, **25**, 455–76.
Goebel, K. (1905). *Organography of Plants*. Oxford: The Clarendon Press.
Green, P. B. (1980). Organogenesis – a biophysical view. *Annual Review of Plant
    Physiology*, **31**, 51–82.
Green, P. B. & Brooks, K. E.(1978). Stem formation from a succulent leaf: its bearing on
    theories of axiation. *American Journal of Botany*, **65**, 13–26.
Green, P. B. & Lang, J. M. (1981). Toward a biophysical theory of organogenesis:
    birefringence observations on regenerating leaves in the succulent, *Graptopetalum
    paraguayense* E. Walther. *Planta*, **151**, 413–26.
Holder, N. (1979). Positional information and pattern formation in plant morphogenesis
    and a mechanism for the involvement of plant hormones. *Journal of Theoretical Biology*,
    77, 195–212.
Inoue, S. & Sato, H. (1967). Cell motility by labile association of molecules. The nature of
    mitotic spindle fibers and their role in chromosome movement. *Journal of General
    Physiology*, **50** (Symposium supplement), 259–92.
Lintilhac, P. M. (1974). Differentiation, organogenesis, and the tectonics of cell wall

orientation. III. Theoretical considerations of cell wall mechanics. *American Journal of Botany*, **61**, 230–7.

Lintilhac, P. M. & Jensen, W. A. (1974). Differentiation, organogenesis, and the tectonics of cell wall orientation. I. Preliminary observations on the development of the ovule in cotton. *American Journal of Botany*, **61**, 129–34.

Lintilhac, P. M. & Vesecky, T. (1980). Mechanical stress and cell wall orientation in plants. I. Photoelastic derivation of principal stresses in a model. With a discussion of the concept of axillarity and the significance of the "arcuate shell zone". *American Journal of Botany*, **67**, 1477–83.

— (1981). Mechanical stress and cell wall orientation in plants. II. The application of controlled directional stress to growing plants; with a discussion on the nature of the wound reaction. *American Journal of Botany*, **68**, 1222–30.

Sachs, T. (1978). Patterned differentiation in plants. *Differentiation*, **12**, 65–73.

Salmon, E. D. (1975). Pressure induced depolymerization of spindle microtubules. II. Thermodynamics of in vivo spindle assembly. *Journal of Cell Biology*, **66**, 114–27.

Sinnott, E. W. (1943). Cell division as a problem of pattern in plant development. *Torreya*, **43**, 29–34.

— (1960). *Plant Morphogenesis*. New York: McGraw Hill.

Wardlaw, C. W. (1955). *Embryogenesis in Plants*. New York: Wiley.

Wolpert, L. (1971). Positional information and pattern formation. *Current Topics in Developmental Biology*, **6**, 183–224.

5

Positional control of development in fungi

DAVID MOORE

Filamentous fungi must be unique in offering, as part of the normal pattern of growth of one organism, the opportunity to study the behaviour of independent cells undergoing vegetative growth as well as the behaviour of cells which are contributing to the structure and development of complex organs and tissues. The free-living microbial cell shows great flexibility and adaptability in responding to its environment but the contact with the environment is so close that the concept of the behaviour of the cell has no meaning unless the whole system of cell plus environment is considered. This is not the case for the cells of a tissue where the organization and metabolism of the cells which constitute the tissue can in large measure determine the environment of individual cells. However, the contribution to a tissue must be viewed as carrying with it some loss of freedom of response – a diminution of the flexibility and adaptability of which the individual cell may be capable – in favour of the concerted action required of the tissue as a whole.

In most organisms the behaviour of independent cells can only be studied under the most artificial conditions by investigation of cell cultures. For the fungi the vegetative growth of 'undifferentiated' cells in the form of a mycelium is a normal part of the life cycle. Yet that same mycelium may give rise to fruiting bodies which are as complex in structure and every bit as massive (some Zambian mushrooms can approach one metre in cap diameter (Piearce, 1981)) as are the organs and tissues of higher organisms. This, of course, is a very considerable experimental advantage. The only other group of organisms which offers the same sorts of experimental attractions are the slime moulds, and even these, as the class Myxomycetes, have been claimed as kindred of the true fungi.

Such uniqueness, though, does bring with it some difficulties. One such has already been encountered by the use of the word 'cell' in the first paragraph. In this essay 'cell' will be used to refer to discrete compartments; whether those compartments are in free cytoplasmic contact with one another, and whether they are uninucleate or multinucleate, will vary with the situation.

The initial emphasis of this essay will be on the growth and structure of the mycelium. The sorts of controls which operate between the constituent cells to determine the pattern of growth of mycelial hyphae will be discussed. The form and structure of the complex organs and tissues will then be illustrated and an attempt made to indicate how the behaviour of mycelial hyphae may be modified to direct morphogenesis of these structures.

Organization of mycelial growth

Control of hyphal growth

Filamentous fungi are well adapted to the colonization of solid substrata. By hyphal extension and regular branching the mycelium can increase in size without disturbing the cell volume/surface area ratio so that metabolite and end-product exchange with the environment can involve translocation over very short distances. Growth of the mycelium, of course, is not haphazard and Bull & Trinci (1977) have identified three mechanisms involved in regulating the growth pattern of undifferentiated mycelia. These are: the regulation of hyphal polarity, the regulation of branch initiation, and the regulation of the spatial distribution of hyphae.

Fungal hyphae are as variable between species as is any other aspect of fungal biology, but generally speaking the hyphal filament, when separated into compartments by cross-walls, has an apical compartment which is perhaps up to ten times the length of the intercalary compartments. The septa which divide hyphae into cells may be complete (imperforate), penetrated by plasmodesmata, or perforated by a large central pore. The latter may be open (and offer little hindrance to the passage of cytoplasmic organelles and nuclei), or protected by a complex cap structure derived from the endoplasmic reticulum (the dolipore septum of basidiomycetes). Septal form and function have recently been reviewed by Gull (1978), and it is worth noting, in relation to problems associated with the communication of signals, the conclusion that septal form may be modified by the hyphal cells on either side of the septum, and may vary according to age, position in the mycelium, or position in the tissues of a differentiated structure.

Growth of the hypha involves the integration of cellular growth processes so as to produce an ordered sequence of events contributing to a duplication cycle which is exactly analogous to the cell cycle of uninucleate cells (Trinci, 1979). Although it is the case that the hypha is a filament composed of numerous cells connected end to end, it is essential to appreciate that hyphal growth is highly polarized, true extension growth being absolutely limited to the hyphal tip, so the whole morphology of the hypha depends on events taking place at its apex (Grove, 1978). It follows from this that the pattern of hyphae in a mycelium, which is largely a consequence of the distribution of hyphal branches, depends on the pattern of formation of the hyphal tips which initiate the branches. It seems now to be generally accepted that the materials necessary for hyphal extension growth are produced at a constant rate (related to the specific growth rate) throughout the mycelium. Under the influence of a mechanism which achieves polarized transport (Trinci, 1978*a*), these materials are transported towards the tip of the growing hypha. Among the materials taking part in this polarized transport are the cytoplasmic vesicles, which are thought to contain wall precursors and the enzymes needed for their insertion into the existing wall, that seem always to be involved in primary wall growth in the hypha (Bartnicki-Garcia, 1973). Trinci (1974, 1978*b*, 1979) has argued that lateral branches are formed at locations where these vesicles (and other components) affect the rigidified wall of the hypha so as to produce a new 'hyphal tip'. What specifies the site of the branch initiation is not entirely clear, though it has been shown that changes in the ion flow pattern accompany branch initiation in *Achlya* (Kropf, Lupa, Caldwell & Harold, 1983). Anything which interferes with the polarized flow of vesicles and other materials towards the established tip would promote a localized accumulation which may initiate a branch. In septate moulds there is often a correlation between septation and branch initiation such that branches arise just behind septa. This implies that the apical flow of vesicles is interrupted by the septum and accumulated vesicles interact with the lateral wall of the hypha. Trinci (1974) showed that a new branch is initiated when the mean volume of cytoplasm per hyphal tip (the hyphal growth unit) exceeds a particular critical value. For a range of fungi the hyphal growth unit increased following spore germination but then exhibited a series of damped oscillations tending towards a constant value (Trinci, 1974). Such constancy demonstrates that over the mycelium as a whole, and not just in single hyphae, the number of branches is regulated in accord with increasing cytoplasmic volume.

Control of hyphal branching

Initiation of a hyphal branch amounts to the initiation of a new gradient of polarity by disturbance of the original gradient. Basically similar events must occur during spore germination when the polarity of the emerging hypha is first established. Most fungal spores swell during germination. In part this is due to imbibition of water but it is also nutrient dependent and involves active 'spherical growth' (Bartnicki-Garcia, Nelson & Cota-Robles, 1968) which can increase the spore diameter by up to three times. During spherical growth the vegetative wall is synthesized beneath the spore wall by the uniform, isotropic, deposition of wall materials (Bartnicki-Garcia, 1973). This non-polarized pattern of wall growth is in some way altered so that in one, or perhaps a few, regions vesicles accumulate and germ tube outgrowths occur which eventually become vegetative hyphae. The nature of the polarizing influence which converts isotropic spherical growth to non-isotropic germ-tube growth is unknown, but germ tubes and hyphae are alike in that their extension growth must be organized by reference to the growth of other germ tubes or hyphae around them.

The mechanism by which this organization is effected has been described as an autotropism – a growth response to a unidirectional stimulus emitted by the same organism or by a separate individual of the same species. Spores express autotropism in the directions of growth of germ tubes produced by neighbouring spores. Three types of interaction may occur: adjacent spores may be neutral with respect to one another, or may show positive, or negative, autotropic behaviour. Emerging germ tubes of paired spores of *Botrytis cinerea* germinating in liquid medium tend to grow towards one another – positive autotropism (Jaffe, 1966), although spores of the same species show neutral autotropism when grown on the surface of solid medium (Robinson, Park & Graham, 1968). Species of *Mucor*, *Rhizopus*, and *Trichoderma* exhibited various degrees of negative autotropism (Robinson *et al.*, 1968). Only a limited number of fungi have been examined and though there are some differences between them, the majority show negative autotropism. Broadly speaking, this is also true for hyphae. Although the early growth in a large spore population may be erratic overall, by the time a young mycelium is established the peripheral hyphae grow in a direction which is diametrically away from the main mass of mycelium. These phenomena have usually been explained by the assumption that they result from a chemotropic response to some (unknown) factor in the environment. Autotropism between spores has generally been studied from the point of view of investigating the relationships between adjacent spores. Hence the distances involved have usually been small – in the region of a few micrometres – and as autotropic effects become less marked with increasing distance, labile chemical species have been postulated to account for the observed effects. Thus, Jaffe (1966) accounted for the positive autotropism exhibited by *Botrytis cinerea* spore pairs (the spores of a pair either touching or being up to 10 μm apart) by postulating '... a diffusible, unstable, locally effective, macromolecular growth stimulator ...' which '... while initially emitted uniformly by each spore, comes to be emitted most rapidly by the very presumptive growing points that it favors' (Jaffe, 1966). It was noted that the spores became negatively autotrophic when the medium was equilibrated with 0.3 to 3% CO_2, and Jaffe (1966) concluded that under these conditions interaction between the spores was dominated by a locally effective growth inhibitor, although the growth stimulator was still thought to be having some effect. A diffusible inhibitor of germ-tube growth was also thought likely to be responsible for the negative autotropic effects described by Robinson *et al.* (1968) between paired spores of *Rhizopus*, *Mucor* and *Trichoderma*. Here again there was a decline in the degree of response as spore pairs separated by greater distances (up to 40 μm in this study) were observed.

Stadler (1952) described experiments in which a dense suspension of spores of *Rhizopus stolonifer* (in agar medium) induced negative autotropism in a less-dense suspension some 2.5 mm away. Arthrospores of *Geotrichum candidum* behaved similarly (Robinson, 1973a). Over these sorts of distances it is unlikely that compounds as labile as that suggested by Jaffe (1966) can have much effect. Although Stadler (1952) was unable to demonstrate any chemotropism caused by cell-free filtrates of dense spore suspensions and therefore suggested that the active metabolite is labile, germinating spores do produce stable metabolites which can inhibit germination of spores of the same species. Spores also produce metabolites which can stimulate germination (Robinson *et al.*, 1968; Robinson, 1973b), and Stadler (1952) detected production of a stable promotor of hyphal extension growth by germinating spore suspensions of *Rhizopus stolonifer*.

There is, therefore, considerable suggestive evidence to the effect that substances may be produced which are able to organize the growth pattern of neighbouring spores or hyphae. The chemical nature of the compounds involved is unknown. Exhaustion of metabolites and production of staling substances by metabolism of the nutrients of the medium have both been proposed as ways in which a gradient of chemotropic activity could be established. Robinson (1973c) has suggested that a gradient of oxygen, in an environment where oxygen is the metabolite most likely to be growth limiting, could account for a variety (but not all) of the autotropic reactions which have been reported. It is envisaged that a gradient in respiratory activity across a spore or hyphal tip could polarize the accumulation of vesicles and thereby determine the direction of germ-tube emergence or growth of the hyphal apex. Further, the position of branch initiation can be correlated with the oxygen concentration gradient in a particular experimental system (Robinson, 1973*b*), so it is feasible that the arrangement of leading and branch hyphae is also influenced by such a gradient. Since oxygen is only sparingly soluble and has a rate of diffusion in water some four orders of magnitude less than that in air, steep oxygen gradients are frequently encountered in nature. Similarly, uptake and accumulation processes in fungi are sufficiently rapid and effective for the cell to act as a 'sink' for many metabolites. It is consequently relatively easy to account for any autotropic response by assuming that growth is directed in one or other direction along these nutrient gradients. Some of the observed tropisms must be based on this sort of response, but there are other observations which imply a degree of specificity which argues against total dependence on nutrient gradients.

If the patterns of hyphal growth and branching were entirely dependent on their immediate environment then mycelial form would be determined rather mechanically. Yet, for one thing, there is plenty of evidence of a genetic control being exercised. An interesting example is the considerable difference between mycelial morphology of monokaryons and dikaryons of *Coprinus cinereus*. The monokaryon is generally comprised of narrow hyphae (4 μm in diameter) which have branches emerging virtually at right angles to the main axis. Dikaryons, as well as having thicker hyphae (7 μm in diameter) which grow much faster than those of the mono-karyon, show a very acute angle of branching and, additionally, have clamp connections. The latter are specialized branches that grow in the opposite direction to their parental hyphae before fusing with the latter (Casselton, 1978). These differences arise as a result of the genotypic change occasioned by the co-existence of two compatible nuclei in the same cytoplasm. Clutterbuck (1978) has reviewed the variety of mutants affecting hyphal morphology, while Lysek (1978) considers another phe-nomenon – mutants causing morphological rhythms. A variety of strains show spontaneous rhythmic alterations in branching patterns which give rise to successive zones in the extension growth of the colony, forming concentric bands. Radially-placed barriers interrupt the regularity of the bands, implying that the periodicity of branching in adjacent hyphae depends on a synchronization of metabolic activities which requires lateral communication.

It seems reasonable to expect that nutrient depletion in the vicinity of a hypha will create gradients which will ensure that other hyphae and hyphal branches grow away from it (negative autotropism). The formation of complex structures, like fruit bodies, involves positive autotropism and very specific systems may therefore have arisen to provide the attractants and responses necessary to secure the growth of hyphae towards one another and across each other's gradients of nutrient depletion. Nothing is known of the phenomena involved in the aggregation of hyphae during the initial stages of organ development. But there are some other behaviour patterns which indicate the sorts of factors which could be involved.

When oidiospores of certain basidiomycetes are spread on solid medium just in front of an advancing mycelium the mycelial hyphae often show a positive tropic response towards the oidia with which the hyphae eventually fuse. This 'homing reaction' is often not very specific but extensive and rapid homing can be followed by vacuolization and death of the homing hyphae. Such a lethal reaction occurs between species which seem on other criteria to be very closely related, and studies of the homing and lethal reactions have been used to distinguish relationships between species of basidiomÿcetes (Kemp, 1975, 1977). Kemp (1970) and Bistis (1970) showed that the oidia produced a growth substance which elicited a positive chemotropic response in the hyphae. It is highly unlikely that such a mechanism would employ a simple metabolite as chemotropic agent since such a strategy would lack the taxonomic specificity which is observed. So the implication is that the process depends on production of a very specific chemical attractant.

Characterized chemo-attractants

Aquatic fungi

A number of lower fungi produce chemo-attractants involved in growth processes leading up to sexual reproduction which can quite properly be described as sex hormones (Bu'lock, 1976). *Allomyces* is an aquatic fungus which produces uniflagellate motile gametes. These are differentiated as male and female although they arise on the same haploid thallus and consequently have the same genotype. The female gametes and gametangia produce a substance called sirenin to which the male gametes show strong chemotaxis. Sirenin is a bicyclic sesquiterpene diol (Fig. 5.1) which is active at concentrations less than 10^{-10} g ml^{-1}. Female gametes are only sluggishly motile but male gametes swim in random smooth arcs interrupted by stops after which the cell swims in a different direction.

Sirenin organizes the direction of swimming by shortening the run between interruptions if the cell moves away from the source of hormone
and diminishing the number of stops if the cell is moving towards the
source. Neither the metabolic origin nor the mode of action of sirenin are
known. Inactivation of the hormone by the male gamete is essential to the
overall activity but it is not known whether this results from enzymic
breakdown or irreversible binding to some component of the cell. This
chemotaxis thus leads to cell contact which is a prelude to plasmogamy.
Bu'lock (1976) suggests that the specific cell adhesion which follows establishment of contact involves special surface polysaccharides. More is
known about cell adhesion in yeasts and this will be discussed later. The
female sex hormone of another water mould, *Achlya*, has also been characterized in some detail. This material, called antheridiol, is a steroid
(Fig. 5.1) the activity of which can be detected by bioassay in 10^{-11} M
solution (Raper, 1952; Barksdale, 1969). The mating sequence reported
by Raper (1966) consisted of the development of antheridial hyphae on
the male, the production of oogonial initials on the female, growth of
antheridial hyphae towards oogonial initials, formation of cross-walls
separating off oogonia and antheridia, and finally, after the two made
contact, the antheridium grew through a lysed portion of the oogonial
wall, after which its own wall was dissolved. Antheridiol is produced
continuously by the female and under the influence of the hormone,
branches on the putative male thallus which might otherwise grow out as
vegetative branches are caused to elongate rapidly and differentiate into
antheridia. The male is also induced to excrete a second hormone, hormone B or oogoniol, and it is in response to hormone B that the female
initiates oogonial differentiation and amplifies antheridiol levels to those
which attract antheridial hyphal growth. There are thus at least two

Fig. 5.1. Structural formulae of some sex hormones of lower fungi.

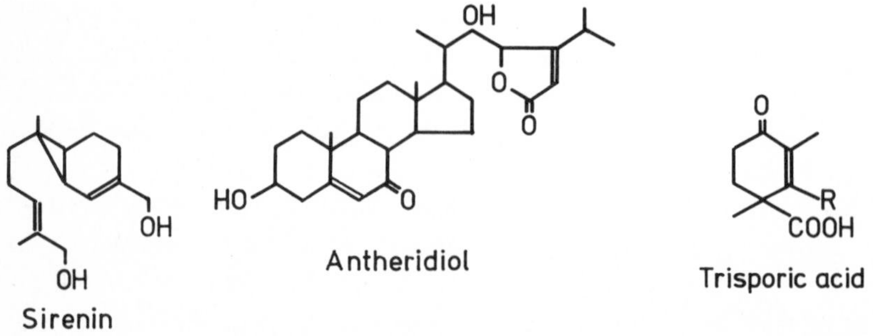

contributors to this hormonal 'conversation'; the female produces antheridiol but takes up very little itself, and does not synthesize hormone B though it does have a receptor for this hormone. On the other hand, the male makes no antheridiol but is sensitive to it, and one of the responses is to produce hormone B to which the male is insensitive. Antheridiol and hormone B are thought to be alternative products of a branched biosynthetic pathway. In the male, antheridiol amplifies that branch of the pathway which leads to hormone B synthesis and also increases respiration, induces breakdown of glucan reserves in the cytoplasm and triggers the *de novo* synthesis of cellulase. These metabolic changes contribute to processes involved in antheridium initiation – including aggregation of vesicles at the sites where initials develop (Mullins & Ellis, 1974). It is likely that broadly similar responses are elicited in the female by hormone B. It is not known how these sterols influence gene regulation but the available data seem to indicate a system very much akin to sterol regulation in animals (Horgen, 1977).

Mucorales

The only other fungi in which the activity of known hormones has been well characterized are some members of the Mucorales. These are filamentous, terrestrial, lower fungi with mycelia typically composed of unbranched coenocytic hyphae. A little way behind the advancing hyphal tips of vegetative mycelia asexual sporangiophores are produced. However, in the vicinity of a mycelium of opposite mating type sporangiophore formation is suppressed and sexual differentiation takes place, involving formation of sexual hyphae (zygophores) which grow towards each other, fusing in pairs to eventually form gametangia. Zygophore formation is determined by trisporic acid (Fig. 5.1); if this chemical is added to pure, unmated, cultures sporangiophore formation ceases and zygophores form instead (Gooday, 1973, 1974*a*; Bu'lock, 1976). Although there are a number of trisporic acid-related compounds, some of them corresponding to intermediates in the pathway, both mating types produce and respond to the same hormone. The trisporic acids are synthesized from β-carotene: the molecule is cleaved to retinal, a C_2 fragment is lost, and then there is a series of oxidations. The complete reaction sequence occurs only when both plus and minus mating types are grown in mixed culture or in an experimental set-up in which they are separated by a membrane permeable to small molecules. Both mating types have the genetic capacity to produce the enzymes of the complete pathway, but the alleles which determine the mating type repress complementary steps in the later stages of trisporic acid synthesis. Thus in plus strains synthesis of enzymes needed to form the 4-keto group is repressed by the MT^+ allele while enzymes involved in forming the 1-carboxylic acid group are repressed by MT^- (Bu'lock, Jones, Quarrie & Winskill, 1973; Bu'lock, 1975). Each mating type thus produces a precursor which only the opposite mating type can convert to trisporic acid. The precursors diffuse between the strains and have the status of prohormones which stimulate trisporic acid synthesis. Early steps in the pathway are repressed to a rate-limiting level by a mechanism which allows activation by trisporic acid. When plus and minus strains come together, therefore, the complementary synthesis of trisporic acid consequent on the co-diffusion of the prohormone precursors leads to derepression of the early part of the pathway and an amplification of overall trisporic acid synthesis. The increasing gradient of prohormone diffusing from each zygophore induces a chemotropic response. The zygophores can grow towards one another from distances of up to two millimetres. When the zygophores make contact they adhere firmly in a way that implies that mating type- and species-specific substances are formed on the zygophore surface. These features are clearly an aspect of the mating type phenotype and are necessary for completion of the mating programme – without adhesion the zygophores continue unproductive extension growth – but the nature of the substances involved is unknown.

Yeasts

Some idea of the sorts of molecules which might be involved in cell to cell contact comes from work with a number of yeast species. Many release diffusible sex hormones ('pheromones') as a prelude to the cell fusion that leads to conjugation. These substances are outside the scope of this discussion apart from noting that they prepare the cells for conjugation and contribute to the recognition of different mating types. However, the major step in the recognition of compatible cell types involves macromolecules on the cell surfaces which cause cells to agglutinate. Some of these are constitutive (i.e., cells agglutinate immediately the different clones are mixed) while others are inducible, the cells only acquiring the ability to agglutinate after growth in mixed culture. In both *Hansenula wingei* and *Saccharomyces cerevisiae* there is evidence that the molecules directly involved in agglutination – the agglutinins – are probably glycoproteins. In *H. wingei* one of the agglutinin components consists of 28 amino acids and about 60 mannose residues. The agglutinins seem to be located on surface filaments external to the cell wall. The function of the agglutinins is to bring cells of opposite mating type together. They do this by virtue of their ability to bind in a complementary manner, the agglutinin of one mating type binding specifically to that produced by the compatible mating type. Following this adhesion of yeast cells by complementary binding, protuberances grow out from the cell walls and when these meet cytoplasmic communication is established by dissolution of the walls. These phenomena (reviewed by Crandall, Egel & Mackay (1977)) are obviously specifically part of the mating process in yeast; yet they clearly demonstrate that fungal cells are capable of producing surface glycoproteins which, by a sort of antigen–antibody reaction, can achieve a very specific adhesion. It is exactly this sort of specific cell binding which one might expect to be part of the cell-to-cell communication which contributes to the construction of differentiated multicellular structures.

The possible role of cyclic-AMP

The aspects of sexual reproduction in a few lower fungi which have just been described represent the only cases (along with examples from related species not referred to here) in which fungal hormonal chemical attractants have been characterized. In a real sense, therefore, they exemplify the only agents of positional signalling in fungi which are known, even if imperfectly. Their service to a discussion centred on morphogenesis and its organization is that such processes, involving as they do the mutual attraction and adhesion of differentiating hyphal branches, illustrate ways in which metabolic signals can be generated and used. I have already described the vegetative hyphal growth form as one which shows a characteristically spreading growth habit; yet all complex fungal structures require that hyphae grow together and this must involve both chemotropism and adhesion. How such concerted growth is initiated and organized in filamentous fungi is completely unknown and the closest approach we have to any indication is provided by the slime mould *Dictyostelium discoideum* (Newell, 1978).

Slime moulds

When the free-living amoebae of *D. discoideum* begin to starve the cell aggregation phase is initiated; this eventually leads to the formation of the multicellular slug (pseudoplasmodium). Aggregation depends on the emission of a chemical – given the generic name of acrasin – which acts as an attractant. In the most widely known case, that of *D. discoideum* itself, the chemical attractant is the nucleotide cyclic-AMP (cAMP). Newell (1978) divides the aggregation process into seven stages. In the first stage the signal is generated. The evidence indicates that signalling is pulsatile rather than forming a continuous gradient. The primary source of the oscillations is enzymic; it may be the adenylate cyclase or the cytochrome chain. The second stage is reception of the signal which seems, at least temporarily, to be the responsibility of receptor molecules on the surface of the amoebae. The characteristics of these receptors are similar to those of some hormone receptors in animals. Thirdly, the receiving cell destroys the incoming signal – via a membrane-bound and a soluble extracellular phosphodiesterase – and, fourthly, relays the signal. There is a delay of about 12 s between receipt of the incoming signal and generation of a new pulse of cAMP, and in part this delay determines the rate of propagation of the signal through the population. The fifth step is the chemotactic response itself. Receipt of a pulse of cAMP causes the amoebae to move towards the signal source for about 100 s, during which time the cells move about two cell diameters. As well as causing the chemotactic response, the pulse of cAMP initiates cellular changes associated with differentiation including the synthesis of cell-surface glycoproteins which are involved in cell adhesion during the final stage – the association of the cells into a multicellular aggregate. Once the aggregate is formed proper progress of the developmental program requires the maintenance of cell contacts; and the consequences of mechanical separation of the cells imply that intercellular signalling is required continuously for completion of development. There is some evidence that cAMP continues to play a role as a morphogen in the aggregated slug, but ammonia has also been claimed to regulate differentiation (Schindler & Sussman, 1977; Sussman, Schindler & Kim, 1977).

Basidiomycetes

Not surprisingly, a great deal of interest has been focused on the likely role of cAMP in controlling differentiation in other organisms. Before discussing one of these cases it is worth emphasizing that this 'magic bullet' does not have the same role in all slime moulds; in *Polysphondylium violaceum* the acrasin (aggregation signal) is a small peptide (Wurster, Pan, Tyan & Bonner, 1976; Bonner, 1977). Nevertheless it is inevitable, considering the widespread degree of interest in cAMP, that attempts have been made to find a developmental role for the compound in filamentous fungi. The most assiduous search has been conducted by Uno and Ishikawa in their work with the basidiomycete *Coprinus cinereus* (using the name *C. macrorhizus*). In a long series of papers these workers have investigated the metabolism of cAMP during fruit-body development in *Coprinus*. Superficially, the developmental problem in a basidiomycete like *Coprinus* is rather similar to that faced by *Dictyostelium*. The spreading growth of the vegetative mycelium can be likened to the migrating slime mould amoebae. The analogue of amoebal aggregation would then be the aggregation of hyphae into the initials of structures such as fruit bodies. Additions of cAMP to preparations of '*Coprinus lagopus*' (= *C. cinereus*) have been reported to accelerate the production of fruit-body primordia (Matthews & Niederpruem, 1972). A fruit-body-inducing substance (FIS) which was able to induce fruiting in certain mutant strains of *C. cinereus* has been extracted from various tissues and identified as cAMP (Uno & Ishikawa, 1973). Although most of the work done by Uno and Ishikawa has been concentrated on the phenomenon of monokaryotic fruiting (which is an abnormality of certain mutant monokaryons) they have included the dikaryon (the normal mycelial origin of fruiting bodies) in sufficient of their work for the picture to be reasonably clear. They have shown that the mycelium accumulates cAMP at the onset of fruiting, the accumulation being first evident at about the time that fruit-body formation is first initiated. The accumulation proceeds to a late primordial stage, but then the amount of cAMP declines as the primordium matures (Uno, Yamaguchi & Ishikawa, 1974). However, supplementation of the medium with cAMP did not induce fruiting in these Japanese strains of *C. cinereus* (Uno & Ishikawa, 1973). Some involvement of cAMP in the phenomenon of catabolite repression has also been demonstrated (Uno & Ishikawa, 1974) and is significant since fruiting is usually initiated as the carbohydrate supply of the medium is becoming exhausted. The nucleotide has also been shown to activate glycogen phosphorylase and inhibit glycogen synthetase (Uno & Ishikawa, 1976, 1978). It is quite clear from the extensive work done by Uno and Ishikawa that cAMP is closely involved in fruit body development in *C. cinereus*; however, despite the amount of work done, the role of the chemical remains obscure. Throughout their analyses Uno and Ishikawa made their extractions from mycelia together with any fruit bodies they may have produced. No attempt was made to separate the two structures. Preliminary attempts to partition the cAMP between the different parts of the organism indicate that in the primordium all of the cAMP is located in the fruit-body cap whereas in more mature fruit bodies it is found in the upper part of the stipe (Darbyshire, 1974). Moreover, when separate analyses are made of the mycelium and its fruiting structures, the cAMP is found to be accumulated in the fruiting bodies and particularly heavy accumulations are observed in the very earliest stages (Milne, 1977). In the fruit body initials cAMP is accumulated to levels up to four times the highest recorded by Uno & Ishikawa (1973). These peaks of cAMP concentration seem to be located in the parts where metabolic changes, especially the mobilization and utilization of reserve materials, occur most rapidly (Moore, Elhiti & Butler, 1979). A particularly interesting point is that, in the first case, the fruit body initial, and subsequently, the cap of the primordium are the recipients of very large quantities of reserve materials channelled to these locations from elsewhere. In the average fruit body about one milligramme of glycogen, first accumulated in the stipe base, is relocated into the cap of the primordium (Moore *et al.*, 1979), while polysaccharide, presumably glycogen too, is mobilized from the mycelium and relocated in the developing fruit body initials (Madelin, 1960). Thus it may be that cAMP serves to organize the directionality of these translocation streams. The evidence does seem to suggest that reserve materials, notably glycogen but probably other compounds as well, are translocated through the hyphae of the mycelium and of the primordium along cAMP gradients. This process, rather than the autotropic growth of the hyphae, may be the broad equivalent of the chemotropic effect which cAMP has on *Dictyostelium* amoebae. *Coprinus* hyphae are unable to migrate like the amoebae but their contents can migrate.

Basidiomycete 'social organization'

This inward translocation of materials towards centres of development highlights a feature of the mycelium which may influence the form and nature of positional control signals. Buller (1931) was the first to describe what he called 'social organization' in the basidiomycete *Coprinus sterquilinus*. He indicated that the numerous hyphal fusions which interconnect the hyphae as they grow through the dung substrate provide the direct routes through which nutrients flow towards the developing fruit. He pointed out that where the substrate has been inoculated with a number of spores the hyphal fusions between the individual mycelia to which they gave rise still allow such nutrient flow and ensure that a fruiting body is still produced on that substrate, although the mycelia may individually have invaded an amount of substrate which is insufficient to provide for formation of a fruit body. In other words, it appears that in these circumstances the mycelia co-operate rather than compete. Such a compound (or 'unit') mycelium, which is genetically heterogeneous but acts as a physiological unit, has been seen as a consequence of the readiness with which hyphal fusions occur (Burnett, 1976). In many ways such a concept views the mycelium like a whole organism in which the reproductive capacity is concentrated in a limited section of the body while being nutritionally supported by the activities of the whole organism. Of course the nuclear heterogeneity sets the situation apart from that evident in higher organisms where the 'division of labour' between cells and tissues is most highly developed. Nevertheless, there is a clear implication that in some way in these compound mycelia the developmental potential of one part is expressed at the expense of the others and this must reflect a degree of organization which requires the dispatch and receipt of some sort of positional control signal.

It must be stated that promiscuous hyphal fusion and the unit mycelium are not universally observed in fungi. Rayner & Todd (1979) review situations in which the opposed relation between mycelia (specifically, between different dikaryons) of the same species is observed – a relationship in which neighbouring mycelia display mutual antagonism. Indeed, these authors provide an authoritative discussion of the reality of the individuality of fungal mycelia. In their view the unit mycelium concept could well only apply in limited circumstances where the mycelia are closely related (a criterion likely to be met in the circumstances envisaged by Buller (1931)). In wood-decaying fungi certainly, and probably in others, Rayner & Todd (1979) believe that the population is made up of dikaryotic individuals analogous to the diploid individuals which make up the populations of higher organisms. If this is the case then it may have implications for the nature and need for signals to pass between neighbouring individuals in the population.

Differentiation of multicellular structures

I have so far discussed events which essentially can be described as involving hyphal interactions within the mycelium. But in the various complex structures produced by the higher fungi in particular, the interactions are much more of the sort that might be anticipated between differentiating cells in organized tissues. The fungi produce many organized structures and only a few of these will be illustrated to give some idea of the level of complexity that these supposedly lower organisms can attain. The selected illustrations are not exhaustive; they happen to be the examples with which I am most familiar, but I believe them to be representative.

Basidiomycete structures

Vegetative organs

The most highly organized structures are the mushroom fruit bodies of the Basidiomycetes, but the vegetative mycelium is capable of elaborating a range of other organs. Among these are sclerotia, strands and rhizomorphs. These represent essentially alternative strategies. The sclerotium is a resting structure with cells specialized to withstand adverse environmental conditions; it tends to be spherical, or at least globose, and fairly small (a few millimetres or less in diameter). Sclerotia generally show some degree of radial symmetry with an outer layer of closely packed rind

Fig. 5.2. Scanning electron micrograph of a broken sclerotium of *Coprinus cinereus*. Harvested sclerotia were frozen in liquid nitrogen and fractured by impact while still frozen. After critical-point drying they were scattered onto an adhesive-covered stud and coated with gold. The image shows the compact internal medulla surrounded by the smooth, thick-walled rind. Scale bar represents 20 μm.

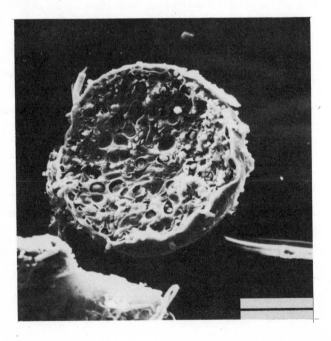

cells which have thickened and pigmented walls surrounding a medullary region of interwoven hyphae (Fig. 5.2). The sclerotium is non-motile; it remains where it is formed to withstand adversity and then germinates when conditions improve. Strands and rhizomorphs, on the other hand, are root-like structures composed of parallel-running hyphae which are interwoven and held together by fusions. Strands increase in diameter by the development of further hyphae from the origin. Rhizomorphs are more highly organized constructions consisting of thousands of hyphae; they are internally differentiated and grow apically from something analogous to an apical meristem. The difference between strands and rhizomorphs may simply be one of degree of co-ordination between the component parts; both are organs of mycelial migration and nutrient translocation. They grow out from colonies on a 'food base' and will grow across non-nutrient surfaces for considerable distances. Low nutrient levels tend to promote strand and rhizomorph initiation, but in *Armillaria mellea* the initiation of a rhizomorph tends to inhibit the formation of further such structures in the vicinity (Garrett, 1953). There is presumably some way in which the growth of one structure controls the initiation of similar structures in the neighbourhood. Garrett (1953) suggested that this is a nutritional phenomenon, the inception of independent rhizomorphs being inhibited by the translocation of nutrients into the existing rhizomorph; but even if this is so there must be some sort of signal which directs the distribution of nutrients. The organization of the rhizomorph is in some way positively maintained. This, too, may be based on a dependence on nutrient gradients since Watkinson (1971a, b) has shown that translocated nutrients leak from strands growing over a nutrient-depleted medium and Garrett (1960) notes that the hyphae composing the strand separate from one another when they grow over a nutrient-rich medium. Both these observations support the idea (Garrett, 1960) that the hyphae of the strand are positively chemotropic towards nutrients, but since the strand grows away from the food base this cannot be true for extension growth of the strand as a whole.

Burnett (1976) identified three factors which are involved in these vegetative aspects of differentiation. These are: synchronization of the behaviour of adjacent hyphae; regulation of the balance between extension growth and branch formation; and the internal control of localized growth through competition for nutrients. Very little is known about the ways these parameters are regulated, but such mechanisms must operate with even greater sophistication in determining the structure of fruit bodies. These, too, are initiated by the localized aggregation of hyphae.

In C. cinereus there is evidence that the same initial hyphal aggregate can serve as the starting point for either a sclerotium or a fruit body (Moore, 1981*a*), but whereas the sclerotium develops a radial symmetry, the fruit-body developmental sequence imposes a polarizing influence on the aggregate, and from a very early stage the shape of the developing mushroom is clearly evident. This is illustrated for *C. cinereus* in Fig. 5.3. Such images

Fig. 5.3. Vertical sections of initials of *C. cinereus* fruit bodies. The structure shown in (*a*) was 0.8 mm tall, that in (*b*), 1.2 mm tall. From Moore *et al.* (1979).

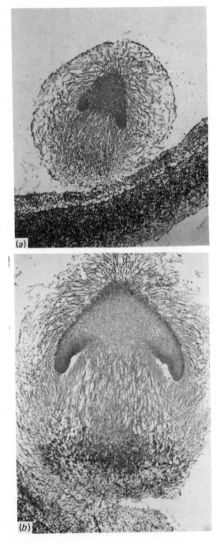

are fairly representative of most gilled mushrooms (Taber, 1966), though the 'bud' or 'button' varies in physical size and rate of development. Further growth of the primordium leads to the delimitation of its various tissues and establishment of the unmistakable mushroom form (Fig. 5.4).

Fig. 5.4. Section of a primordium of *C. cinereus* which was 15 mm tall. The figure is a photograph of a montage of light micrographs. From Moore *et al.* (1979).

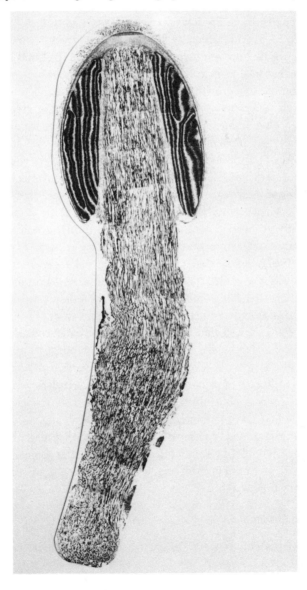

Mushroom development

Development of such a highly organized structure can proceed only under the most rigid control, but we are almost completely ignorant of the means by which such control may be exercised. For nearly 150 years, though, efforts have been made to understand the regulation of mushroom growth by using experiments involving surgical removal of portions of the developing fruit body. A number of such studies have been carried out over the years. These are reviewed by Gruen (1963, 1967), Gruen & Wu (1972*a*, *b*) and, especially, Burnett (1976). There have been consistent indications that growth factors produced by the fruit-body cap influence, perhaps control, growth of the stipe. Total removal of the cap leads to cessation of stipe growth. When segments of cap are left attached to the stipe, the latter shows a growth curvature with the greatest extension growth being immediately beneath the remaining sector of cap. The source of the implied growth factor seems to be the gills themselves. Removal of cap tissues other than the gills had little effect on growth. Removal of the gills, leaving the rest of the cap tissue intact, was followed by a decline in growth. The gills also seem to promote growth of the cap since if a small amount of gill tissue is left intact at the outer edge it is able to safeguard development of the cap even though the amount of tissue removed is sufficient seriously to reduce stipe growth. Thus both Gruen (1963, 1967) and Borriss (1934), working with *Agaricus* and *Coprinus* respectively, concluded that the gill lamellae were the origin of controlling factors for stipe elongation and cap expansion. Some qualification of this generalization may be necessary in individual cases. For example, Gooday (1974*b*) showed that later stages of stipe extension in *Coprinus* were quite independent of the cap, confirming the report of Borriss (1934) that the influence of the cap in this organism is restricted to the earliest (but formative) stages of primordium growth. Nevertheless, it is clear that the gills do exert a controlling influence. There have been claims for the isolation of growth-regulating substances although many of these remain to be substantiated. Attempts to identify the alleged growth promoters have revealed mixtures of amino acids including, especially, derivatives of glutamate, but until more extensive work has been done these suggestions can carry little weight.

Pattern formation in the fruit body cap

Most of the studies done on growth regulation in the mushroom fruit

Fig. 5.5. Structure of the gill surfaces (hymenia) of primordia similar to that shown in Fig. 5.4, as revealed by scanning (*a*) and transmission (*b*) electron microscopy. Scale bars represent 10 μm. From Moore *et al.* (1979).

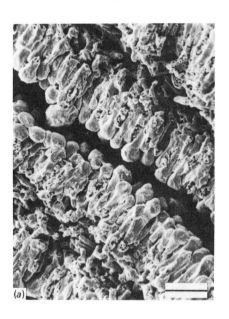

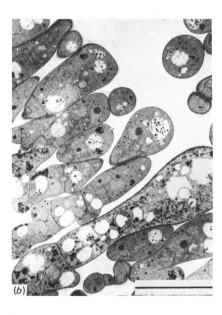

body have concentrated on growth of the stipe, yet they have revealed
that some factors do influence cap expansion. It is the cap which displays
the greatest morphological, and consequently morphogenetic, complexity.
Burnett (1976) has noted that the work which has been done points to the
regulation of such features as gill initiation, gill spacing, cell morphology
and cell packing. The cell differentiation which occurs can be illustrated

Fig. 5.6. Structure of the hymenium of a mature *C. cinereus* fruit body as revealed
by scanning (*a*) and transmission (*b*) electron microscopy. Scale bars represent
10 μm. Note the close-packed pavement of box-like paraphyses and the regularly
interspersed club-shaped basidia. From Moore *et al.* (1979).

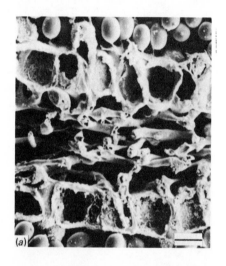

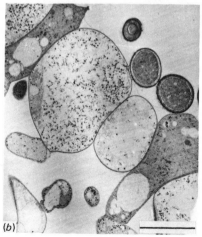

by reference to *Coprinus*. Early in development the gill surface (hymenium) is made up of a loosely-packed array of essentially undifferentiated hyphal tips (Fig. 5.5). By the time the cap has matured three morphologically distinct cell types have been produced, for at maturity the hymenium consists of a pavement of highly inflated paraphyses within which are embedded a great many basidia and a scattering of cystidia (Figs. 5.6,

Fig. 5.7. (*a*), Scanning electron micrograph of a torn gill surface of *C. cinereus* showing the large bulbous cystidia scattered in the regular carpet of spore-bearing basidia. Scale bar represents 200 μm. (*b*), Edge of the cut gill surface of a mature hymenium showing the point of emergence of a cystidium. Scale bar represents 10 μm. From Moore *et al.* (1979).

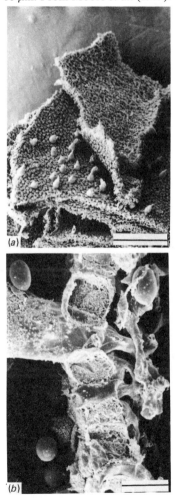

5.7). The development of these structures could provide a classic example of a morphogenetic field, the more so since the gill plate grows in two directions (inwards, towards the stipe, and along the length of the cap as the cap increases in diameter) which are essentially at right angles and which could contribute to a co-ordinating system to which cell differentiation could be referenced. Models have been proposed which account for the development of such morphologies by assuming that the progressive distribution of a substance – the morphogen – is the factor which regulates cell differentiation. The models have been most effectively analysed by Meinhardt & Gierer (1974) who show how two-dimensional patterns closely similar to those observed in the basidiomycete hymenium may be generated, based on activators and inhibitors capable of diffusing through the tissues. However, the developing mushroom is faced with a problem which may be of a different order of difficulty to that met by higher plants and animals. Models of morphogenetic processes based on distribution of morphogens are very dependent on adequate communications within the tissue, communication which must extend over many cell diameters. Higher animals and higher plants are well provided – via cell processes, gap junctions, plasmodesmata and the like – with avenues for such communication, but the developing basidiomycete hymenium is an array of separate cells (Fig. 5.5). There is no evidence for any lateral cytoplasmic contact between neighbouring cells. Indeed, the electron micrographs show considerable space between hymenial cells of primordial tissues at about the time that the morphogenetic pre-patterning might be expected to be taking place. Of course, the cells of the hymenium are branches from the subhymenial hyphae, but any suggestion that communication of morphogens takes place through the subhymenium must account for the fact that adjacent branches from the same subhymenial hypha can have different morphogenetic fates (Fig. 5.8). Models involving gaseous or volatile morphogens could clearly be constructed, and might have some basis in known metabolic events (Ewaze, Moore & Stewart, 1978; Moore 1981*b*), but it is a sad fact that we are largely ignorant of the detailed aspects of tissue construction – the relationships between adjacent cells are known in only the vaguest way – and the knowledge we do have about the possible effects of growth factors is woefully inadequate. Nevertheless, I believe the pursuit of this knowledge is worthwhile. Understanding of the ways in which these lowly organisms have solved the organizational problems associated with their morphogenesis could tell us a great deal about the more highly-regarded candidates for developmental studies.

Fig. 5.8. Transmission electron micrograph of the mature gill hymenium of *C. cinereus* showing a basidium and paraphysis arising as branches from the same subhymenial hypha. Scale bar represents 5 μm.

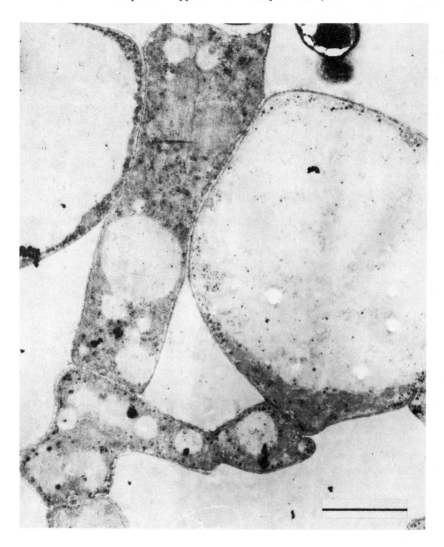

References

Barksdale, A. W. (1969). Sexual hormones of *Achlya* and other fungi. *Science*, **166**, 831–7.
Bartnicki-Garcia, S. (1973). Fundamental aspects of hyphal morphogenesis. *Symposia of the Society for General Microbiology*, **23**, 245–67.
Bartnicki-Garcia, S., Nelson, L. & Cota-Robles, E. (1968). A novel apical corpuscle in hyphae of *Mucor rouxii*. *Journal of Bacteriology*, **95**, 2399–402.
Bistis, G. N. (1970). Dikaryotisation in *Clitocybe truncicola*. *Mycologia*, **62**, 911–23.
Bonner, J. T. (1977). Some aspects of chemotaxis using the cellular slime moulds as an example. *Mycologia*, **69**, 443–59.
Borriss, H. (1934). Beiträge zur Wachstums und Entwicklungsphysiologie der Fruchtkörper von *Coprinus lagopus*. *Planta*, **22**, 28–69.
Bull, A. T. & Trinci, A. P. J. (1977). The physiology and metabolic control of fungal growth. *Advances in Microbial Physiology*, **15**, 1–84.
Buller, A. H. R. (1931). *Researches on Fungi*, vol. 4. London: Longman, Green & Co.
Bu'lock, J. D. (1975). Cascade expression of the mating type locus in Mucorales. *Proceedings of the 1974 Symposium on the Genetics of Industrial Micro-organisms.* London: Academic Press.
— (1976). Hormones in fungi. In *The Filamentous Fungi*, vol. 2, *Biosynthesis and Metabolism*, ed. J. E. Smith & D. R. Berry, pp. 345–68. London: Edward Arnold.
Bu'lock, J. D., Jones, B. E., Quarrie, S. & Winskill, N. (1973). The biochemical basis of sexuality in Mucorales. *Die Naturwissenschaften*, **60**, 550–1.
Burnett, J. H. (1976). *Fundamentals of Mycology*, 2nd edn. London: Edward Arnold.
Casselton, L. A. (1978). Dikaryon formation in higher basidiomycetes. In *The Filamentous Fungi*, vol. 3, *Developmental Mycology*, ed. J. E. Smith & D. R. Berry, pp. 275–97. London: Edward Arnold.
Clutterbuck, A. J. (1978). Genetics of vegetative growth and asexual reproduction. In *The Filamentous Fungi*, vol. 3, *Developmental Mycology*, ed. J. E. Smith & D. R. Berry, pp. 240–56. London: Edward Arnold.
Crandall, M., Egel, R. & Mackay, V. L. (1977). Physiology of mating in three yeasts. *Advances in Microbial Physiology*, **15**, 307–98.
Darbyshire, J. (1974). Developmental studies on *Coprinus lagopus* (*sensu* Buller). *Ph.D. Thesis, University of Manchester.*
Ewaze, J. O., Moore, D. & Stewart, G. R. (1978). Co-ordinate regulation of enzymes involved in ornithine metabolism and its relation to sporophore morphogenesis in *Coprinus cinereus*. *Journal of General Microbiology*, **107**, 343–57.
Garrett, S. D. (1953). Rhizomorph behaviour in *Armillaria mellea* (Vahl) Quél. I. Factors controlling rhizomorph initiation by *A. mellea* in pure culture. *Annals of Botany* **17**, 63–79.
— (1960). Rhizomorph behaviour in *Armillaria mellea* (Fr.) Quél. III. Saprophytic colonization of woody substrates in soil. *Annals of Botany*, **24**, 275–85.
Gooday, G. W. (1973). Differentiation in the Mucorales. *Symposia of the Society for General Microbiology*, **23**, 269–94.
— (1974a). Fungal sex hormones. *Annual Review of Biochemistry*, **43**, 35–49.
— (1974b). Control of development of excised fruit bodies and stipes of *Coprinus cinereus*. *Transactions of the British Mycological Society*, **62**, 391–9.

Grove, S. N. (1978). The cytology of hyphal tip growth. In *The Filamentous Fungi*, vol. 3, *Developmental Mycology*, ed. J. E. Smith & D. R. Berry, pp. 28–50. London: Edward Arnold.

Gruen, H. E. (1963). Endogenous growth regulation in carpophores of *Agaricus bisporus*. *Plant Physiology*, **38**, 652–66.

— (1967). Growth regulation in fruit bodies of *Agaricus bisporus*. *Mushroom Science*, **6**, 103–20.

Gruen, H. E. & Wu, S. (1972a). Promotion of stipe elongation in isolated *Flammulina velutipes* fruit bodies by carbohydrates, natural extracts, and amino acids. *Canadian Journal of Botany*, **50**, 803–18.

— (1972b). Dependence of fruit body elongation on the mycelium in *Flammulina velutipes*. *Mycologia*, **64**, 995–1007.

Gull, K. (1978). Form and function of septa in filamentous fungi. In *The Filamentous Fungi*, vol. 3, *Developmental Mycology*, ed. J. E. Smith & D. R. Berry, pp. 78–93. London: Edward Arnold.

Horgen, P. A. (1977). Steroid induction of differentiation: *Achlya* as a model system. In *Eucaryotic Microbes as Model Developmental Systems*, ed. D. H. O'Day & P. A. Horgen, pp. 272–93. New York: Marcel Dekker.

Jaffe, L. F. (1966). On autotropism in *Botrytis*: technique and control by pH. *Plant Physiology*, **41**, 303–6.

Kemp, R. F. O. (1970). Interspecific sterility in *Coprinus bisporus*, *C. congregatus* and other basidiomycetes. *Transactions of the British Mycological Society*, **54**, 488–9.

— (1975). Breeding biology of *Coprinus* species in the section Lanatuli. *Transactions of the British Mycological Society*, **65**, 375–88.

— (1977). Oidial.homing and the taxonomy and speciation of basidiomycetes with special reference to the genus *Coprinus*. In *The Species Concept in Hymenomycetes*, ed. H. Clemençon, pp. 259–74. Vaduz: J. Cramer.

Kropf. D. L., Lupa, M. D. A., Caldwell, J. H. & Harold, F. M. (1983). Cell polarity: endogenous ion currents precede and predict branching in the water mold, *Achlya*. *Science*, **220**, 1385–7.

Lysek, G. (1978). Circadian rhythms. In *The Filamentous Fungi*, vol. 3, *Developmental Mycology*, ed. J. E. Smith & D. R. Berry, pp. 376–88. London: Edward Arnold.

Madelin, M. F. (1960). Visible changes in the vegetative mycelium of *Coprinus lagopus* Fr. at the time of fruiting. *Transactions of the British Mycological Society*, **43**, 105–10.

Matthews, T. R. & Niederpruem, D. J. (1972). Differentiation in *Coprinus lagopus*. I. Control of fruiting and cytology of initial events. *Archives of Microbiology*, **87**, 257–68.

Meinhardt, H. & Gierer, A. (1974). Applications of a theory of biological pattern formation based on lateral inhibition. *Journal of Cell Science*, **15**, 321–46.

Milne, D. P. (1977). Differentiation in *Coprinus*. *Ph.D. Thesis, University of Manchester*.

Moore, D. (1981a). Developmental genetics of *Coprinus cinereus*: genetic evidence that carpophores and sclerotia share a common pathway of initiation. *Current Genetics*, **3**, 145–50.

— (1981b). Evidence that the NADP-linked glutamate dehydrogenase of *Coprinus cinereus* is regulated by acetyl-CoA and ammonium levels. *Biochimica et Biophysica Acta*, **661**, 247–54.

Moore, D., Elhiti, M. M. Y. & Butler, R. D. (1979). Morphogenesis of the carpophore of *Coprinus cinereus*. *The New Phytologist*, **83**, 695–722.

Mullins, J. T. & Ellis, E. A. (1974). Sexual morphogenesis in *Achlya*: ultrastructural basis for the hormonal induction of antheridial hyphae. *Proceedings of the National Academy of Sciences of the USA*, **71**, 1347–50.

Newell, P. C. (1978). Cellular communication during aggregation of *Dictyostelium*. *Journal of General Microbiology*, **104**, 1–13.

Pearce, G. D. (1981). Zambian mushrooms – custom and folklore. *Bulletin of the British Mycological Society*, **15**, 139–42.

Raper, J. R. (1952). Chemical regulation of sexual processes in the Thallophytes. *Botanical Reviews*, **18**, 447–545.

— (1966). Life cycles, basic patterns of sexuality and sexual mechanisms. In *The Fungi*, vol. 2, ed. G. C. Ainsworth & A. S. Sussman, pp. 473–511. London: Academic Press.

Rayner, A. D. M. & Todd, N. K. (1979). Population and community structure and dynamics of fungi in decaying wood. *Advances in Botanical Research*, **7**, 333–420.

Robinson, P. M. (1973a). Autotropism in fungal spores and hyphae. *Botanical Reviews*, **39**, 367–84.

— (1973b). Chemotropism in fungi. *Transactions of the British Mycological Society*, **61**, 303–13.

— (1973c). Oxygen – positive chemotropic factor for fungi? *The New Phytologist*, **72**, 1349–56.

Robinson, P. M., Park, D. & Graham, T. A. (1968). Autotropism in fungi. *Journal of Experimental Botany*, **19**, 125–34.

Schindler, J. & Sussman, M. (1977). Ammonia determines the choice of morphogenetic pathways in *Dictyostelium discoideum*. *Journal of Molecular Biology*, **116**, 161–9.

Stadler, D. R. (1952). Chemotropism in *Rhizopus nigricans*: the staling reaction. *Journal of Cellular and Comparative Physiology*, **39**, 449–74.

Sussman, M., Schindler, J. & Kim, H. (1977). Toward a biochemical definition of the morphogenetic fields in *D. discoideum*. In *Development and Differentiation in the Cellular Slime Moulds*, ed. P. Cappuccinelli & J. M. Ashworth, pp. 31–50. Amsterdam: Elsevier/North-Holland Biomedical Press.

Taber, W. A. (1966). Morphogenesis in basidiomycetes. In *The Fungi*, vol. 2, ed. G. C. Ainsworth & A. S. Sussman, pp. 387–412. London: Academic Press.

Trinci, A. P. J. (1974). A study of the kinetics of hyphal extension and branch initiation of fungal mycelia. *Journal of General Microbiology*, **81**, 225–36.

— (1978a). Wall and hyphal growth. *Science Progress*, **65**, 75–99.

— (1978b). The duplication cycle and vegetative development in moulds. In *The Filamentous Fungi*, vol. 3, *Developmental Mycology*, ed. J. E. Smith & D. R. Berry, pp. 132–63. London: Edward Arnold.

— (1979). The duplication cycle and branching in fungi. In *Fungal Walls and Hyphal Growth*, ed. J. H. Burnett & A. P. J. Trinci, pp. 319–58. London: Cambridge University Press.

Uno, I. & Ishikawa, T. (1973). Purification and identification of the fruiting-inducing substances in *Coprinus macrorhizus*. *Journal of Bacteriology*, **113**, 1240–8.

— (1974). Effect of glucose on the fruiting body formation and adenosine 3′, 5′-cyclic monophosphate levels in *Coprinus macrorhizus*. *Journal of Bacteriology*, **120**, 96–100.

— (1976). Effect of cyclic AMP on glycogen phosphorylase in *Coprinus macrorhizus*. *Biochimica et Biophysica Acta*, **452**, 112–20.

— (1978). Effect of cyclic AMP on glycogen synthetase in *Coprinus macrorhizus*. *Journal of General and Applied Microbiology*, **24**, 193–7.

Uno, I., Yamaguchi, M. & Ishikawa, T. (1974). The effect of light on fruiting body formation and adenosine 3':5'-cyclic monophosphate metabolism in *Coprinus macrorhizus*. *Proceedings of the National Academy of Sciences of the USA*, **71**, 479–83.

Watkinson, S. C. (1971*a*). The mechanism of mycelial strand induction in *Serpula lacrimans*: a possible effect of nutrient distribution. *The New Phytologist*, **70**, 1079–88.

— (1971*b*). Phosphorus translocation in the stranded and unstranded mycelium of *Serpula lacrimans*. *Transactions of the British Mycological Society*, **57**, 535–9.

Wurster, B., Pan, P., Tyan, G. G. & Bonner, J. T. (1976). Preliminary characterization of the acrasin of the cellular slime mould *Polysphondylium violaceum*. *Proceedings of the National Academy of Sciences of the USA*, **73**, 795–9.

6

Positional control of development in algae

SUSAN DRURY WAALAND

In studies of development one would like to know what controls the developmental fate of cells. In vascular plants, the position of a cell within a tissue appears to play an important role in determining which developmental pathway it will follow. However, it is often difficult to follow individual cells through their development in complex tissues. In many algae, tissues are less complex and individual cells more accessible. In filamentous or unicellular algae, one can easily manipulate single cells. Such algae are particularly well-suited for studies of positional gradients within an individual cell. Through such studies one hopes to find the factors which regulate the developmental fate of daughter cells following division, or of the two ends of a single, large cell. This chapter will concentrate on the positional control of development in three types of algal cells: the germinating zygotes of fucoid brown algae; the growing cells of a green alga, *Acetabularia*, and the regenerating shoot cells of a red alga, *Griffithsia*. These three kinds of cells give examples of three aspects of positional regulation of development within cells: the development of positional gradients in initially apolar cells in fucoid zygotes; the molecular basis of positional information within cells in *Acetabularia*; and hormonal modification of positional information in *Griffithsia*. Evidence for tissue polarity and correlative control of development in multicellular algae will also be discussed.

Development of positional gradients in apolar cells

One aspect of the positional control of development is the establishment of a gradient of positional information within a cell, a filament, or a tissue. To investigate this question it is useful to look at how differences in developmental potential arise within single cells. In the germination of algal spores and zygotes, we find many examples of cells which are initially apolar and which develop an intracellular asymmetry that leads

to the production of very different daughter cells. The control of the
development of positional differences in single cells has been investigated
in detail using zygotes of the fucoid brown algae *Fucus* and *Pelvetia* (Jaffe,
1968; Jaffe & Nuccitelli, 1977; Quatrano, 1978; Quatrano, Brawley &
Hogsett, 1979).

In the fucoid algae, naked gametes are shed and fertilization takes place
outside the parent plants. After fertilization, the first step in zygote development is the production of a cell wall by the naked zygote. At this stage the
zygote is apparently apolar; no morphological or biochemical differences
have been detected within it (Fig. 6.1). As development proceeds, cytoplasmic polarity becomes visible; one side of the zygote begins to bulge out to
form a rhizoid protuberance. As the rhizoid outgrowth continues to enlarge, a refractile zone becomes obvious at its tip. When the nucleus divides,
the mitotic spindle is oriented parallel to the rhizoid axis; cytokinesis
separates the zygote into two cells: a rhizoid cell gives rise to the holdfast of
the plant while the thallus cell is the progenitor of the upright, photosynthetic portion of the plant. Thus, during zygote germination, a symmetrical
cell develops cytoplasmic asymmetry which subsequently leads to the
formation of two distinct cell types with different developmental fates.

Fig. 6.1. Zygote germination in *Fucus*. A, An apolar zygote; B, 12–14 h after
fertilization a rhizoid protuberance begins to form; C, 18–20 h after fertilization
cytokinesis has occurred to form a thallus cell and a rhizoid cell. rp, rhizoid
protuberance; tc, thallus cell; rc, rhizoid cell. (Modified after Quatrano *et al.*,
1979.)

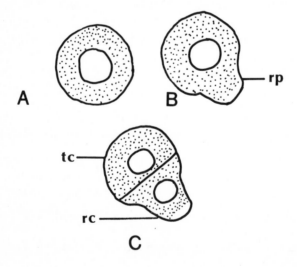

In order to understand how asymmetry develops within these zygotes, we need to look in greater detail at the events leading to the formation of a rhizoid protuberance. First of all, the site of rhizoid formation can be determined by a number of environmental factors including light, gradients in ionic concentration and pH, and zygote density (Jaffe, 1968). Rhizoids form on the darkest side of a zygote; towards the high side of concentration gradients of calcium, potassium and hydrogen ions; and towards the positive pole of an externally-applied electrical gradient. Thus, exposure of zygotes to an environmental stimulus such as unilateral light, causes formation of rhizoids in a predictable location. Unilateral light stimuli have often been used to produce zygotes with a predictable orientation; zygotes of *Fucus* are sensitive to unilateral light from the time that they complete the synthesis of their first cell wall until just before their rhizoid outgrowths appear (Quatrano, 1972). During the period when zygotes are sensitive to a light stimulus, the rhizoid pole is labile and can be re-oriented by a second directional stimulus.

Some of the first visible signs of cytoplasmic asymmetry in zygotes are the appearance of a thin refractile area just under the plasma membrane at the presumptive rhizoid pole (Nuccitelli, 1978), the asymmetric secretion of an adhesive, jelly-like substance at this pole (Schröter, 1978) and indications of increased fusion of vesicles with the plasma membrane in this region (Peng & Jaffe, 1976). These events all appear to occur during the time when the rhizoid pole is still labile. At about the time the rhizoid pole becomes permanently fixed, further cytoplasmic asymmetry can be seen. An accumulation of swollen Golgi bodies, vesicles, and other organelles is seen on the side of the nucleus which faces the rhizoid pole (Quatrano, 1972). A fucose polymer, fucoidin, is sulphated within the Golgi complex; after this sulphation, vesicles containing fucoidin migrate to the rhizoid pole and fuse with the plasma membrane to release fucoidin into the rhizoid wall (Quatrano, 1974; Crayton, Wilson & Quatrano, 1974; Callow, Coughlan & Evans, 1978; Brawley & Quatrano, 1979).

These results show that the development of positional differences within the fucoid zygote occurs in a series of steps. Many of the steps appear to be related to the development of a centre of tip growth at the rhizoid pole. There is a still-earlier indication of the development of a rhizoid pole within the zygote. As mentioned above, when an external current is passed through zygotes, the rhizoid pole is localized towards the anode. Jaffe (1966) showed that when a series of zygotes was arranged in a row, it was possible to detect a self-generated current running through the zygotes. Subsequently Jaffe and his co-workers developed an ultrasensitive vibrating probe electrode which can detect the current fluxes around individual zygotes (Jaffe & Nuccitelli, 1974; Nuccitelli & Jaffe, 1974). They have shown that zygotes of *Fucus* and *Pelvetia* have a self-generated current which enters at the rhizoid pole and leaves at the opposite side of the cell. This current appears to be carried by the movement of calcium ions which enter at the rhizoid pole and exit at the thallus side of the cell (Jaffe & Nuccitelli, 1977; Jaffe, 1979). Nuccitelli (1978) found that *Pelvetia* zygotes, which have been given a light stimulus, have patches of inflowing current which are, at first, randomly positioned in the cell membrane around the zygote; these patches then become localized on the dark side of the zygotes. The localization of current precedes the formation of a refractile zone at the rhizoid pole, the first morphological evidence of polarity. If a second light stimulus is given from another direction, the site of inflowing current moves to the newly determined rhizoid pole and a new refractile zone then appears in this region. Jaffe and co-workers (Jaffe, 1968; Jaffe, Robinson & Nuccitelli, 1974) have proposed that this self-generated current causes the cytoplasmic localization of rhizoid-specific components by a self-electrophoresis of charged molecules within the cytoplasm and the cell membrane. This hypothesis is supported by the observation that localized secretion of fucoidin occurs only when the molecule is sulphated and thus acquires a negative charge (Crayton *et al.,* 1974; Hogsett & Quatrano, 1978; Quatrano *et al.,* 1979).

To understand fully the relationship between ion currents and the development of positional differences within cells, one would like to know what is responsible for the localization, to specific positions in the plasma membrane, of the ion channels and pumps responsible for the transcellular current. In animal cells, microfilaments have been shown to function in the movement of proteins within membranes and in the anchoring of proteins in a specific region of the membrane (cf. Quatrano *et al.,* 1979). In zygotes of both *Pelvetia* (Nelson & Jaffe, 1973) and *Fucus* (Quatrano, 1973), the drug cytochalasin B (CB) inhibits the light-stimulated localization of the rhizoid pole; it does not disrupt an already localized pole but will extend the time during which the pole is sensitive to reorienting light stimuli. CB has been shown to disrupt microfilaments in many kinds of cells. It has been suggested that microfilaments may direct the movement and/or the anchoring of ion channels in the cell membrane on the dark side of the zygotes (Quatrano, 1978; Quatrano *et al.,* 1979; Jaffe & Nuccitelli, 1977). If this is the case, CB should prevent the light-induced localization of inflowing current to the dark side of the zygote; the effect of CB on localization of inflowing current has not been tested. However, in

Fucus, when CB or cytochalasin D (CD) is used to prevent zygote polarization, later events associated with the development of a rhizoidal pole are inhibited. For example, in the presence of CB the secretion of rhizoid-specific, sulphated fucoidin no longer is localized to the rhizoid pole, but occurs randomly around the zygote wall (Novotny & Forman, 1974; Brawley & Quatrano, 1979).

In the germination of fucoid zygotes we find a polarization of initially uniform cytoplasm into rhizoid and thallus components and the initiation of a tip-growing rhizoid cell. The polarization and the establishment of a tip-growth centre are correlated with the development of a localized in-flowing current which is carried, in part, by an influx of calcium ions at the rhizoid pole and a pumping out of calcium ions from the thallus region. Microfilaments also appear to be important for the localization of the rhizoid pole, for the movement of rhizoid specific substances to this pole, and for the localized secretion of cell wall precursors. Locally high concentrations of ions such as calcium in the rhizoid half of the zygote may also serve a regulatory function in these processes.

Ion currents also appear to be related to the continued growth of the rhizoid of two-celled embryos; fluctuations in these currents have been correlated with fluctuations in the rate of rhizoid elongation (Nuccitelli & Jaffe, 1974). Ion currents have been shown to occur in a number of other plant cells which grow by tip growth (Weisenseel & Kicherer, 1981). In cells which elongate by tip growth, growth is confined to the apical dome of an apical cell (Green, 1969). The tips of such cells usually have a refractile zone which is filled with vesicles, and vesicle fusion with the plasma membrane appears to be concentrated in this zone (Sievers & Schnepf, 1981). These vesicles are thought to contain cell wall precursors and possibly wall-loosening enzymes (Bartnicki-Garcia, 1973).

Ion currents have been shown to be correlated with the establishment of centres of tip growth in other algae (Weisenseel & Kicherer, 1981). Filaments of the coenocytic, yellow-green alga, *Vaucheria* elongate by tip growth (Kataoka, 1975). In these filaments, unilateral light has been shown to induce the formation of branches by inducing the formation of new regions of tip growth (Kataoka, 1975; Åberg, 1978). A localized efflux of current followed by current influx at the presumptive site of the new growing point has been shown to precede the initiation of branch growth (Weisenseel & Kicherer, 1981).

Localized ion fluxes may also play a role in the development of *Micrasterias*, a placoderm desmid. In these unicellular algae, precise regulation of the sites of tip growth is important for the development of the cells, each of which consists of two identical halves or semi-cells connected by an isthmus. The wall of each semi-cell has a genetically determined pattern of ornamentation with several highly-lobed wings. When cells divide, the plane of division runs perpendicularly through the isthmus so that each daughter cell has one mature semi-cell and a newly formed septum (Fig. 6.2). The septum of each cell grows to form a semi-cell which is identical to its mature semi-cell. The literature on the control of semi-cell development by the nucleus and by the cytoplasm has recently been reviewed in detail (Kiermayer, 1981; Kallio & Lehtonen, 1981). The new semi-cell is produced by a series of precisely localized centres of tip growth (Lacalli, 1975a, b). At first, three centres form to initiate the

Fig. 6.2. Morphogenesis in *Micrasterias*. A, A single cell consisting of two identical semi-cells which are connected by an isthmus; B, just after cytokinesis each daughter cell has one mature semi-cell from the parent cell and one newly-formed septum; C, development of new semi-cells has proceeded to the five-lobed stage. Further tip growth and division of growing centres will result in the production of a semi-cell identical to the mature semi-cell. (Modified after Pickett-Heaps, 1975.)

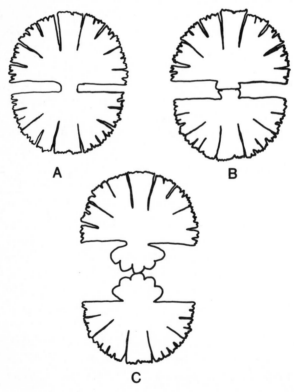

growth of the three wings of the semi-cell; later these centres elongate and divide repeatedly to form the 33 or more lobes characteristic of the mature cell.

Several experiments suggest that localized ion fluxes and microfilaments may be important in the initiation of centres of tip growth in these cells. First, externally applied currents can change the direction of lobe growth; when developing semi-cells are placed in an electrical field, the elongating lobes tended to grow towards the cathode (Brower & McIntosh, 1980; Brower & Giddings, 1980). Second, as in fucoid zygotes, there appears to be a locally high concentration of calcium in growing lobe tips (Meindl, 1982). Finally, if developing semi-cells are placed in CB for one hour, cytoplasmic streaming stops, growth ceases, and wall deposition is inhibited; growth is not re-established after the drug is removed (Tippit & Pickett-Heaps, 1974). Thus it appears that microfilaments may also be important for localizing centres of tip growth in *Micrasterias* cells; in these cells, centres of tip growth do not appear to reorganize once they have been disrupted. From these observations one might predict that inflowing currents would be found at the tips of elongating lobes of *Micrasterias* semi-cells and that changes in the location of currents might precede lobe branching; to date, however, ion currents have not been measured in *Micrasterias* cells.

Thus, localized influxes of ions and self-generated currents appear to play an important role in morphogenesis in a number of algal cells. They are correlated with the establishment and maintenance of centres of tip growth. They may also regulate the distribution of factors which control developmental potential within single cells.

Intracellular gradients and positional control

A second aspect of the positional control of development of single cells is the consideration of the macromolecular basis for gradients in developmental potential within growing cells. Experiments on the control of morphogenesis in *Acetabularia*, a large uninucleate, unicellular green alga, have yielded important information about this question (Hämmerling, 1953, 1963; Bonotto, Lurquin & Mazza, 1976; Schweiger & Berger, 1979, 1981). The life cycle of an *Acetabularia* cell is summarized in Fig. 6.3. A germinating zygote produces a polarized tubular cell. At the base of the cell, where the single nucleus resides, a clump of branched rhizoid outgrowths is formed; the apex of the filament grows by tip growth to form a tubular stalk. Periodically, the tip stops growing, flattens, and

forms a whorl of fine, branched hairs; then the growing tip reorganizes
and elongation resumes, eventually forming a stalk 6–10 cm in length. At
maturity, the stalk tip forms a species-specific cap which is a whorl of
compartments or rays. When the cap reaches mature size, the large primary nucleus, still at the base of the cell, divides many times to form
hundreds of secondary nuclei. The secondary nuclei are carried, by cytoplasmic streaming, to the cap and distributed into the rays. Cyst walls
form around each nucleus. The cysts are shed at maturity; biflagellate
gametes are produced by each cyst. Fusion of gametes to form a new
zygote completes the life cycle.

The aspects of this life cycle which are of interest here are the control of
polar growth and development. The classic work on this aspect of the
development of *Acetabularia* was done by Hämmerling and is summarized in two reviews by him (Hämmerling, 1953, 1963). He found that if

Fig. 6.3. Life cycle of *Acetabularia*. A, Recently settled zygote; B, a growing
uninucleate plant with rhizoid and stalk; C, a uninucleate plant with mature cap;
D, multiple divisions of the primary nucleus result in the production of many
secondary nuclei which are carried to the cap; E, an individual cap ray containing
uninucleate cysts; F, a germinating cyst releasing biflagellate gametes; G, gamete
fusion. r, rhizoid; s, stalk; pn, primary nucleus; cr, cap ray; sn, secondary nuclei;
cy, cyst; g, gamete.

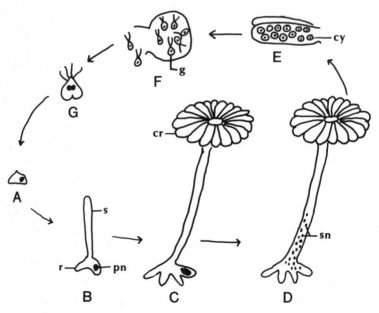

Acetabularia cells are cut in half to produce an anucleate stalk and a nucleate base, the nucleate basal portion regenerates a new stalk which resumes growth and development. More remarkably, the anucleate stalk also grows, producing several whorls of hairs and ultimately a species-specific cap; thus the anucleate stalk appears to contain morphogenetic substances (MGSs) which can direct growth and morphogenesis. These MGSs are produced by the nucleus and exported to the stalk, where they are present throughout most of the period of elongation. Within the anucleate stalk there appears to be a gradient in cap-forming MGS. If this portion is cut into three segments, the apical-most segment grows and produces a cap. The lower two segments grow but do not usually produce caps. The nucleus appears to be able to regulate the polarity of the stalk. When a rhizoid segment containing a nucleus is grafted onto either the apical or basal end of an anucleate stalk segment, the end nearest the nucleus forms rhizoidal outgrowths and the opposite end forms the growing stalk tip. Thus, the nucleus appears to be the source not only of stalk-forming MGS but also of MGS for rhizoid formation. The nucleus, in some way, influences the distribution of both substances.

Subsequent research strongly supports the idea that the MGSs are messenger RNA (mRNA) molecules, possibly in the form of ribonucleo-protein particles (Brachet, 1968; Bonotto *et al.*, 1976; Schweiger & Berger, 1981). The patterns of proteins synthesized at different stages of morphogenesis are remarkably similar in nucleate and anucleate cells (cf. Bonotto *et al.*, 1976). In the absence of a nucleus, anucleate cells have the ability not only to synthesize new nuclear-encoded proteins, but also to initiate the synthesis of stage-specific proteins at the appropriate time during development (Ceron & Johnson, 1971; Schweiger, Apel & Kloppstech, 1972). Zetsche and co-workers (Zetsche, Grieninger & Anders, 1970; Brändle & Zetsche, 1971) studied the effect of inhibitors of protein and RNA synthesis on the synthesis of UDPG-pyrophosphorylase, an enzyme which appears just before cap formation in nucleate and anucleate cells. Inhibitors of protein synthesis, such as cycloheximide and puromycin, prevent the appearance of this enzyme. Inhibitors of nuclear and organellar RNA synthesis do not prevent the synthesis of the enzyme. These data support the hypothesis that anucleate cells contain the mRNA needed for enzyme synthesis, and that increases in enzyme activity require protein synthesis. They also found a gradient in the synthesis of UDPG-pyrophosphorylase; more enzyme was produced at the tip of the growing cell than at its base. Thus, in *Acetabularia* a gradient in morphogenetic substances, most likely mRNA molecules, appears to control the developmental potential of different parts of the cell. Gradients in mRNA may lead to gradients in the concentrations of proteins synthesized on that RNA. These gradients in enzyme activity may in turn lead to gradients in morphogenesis.

One might now ask what is responsible for the generation of the gradient in mRNA in these cells. Intracellular mechanisms for the polar distribution of morphogenetic substances and of organelles such as chloroplasts have been investigated. There is active cytoplasmic streaming in *Acetabularia* cells which might carry substances from the nucleus to the tip of the stalk and to the basal rhizoid. Koop & Kiermayer (1980) have shown that there are numerous channels of streaming, and that the rates and direction of movement within channels can vary. Both microfilaments and microtubules appear to be important for the maintenance of streaming.

In *Acetabularia*, as in fucoid zygotes and filaments of *Vaucheria*, transcellular currents are present during development. When basal anucleate stalk segments regenerate growing tips, a number of currents can be shown to traverse the cell (Novák & Bentrup, 1972; Bentrup, 1977). In an externally-applied current gradient, caps tend to form at the end of the cell which faces the anode. Self-generated currents may be important in the initiation of tip growth in *Acetabularia* and in setting up a gradient in cap-forming substances.

Hormones and positional control

A third aspect of the role of positional gradients in controlling the development of cells is the modification of gradients by hormones. Regenerating cells of *Griffithsia*, a marine red alga, show a polarity which can be modified by an endogenous hormone. Development and its control have been studied in detail in *Griffithsia pacifica* (Duffield, Waaland & Cleland, 1972; Waaland & Cleland, 1972, 1974).

Plants of *G. pacifica* grow as simple, branched filaments with large, multinucleate cells. There are basically two types of vegetative cells in a plant. The upright shoot portion of the plant has club-shaped, deeply-pigmented shoot cells; the prostrate rhizoids have pale, elongate, adhesive rhizoid cells. In both shoot filaments and rhizoids, new cells are added to the filament by division of the apical cell. Subapical cells do not divide anticlinally although they may bud to produce branches. In addition to the morphological differences between shoot cells and rhizoidal cells, these two types of cells exhibit other differences in development. In a rhizoid, elongation is confined to the tip of the apical cell; subapical cells do not grow in length. In a shoot filament, intercalary cells elongate. Elongation of an intercalary shoot cell takes place in narrow (20 μm) bands at the top and at the bottom of the cell (Waaland, Waaland & Cleland, 1972; Waaland & Waaland, 1975). Rhizoids and shoots also differ in their response to unilateral light; rhizoids are negatively phototropic and shoot filaments are positively phototropic (Waaland, Nehlsen & Waaland, 1977).

While intercalary cells normally do not divide, a cell can be induced to divide by removing the cells above or below it. In fact, if a single shoot cell is excised from a filament, it will divide twice in the first 24 h producing a new shoot apical cell from its apex and a new rhizoidal cell from its base (Fig. 6.4) (Duffield *et al.*, 1972). The polarity of regeneration is quite strong. It does not seem to be changed by environmental factors such as light and normal gravity. Thus, in *Griffithsia* cells, we see a striking example of positional control of development. From the two ends of a single cell are produced two morphologically, physiologically, and biochemically different cells. Branch formation on intercalary shoot cells also appears to be regulated by position. Intercalary shoot cells only produce branches near their apical ends. The position of lateral rhizoids does not

Fig. 6.4. Development in *Griffithsia pacifica*. A, a single, isolated shoot cell excised by slicing through adjacent cells with a razor blade; B, after 24 h the shoot cell has produced a new shoot apical cell at its apex and a rhizoid at its base; C, after 48 h, two more shoot cells have been produced by division of the apical cell. a, apex; b, base; sac, shoot apical cell; r, rhizoid.

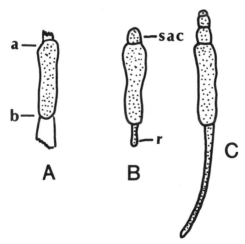

appear to be localized as precisely; lateral rhizoids may be formed near either the apex or the base of an older intercalary cell.

There are no obvious apico-basal differences in the distribution of organelles within shoot cells which might explain the differences in developmental potential; the distribution of chloroplasts, nuclei, Golgi bodies, and other organelles is uniform. When division is induced by removal of adjacent cells, there is a movement of organelles to the ends of a cell and the end walls of the cell bulge out. A regenerating rhizoid can be distinguished from a regenerating shoot cell before cytokinesis cuts them off from the parent cell. Two external factors have been shown to affect the positional gradient in *Griffithsia* cells. Schechter (1934), working with *Griffithsia globulifera* (= *G. bornetiana*) found that when *Griffithsia* filaments are placed in an electrical field, rhizoids tend to form toward the anode regardless of the orientation of the filament; shoot formation is not affected by current. He also found that if the cytoplasm in *Griffithsia* filaments is centrifuged to the base of cells regeneration patterns are also changed; shoots are produced at the basal end of the filament and branches are formed near the bases of intercalary cells. Rhizoid formation is not affected by centrifugation. We have found that when intercalary filaments or single cells of *Griffithsia pacifica* are centrifuged with their bases in a centripetal position for 24 h at low speed, about 50% regenerate shoot apical cells at their bases; of these, 50–90% also produce shoot cells at their exposed apices (unpublished observation). Thus, the polarity of cells has not been completely reversed. If cells which regenerated shoot cells at their bases are re-isolated, they again produce shoot cells at their bases. Thus, it is possible to disturb permanently the polar distribution of positional determinants by electrical fields or low-speed centrifugation.

The events discussed above are those which take place during normal regeneration, when the cells either above or below a given cell are completely removed. If an intercalary cell is killed, but not severed, so that its cell wall holds together the cells above and below it, a different series of developmental events, 'cell repair', occurs (Waaland & Cleland, 1972). In *G. pacifica*, about 5 h after a cell is killed, the cell above the dead cell regenerates a rhizoid, as it would have had the filament been severed. However, the cell below the dead cell divides at 5–7 h, 2–3 times sooner than it would have during regeneration; the cell which is produced looks and grows like a rhizoid. This cell has been called a 'repair shoot cell'. The new rhizoid and the repair shoot grow towards each other through the lumen of the dead cell (Fig. 6.5). As they approach each other the repair shoot is attracted to the rhizoid; when they meet, they fuse to form a single cell. This cell expands laterally to fill the wall of the dead cell. It becomes an intercalary shoot cell, replacing the dead cell. The entire process from puncture to production of a new intercalary shoot cell takes about 24 h. Thus, during cell repair there appear to be several changes in the normal pattern of development. First a shoot apex regenerates a cell which strongly resembles a rhizoid in morphology and growth pattern. Then two vegetative cells fuse. And finally the product of the fusion between a rhizoid and a rhizoid-like repair shoot cell becomes a shoot cell. The process of cell repair occurs in many species of *Griffithsia* and in certain other species in the Ceramiaceae (Höfler, 1934; L'Hardy-Halos, 1971a; Whittick & South, 1972; Waaland, 1975, 1978).

In *Griffithsia*, the process of cell repair has been shown to be co-ordinated by a species-specific hormone called 'rhodomorphin' (Waaland, 1975; Waaland & Watson, 1980; Watson & Waaland, 1983). During cell repair this hormone is produced by the repair rhizoid; it is also produced by other growing rhizoid cells. Rhodomorphin from *G. pacifica* has been isolated and partially purified (Watson & Waaland, 1983); it appears to be a glycoprotein with a molecular weight of 14–15 000.

Repair shoot cells resemble rhizoids both in their morphology and in

Fig. 6.5. Cell repair in *Griffithsia pacifica*. A cell (cw) was killed by puncturing 12 h before these photographs were taken. A, Low-power view (scale bar = 0.5 mm); B, close-up of the cell being repaired (scale bar = 0.1 mm). Note difference in morphology between the normal shoot apical cell (ns) and the repair shoot cell (rs). cw, cell wall of the dead cell; rr, repair rhizoid; rs, repair shoot cell.

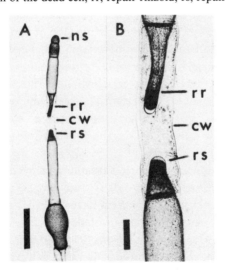

their rapid tip growth. Does rhodomorphin, in addition to inducing cell division, change the positional information within a shoot cell so that the apical end of a cell regenerates a rhizoid instead of a shoot cell? Observations of normal cell repair seem to indicate that repair shoot cells are shoot cells which have been induced to grow like rhizoids. During cell repair, a repair shoot maintains its rhizoid-like behaviour only when rhodomorphin is present. If repairing cells are separated before they begin to fuse, the repair shoot cell stops elongating; it then widens laterally and divides to produce a normal shoot apical cell at its apex (Waaland & Cleland, 1972). Thus, although a repair shoot acts like a rhizoid in the presence of rhodomorphin, it is not a rhizoid cell. If decapitated filaments are placed in a solution of partially purified rhodomorphin they will produce repair shoot cells. As long as they are in hormone solution, these repair shoots grow like rhizoids. When, after 3–4 days, repair shoots are removed from hormone solution, they stop elongating and regenerate shoot apical cells at their apices (Waaland & Watson, 1980). However, if the repair shoots remain in rhodomorphin for as long as 4–5 days, they will not revert to shoot cell behaviour when removed from hormone solution (unpublished observation). They are permanently committed to a rhizoidal differentiation state. Thus, it appears that rhodomorphin may induce a gradual change in the developmental commitment of repair shoot cells; this change requires an extended, continuous exposure to the hormone.

In *Griffithsia*, distinct positional differences in developmental potential can be seen within individual cells. The internal factors which regulate the expression of position are not known. Rhodomorphin, an endogenous, species-specific hormone, changes the expression of developmental potential when it induces the formation of a new type of cell, the repair shoot cell, whose behaviour is intermediate between normal shoot cells and rhizoidal cells. The repair shoot cells grow like rhizoids, but retain the ability to revert to a shoot cell pattern of growth and division.

A final example of positional gradients in single cells can be found in the giant coenocytic cells of certain green algae. *Caulerpa prolifera* is an example of such an alga. Plants of *C. prolifera* have a prostrate 'rhizome' from which are produced upright blades and downward-growing rhizoid clusters; rhizome, blades and rhizoids are interconnected parts of a single coenocytic cell. Jacobs and co-workers have described the characteristics of rhizome growth and the pattern of blade and rhizoid formation in *C. prolifera* (Jacobs, 1970; Jacobs & Olson, 1980). To date no morphological or chemical gradients have been found in these cells which might help to explain how this striking intracellular differentiation is controlled (Brennan & Jacobs, 1980). In fact, rapid cytoplasmic streaming appears to move cytoplasmic contents between blades, rhizome and rhizoids. Recently, Jacobs & Olson (1980) have studied the effect of gravity on the production of rhizoids and blades by the rhizome. Normally, rhizoids are initiated on the bottom of a rhizome and blades are initiated on the top. If the growing rhizome tip is rotated so that the side which faced downwards is now up, the pattern of initiation immediately switches. Rhizoids are now produced on the now-bottom side and blades on the now-top. Thus, there appear to be gravity sensors within the cell which can regulate the position of intracellular differentiation.

This chapter has emphasized the development and maintenance of positional gradients within single cells. In the three algal systems discussed in detail above, we have approached the question of the positional control of development within cells from three levels: control within the cytoplasm by gradients in macromolecules; control at the cell membrane; and intercellular control. In *Acetabularia* we saw that intracellular differences in developmental potential may be controlled by cytoplasmic gradients in the distribution of specific mRNA molecules. From studies of *Fucus* and *Pelvetia*, we found that the cell membrane, by regulating localized ion fluxes and transcellular currents, may be important in establishing and maintaining intracellular gradients. Finally, in *Griffithsia* we saw that intercellular communication, mediated by hormones, may modify patterns of localized growth and cause the production of specialized tip-growing cells.

Polarity and correlative control of development have also been observed within entire filaments and in tissues of more complex multicellular algae (Buggeln, 1981). For example, L'Hardy-Halos (1971b) has found that decapitation of filaments of the red alga *Antithamnion* stimulates branch initiation and lateral branch growth of these filaments. Moss (1965, 1967, 1970) has found evidence for apical dominance in thalli of the fucoid brown algae *Fucus* and *Ascophyllum*. In these algae the apical meristem has a single apical cell subtended by a promeristem. Removal of, or damage to, the apical cell destroys the meristematic activity of the promeristem. At the same time branch formation is stimulated. In regenerating internodes of *Ascophyllum*, the presence of blade primordia influences the pattern of regeneration of the node.

Perrone & Felicini (1972, 1974, 1981) have studied regeneration of excised sections of blades of the marine red alga *Petroglossum nicaeense* (= *Schottera nicaeensis*). This alga has a multicellular prostrate, cylindrical axis from which upright, multicellular, flattened blades are formed. If pieces of a blade are excised, they regenerate in a polar fashion producing blade initials from their apical ends and cylindrical outgrowths from their bases (Perrone & Felicini, 1972, 1974). The production of cylindrical outgrowths appears to depend on the previous initiation of blade initials (Perrone-Pesola & Felicini, 1981). If blade initials are removed repeatedly from the apical end of an excised segment before its basal end begins to regenerate, no cylindrical outgrowths are formed. If blade initials are removed from the apical end after the initiation of cylindrical outgrowths at the base, these outgrowths develop into blades instead of cylindrical axes. In this case cylindrical outgrowths are subsequently formed from the originally apical end of the segment. Thus, blades appear to control, in some way, the initiation and differentiation of prostrate axes. The way in which this control is maintained is not understood.

In multicellular algae there is evidence for positional differences along axes, and for the control by one part of the plant of development in another part of the same plant. Such controls must also operate in regulating the development of more complex algal thalli such as those of the giant kelps *Nerocystis* and *Macrocystis* which attain very large sizes and in which cellular differentiation is nearly as complex as that of vascular plants. If co-ordination of development within the thalli of algae is regulated in a manner similar to that of land plants, one would predict that diffusible hormones are involved in this process. To date, however, there has been no conclusive demonstration of endogenous hormonal co-ordination of growth and differentiation in complex algae (Buggeln, 1981). These algae should provide a rich source of material for future studies of the mechanism of positional control of the development of whole organisms.

References

Åberg, H. (1978). Light and branch formation in the alga *Vaucheria dichotoma* (Xanthophyceae). *Physiologia Plantarum*, **44**, 224–30.
Bartnicki-Garcia, S. (1973). Fundamental aspects of hyphal morphogenesis. *Symposia of the Society for General Microbiology*, **23**, 245–67.
Bentrup, F. W. (1977). Electrical events during apex regeneration in *Acetabularia mediterranea*. In *Progress in Acetabularia Research*, ed. C. L. F. Woodcock, pp. 249–54. New York & London: Academic Press.
Bonotto, S., Lurquin, P. & Mazza, A. (1976). Recent advances in research on the marine alga *Acetabularia*. *Advances in Marine Biology*, **14**, 123–250.
Brachet, J. (1968). Synthesis of macromolecules and morphogenesis in *Acetabularia*. *Current Topics in Developmental Biology*, **3**, 1–36.

Brändle, E. & Zetsche, K. (1971). Die Wirkung von Rifampicin auf die RNA- und Protein Synthese sowie die Morphogenese und den Chlorophyllgehalt kernhaltiger und kernloser *Acetabularia*-Zellen. *Planta*, **99**, 46–55.

Brawley, S. H. & Quatrano, R. S. (1979). Sulfation of fucoidin in *Fucus* embryos IV. Autoradiographic investigations of fucoidin sulfation and secretion during differentiation and the effect of cytochalasin treatment. *Developmental Biology*, **73**, 193–205.

Brennan, T. & Jacobs, W. P. (1980). Polarity and the movement of [^{14}C]indol-3-yl acetic acid in the coenocyte, *Caulerpa prolifera*. *Annals of Botany*, **46**, 129–31.

Brower, D. L. & McIntosh, J. R. (1980). The effects of applied electric fields on *Micrasterias*. I. Morphogenesis and the pattern of cell wall deposition. *Journal of Cell Science*, **42**, 261–77.

Brower, D. L. & Giddings, T. H. (1980). The effects of applied electric fields on *Micrasterias*. II. The distributions of cytoplasmic and plasma membrane components. *Journal of Cell Science*, **42**, 279–90.

Buggeln, R. G. (1981). Morphogenesis and growth regulators. In *The Biology of Seaweeds*, ed. C. S. Lobban & M. J. Wynne, pp. 627–59. Oxford: Blackwell Scientific Publications.

Callow, M. E., Coughlan, E. J. & Evans, L. V. (1978). The role of Golgi bodies in polysaccharide sulphation in *Fucus* zygotes. *Journal of Cell Science*, **32**, 337–56.

Ceron, G. & Johnson, E. M. (1971). Control of protein synthesis during the development of *Acetabularia*. *Journal of Embryology and Experimental Morphology*, **26**, 323–38.

Crayton, M. A., Wilson, E. & Quatrano, R. S. (1974). Sulfation of fucoidin in *Fucus* embryos. II. Separation from initiation of polar growth. *Developmental Biology*, **39**, 164–7.

Duffield, E. C. ·S., Waaland, S. D. & Cleland, R. (1972). Morphogenesis in the red alga, *Griffithsia pacifica*: regeneration from single cells. *Planta*, **105**, 185–95.

Green, P. B. (1969). Cell morphogenesis. *Annual Review of Plant Physiology*, **20**, 365–94.

Hämmerling, J. (1953). Nucleocytoplasmic relationships in the development of *Acetabularia*. *International Review of Cytology*, **2**, 475–98.

– (1963). Nucleocytoplasmic interactions in *Acetabularia* and other cells. *Annual Review of Plant Physiology*, **14**, 65–92.

Höfler, K. (1934). Regenerationsvorgänge bei *Griffithsia Schousboei*. *Flora*, **27**, 331–44.

Hogsett, W. S. & Quatrano, R. S. (1978). Sulfation of fucoidin in *Fucus* embryos. III. Required for localization in the rhizoid wall. *Journal of Cell Biology*, **78**, 866–73.

Jacobs, W. P. (1970). Development and regeneration of the algal giant coenocyte *Caulerpa*. *Annals of the New York Academy of Sciences*, **175**, 732–48.

Jacobs, W. P. & Olson, J. (1980). Developmental changes in the algal coenocyte *Caulerpa prolifera* (Siphonales) after inversion with respect to gravity. *American Journal of Botany*, **67**, 141–6.

Jaffe, L. F. (1966). Electrical currents through the developing *Fucus* egg. *Proceedings of the National Academy of Sciences of the USA*, **56**, 1102–9.

– (1968). Localization in the developing *Fucus* egg and the general role of localizing currents. *Advances in Morphogenesis*, **7**, 295–328.

– (1979). Control of development by ionic currents. In *Membrane Transduction Mechanisms*, ed. R. A. Cone & J. E. Dowling, pp. 199–231. New York: Raven Press.

Jaffe, L. F. & Nuccitelli, R. (1974). An ultrasensitive vibrating probe for measuring steady extracellular currents. *Journal of Cell Biology*, **63**, 614–28.

– (1977). Electrical controls of development. *Annual Review of Biophysics and Bioengineering,* **6,** 445–76.

Jaffe, L. F., Robinson K. R. & Nuccitelli, R. (1974). Local cation entry and self-electrophoresis as an intracellular localization mechanism. *Annals of the New York Academy of Science,* **238,** 372–9.

Kallio, P. & Lehtonen, J. (1981). Nuclear control of morphogenesis in *Micrasterias.* In *Cytomorphogenesis in Plants,* ed. O. Kiermayer, pp. 191–213. Wien: Springer-Verlag.

Kataoka, H. (1975). Phototropism in *Vaucheria geminata.* II. The mechanism of bending and branching. *Plant and Cell Physiology,* **16,** 439–48.

Kiermayer, O. (1981). Cytoplasmic basis of morphogenesis in *Micrasterias.* In *Cytomorphogenesis in Plants,* ed. O. Kiermayer, pp. 147–89. Wien: Springer-Verlag.

Koop, H.-U. & Kiermayer, O. (1980). Protoplasmic streaming in the giant unicellular alga *Acetabularia mediterranea.* I. Formation of intracellular transport systems in the course of cell differentiation. *Protoplasma,* **102,** 147–66.

Lacalli, T. C. (1975a). Morphogenesis in *Micrasterias.* I. Tip growth. *Journal of Embryology and Experimental Morphology,* **33,** 95–115.

– (1975b). Morphogenesis in *Micrasterias.* II. Patterns of morphogenesis. *Journal of Embryology and Experimental Morphology,* **33,** 117–26.

L'Hardy-Halos, M.-T. (1971a). Recherches sur les Céramiacées (Rhodophycées-Céramiales) et leur morphogénèse. II. Les modalités de la croissance et les remaniements cellulaires. *Revue Générale de Botanique,* **78,** 201–56.

– (1971b). Recherches sur les Céramiacées (Rhodophycées-Céramiales) et leur morphogénèse. III. Observations et recherches expérimentales sur la polarité cellulaire et la hiérarchisation des éléments de la fronde. *Revue Générale de Botanique,* **78,** 201–56.

Meindl, U. (1982). Local accumulation of membrane-associated calcium according to cell pattern formation in *Micrasterias denticulata,* visualized by chlorotetracycline fluorescence. *Protoplasma,* **110,** 143–6.

Moss, B. (1965). Apical dominance in *Fucus vesiculosus. The New Phytologist,* **64,** 387–92.

– (1967). The apical meristem of *Fucus. The New Phytologist,* **66,** 67–74.

– (1970). Meristems and growth control in *Ascophyllum nodosum. The New Phytologist,* **69,** 253–60.

Nelson, D. R. & Jaffe, L. F. (1973). Cells without cytoplasmic movements respond to cytochalasin. *Developmental Biology,* **30,** 206–8.

Novák, B. & Bentrup, F. W. (1972). An electrophysiological study of regeneration in *Acetabularia mediterranea. Planta,* **108,** 227–44.

Novotny, A. M. & Forman, M. (1974). The relationship between changes in cell wall composition and the establishment of polarity in *Fucus* embryos. *Developmental Biology,* **40,** 162–73.

Nuccitelli, R. (1978). Oöplasmic segregation and secretion in the *Pelvetia* egg is accompanied by a membrane-generated electrical current. *Developmental Biology,* **62,** 13–33.

Nuccitelli, R. & Jaffe, L. F. (1974). Spontaneous current pulses through developing fucoid eggs. *Proceedings of the National Academy of Science of the USA,* **71,** 4855–9.

Peng, H. B. & Jaffe, L. F. (1976). Cell wall formation in *Pelvetia* embryos: a freeze-fracture study. *Planta,* **133,** 57–71.

Perrone, C. & Felicini, G. P. (1972). Sur les bourgeons adventifs de *Petroglossum nicaeense* (Duby) Schotter (Rhodophycées, Gigartinales) en culture. *Phycologia,* **11,** 87–95.

– (1974). Dominance apicale et morphogénèse chez *Petroglossum nicaeense* (Duby) Schotter (Rhodophyceae). *Phycologia*, **13**, 187–94.

Perrone-Pesola, C. & Felicini, G. P. (1981). Polarité dans la fronde de *Schottera nicaeensis* (Phyllophoracées). *Phycologia*, **20**, 142–6.

Pickett-Heaps, J. D. (1975). *Green Algae: Structure, Reproduction, and Evolution in Selected Genera*. Sunderland, Mass.: Sinauer Associates.

Quatrano, R. S. (1972). An ultrastructural study of the determined site of rhizoid formation in *Fucus* zygotes. *Experimental Cell Research*, **70**, 1–12.

– (1973). Separation of processes associated with differentiation of two-celled *Fucus* embryos. *Developmental Biology*, **30**, 209–13.

– (1974). Developmental biology: development in marine organisms. In *Experimental Marine Biology*, ed. R. Mariscal, pp. 303–46. New York: Academic Press.

– (1978). Developmental of cell polarity. *Annual Review of Plant Physiology*, **29**, 487–510.

Quatrano, R. S., Brawley, S. H. & Hogsett, W. E. (1979). The control of the polar deposition of a sulfated polysaccharide in *Fucus* zygotes. In *Determinants of Spatial Organization*. ed. S. Subtelny & I. R. Konigsberg, pp. 77–96. New York: Academic Press.

Schechter, V. (1934). Electrical control of rhizoid formation in the red alga, *Griffithsia bornetiana*. *Journal of General Physiology*, **18**, 1–21.

Schröter, K. (1978). Asymmetrical jelly secretion of zygotes of *Pelvetia* and *Fucus*: an early polarization event. *Planta*, **140**, 69–73.

Schweiger, H. G., Apel, K. & Kloppstech, K. (1972). Source of genetic information of chloroplast proteins in *Acetabularia*. *Advances in Biosciences*, **8**, 249–62.

Schweiger, H. G. & Berger, S. (1979). Nucleocytoplasmic interrelationships in *Acetabularia* and some other Dasycladaceae. *International Review of Cytology, Supplement* **9**, 11–44.

– (1981). Pattern formation in *Acetabularia*. In *Cytomorphogenesis in Plants*, ed. O. Kiermayer, pp. 119–45. Wien: Springer-Verlag.

Sievers, A. & Schnepf, E. (1981). Morphogenesis and polarity of tubular cells with tip growth. In *Cytomorphogenesis in Plants*, ed. O. Kiermayer, pp. 265–99. Wien: Springer-Verlag.

Tippit, D. H. & Pickett-Heaps, J. D. (1974). Experimental investigations into morphogenesis in *Miscrasterias*. *Protoplasma*, **81**, 271–96.

Waaland, S. D. (1975). Evidence for a cell fusion hormone in red algae. *Protoplasma*, **86**, 253–61.

– (1978). Parasexually produced hybrids between male and female plants of *Griffithsia tenuis* C. Agardh, a red alga. *Planta*, **138**, 65–8.

Waaland, S. D. & Cleland, R. (1972). Development in the red alga, *Griffithsia pacifica*: control by internal and external factors. *Planta*, **105**, 196–204.

– (1974). Cell repair through cell fusion in the red alga *Griffithsia pacifica*. *Protoplasma*, **79**, 185–96.

Waaland, S. D., Nehlsen, W. & Waaland, J. R. (1977). Phototropism in a red alga *Griffithsia pacifica*. *Plant and Cell Physiology*, **18**, 603–12.

Waaland, S. D. & Waaland, J. R. (1975). Analysis of cell elongation in red algae by fluorescent labelling. *Planta*, **126**, 127–38.

Waaland, S. D., Waaland, J. R. & Cleland, R. (1972). A new pattern of plant cell elongation: bipolar band growth. *Journal of Cell Biology*, **54**, 184–90.

Waaland, S. D. & Watson, B. A. (1980). Isolation of a cell-fusion hormone from *Griffithsia pacifica* Kylin, a red alga. *Planta*, **149**, 493–7.

Watson, B. A. & Waaland, S. D. (1983). Partial purification and characterization of a cell fusion hormone from *Griffithsia pacifica*, a red alga. *Plant Physiology*, **71**, 327–32.

Weisenseel, M. H. & Kicherer, R. M. (1981). Ionic currents as control mechanisms in cytomorphogenesis. In *Cytomorphogenesis in Plants*, ed. O. Kiermayer, pp. 379–99. Wien: Springer-Verlag.

Whittick, A. & South, G. R. (1972). *Olpidiopsis antithamnionis* n. sp. (Oomycetes, Olpidiopsidaceae), a parasite of *Antithamnion floccosum* (O. F. Müll.) Kleen from Newfoundland. *Archiv für Mikrobiologie*, **82**, 353–60.

Zetsche, K., Grieninger, G. E. & Anders, J. (1970). Regulation of enzyme activity during morphogenesis of nucleate and anucleate cells of *Acetabularia*. In *Biology of Acetabularia*, ed. J. Brachet & S. Bonotto, pp. 87–110. New York: Academic Press.

7

Cell pattern and differentiation in bryophytes*

MARTIN BOPP

In contrast to higher plants, cell pattern in mosses and liverworts usually arises from a series of very regular cell divisions. Such a regularity, at least for a certain number of divisions, is always found if one single apical cell exists, as has been shown in the formation of roots in the fern *Azolla* (Gunning, 1981). Therefore, to understand the background of pattern of cells, tissues, or organs in Bryophytes the sequence of equal and unequal cell divisions has to be considered, and one has to ask whether such sequences may be sufficient to explain the final cell pattern or whether deviations from the regular pattern may indicate components of positional information other than the cell divisions. Such may be the case also for a newly-formed cell wall which has different positions according to its 'history' and which cannot be explained purely by the geometric relationship of the mother cells. For this reason we can expect in mosses a predetermination of cell differentiation according to both the division sequence and the 'positional information' specified by factors external to the cell (Wolpert, 1971).

In mosses and most liverworts development starts from a single apical cell. In the shoots of moss such cells are, with only one exception, three sided. In the leaves they are always two sided, producing daughter cells in two directions. In liverworts the system of apical cells is much more variable, and in shoots and thalli at least four different types of apical cells are described (Crandall-Stotler, 1981), namely pyramidal-tetrahedral, cuneate, lenticular and hemidiscoidal. These types can be attributed to the major taxonomic levels even if a remarkable number of exceptions exists. Besides these clearly defined unicellular structures, multicellular meristems are found, as in *Riella* or *Riccia*, but in most cases these also can be reduced to a cytologically-distinct, centrally-located notch cell (Crandall, 1969, and Fig. 7.1). In the leaves of hepatics, development, which starts with

---

*In memoriam Dr Charles Hébant, Montpellier.

one single cell that divides early into two identical cells side by side, is also much less uniform than in mosses and, therefore, illustrates important features useful for studying the cell pattern and its influence on leaf shape.

The two one-sided apical cells in the leaf of liverworts, and the two-sided apical cells in moss leaves, are the starting points for a two-dimensional cell pattern which will be completed by further divisions. In a later stage of leaf development this single-plane pattern can proceed to a three-dimensional structure produced completely independently from the former. In a fundamentally different fashion the tetrahedral, three-faced apical cells form the three-dimensional structures of a shoot and its leaves and branches. These elements are the immediate derivatives (or merophytes) of the apical cell and the products of these merophytes are formed by a programmed sequence of formative divisions which introduce tangential and radial walls (Gunning, 1981). Whereas the different types of cells can be determined by the sequence of cell division, the final shape of a leaf, a shoot, or any other organ, is the consequence not only of divisions but also of the specific wall elongation between divisions (Green, 1980) and, to a greater extent, to the wall growth that occurs after divisions have ceased and the final number of cells has been attained.

Two-dimensional patterns in moss leaves

Divisions of the apical cell

As mentioned, the final cell pattern of a leaf (which contains only few cell-types) is a consequence of cell division and cell elongation proceeding in

Fig. 7.1. Meristem of the liverwort *Riccia glauca*. (a) A very young stage, (b) enlarged at a later stage (after Hagemann, 1978).

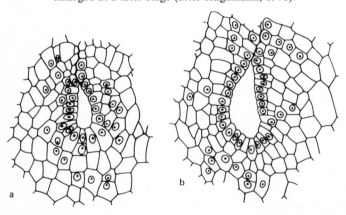

harmony. Even if the divisions are not restricted to the apical cell, which is the general rule, the origin of the pattern can be attributed to the regular divisions of this cell in a fixed manner such that each consecutive division involves a rotation of the spindle axis through about 90° (or less if the apical cell is more rhomboidal). The shape of the apical cell is more or less correlated to the form of the young leaf (compare Figs. 7.2 and 7.3). The result of the regular change of division planes is the equal distribution of daughter cells to either side of the apical cell, forming two symmetrical halves. With this type of division the cells obtain a first determination from the axis of the spindle because the direction of cell elongation is fixed and with this also the next division steps, which are perpendicular to the original first wall.

Depending on the species, this type of division persists so that the descendants of the apical cell produce strips of cells (Fig. 7.2). Another possibility is that after a first division of the descendants a second division, perpendicular to the first, produces quadrants and octants of uniform cells by a sequence of division and elongation in which the greatest elongation takes place in the newly-formed cell wall and the wall parallel to it. The next division is then perpendicular to the largest wall, and so on (Fig. 7.3; Schnepf, 1973; Frey, 1970). This means that between two divisions the direction always changes about 90°. For our purpose it is not important whether the descendants of the apical cell or the apical cell itself show the higher division rate, as the difference has no influence on the final cell pattern. Because quadrant formation also takes place after a while in the first type of division pattern, at a certain stage of development

Fig. 7.2. Tip of a young leaf of *Sphagnum palustre* (redrawn from Sych, 1982).

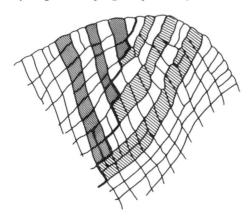

each whole young leaf consists of identical cells (of different age of course) belonging to two symmetric 'families', the 'right-hand family' and the 'left-hand family'. Normally all cells of a quadrant have the same division rate (Schnepf, 1973), whereas between quadrants development is not fully synchronized. The regularity and the equal status of the cells, therefore, seems to disappear.

Pattern formation by cell elongation

In all moss leaves the apical cell is the first to cease to divide, and it starts to differentiate. At the same time the cells of the leaf base are still dividing. Differentiation is associated with cell elongation and the final cell-net depends on the relative amount of elongation of the different longitudinal and transverse cell walls.

As may be seen in Figs. 7.3 and 7.4, the main elongating cell walls are predetermined by the last cell division. The new cell walls of neighbouring cells do not align directly with each other (so as to form a cross), but are always a little bit translocated forming a double Y. Each cell is, therefore, surrounded by six neighbours, four of them with a long common wall and two with a very short one. This type of pattern can be realized if the spindle axis is always slightly inclined away from the long axis of the dividing cell (see below). The short walls of the cells are where the main elongation takes place (Fig. 7.4) while the walls which are longer immediately after division form only the more-or-less triangular ends of the cells. Depending on the elongation rate, either a parenchymatous (as in many acrocarpous mosses) or a prosenchymatous cell-net (in most pleurocarpous mosses) is formed (Frey, 1970, 1981). During the formation of such cell-nets, the alternative structure can appear in some species as an intermediate pattern (as in the Thuidiaceae) which then goes over to the final parenchymatous or prosenchymatous pattern.

To form this cell-net no further positional information besides the division sequence is necessary, with the following few exceptions: (a) the apical cell which initiates division and differentiation may indicate a time sequence; (b) the edge cells, characterized by their exceptional position, are surrounded by only four rather than six cells; and finally (c) the nerve cells forming the mid-line of a leaf, and occasionally the basal cells. Differentiation of the cells of the nerves begins at the leaf tip; therefore, the information for this differentiation could come from the already-differentiated apical cell. The special situation of the apical cell and the edge cells is illustrated by very thick cell walls and a narrow internal cell lumen.

Although the spatial situation of these cells is very prominent, if there are two or more rows of identical differentiated edge cells the differentiation in the second and third row must be induced independently of the division pattern by a supplementary positional signal.

Development of the Sphagnum *leaf*

A quite different development from leaves with identical leaf-blade cells is found in the highly specialized leaves of *Sphagnum*. The cell pattern of

Fig. 7.3. *Neckera crispa* (moss), development of the cell net in a very young leaf (after Frey, 1970).

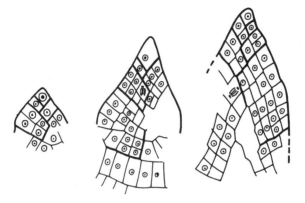

Fig. 7.4. *Neckera crispa*, cell-wall growth forming the prosenchymatous cell net in the leaf (arrows at the main growth area) (after Frey, 1970).

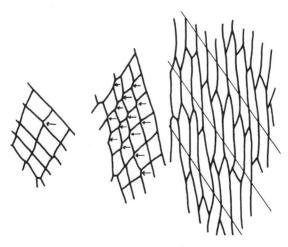

this genus is characterized by two different cell-types in the mature leaf: small chloroplast-rich 'chlorophyll cells' and large empty 'hyaline cells'. As long as identical members of a structure (leaf cells) are produced, all cell divisions are equal and it is not necessary to distinguish between the two daughter cells of a division, to count the number of cell divisions, or to introduce other factors. However, as soon as different types of cells are produced, unequal (asymmetric) cell division must be involved. This is clearly demonstrated in the leaf of *Sphagnum* (Zepf, 1952). Formation of the new cell pattern, with two different cell-types, starts, as expected, at the leaf tip with a second wave of unequal cell divisions.

After the first unequal cell division, separating a smaller plasma-dense cell from a larger one, a second division takes place at right angles to the first, forming a triplet of cells. Two cells are smaller and one is larger (Fig. 7.5). For the positioning of the new cell wall, it is important that in both divisions the spindle is always inclined against the axis of the cell so that the walls of two neighbouring cells form a double Y with the original wall, as we have mentioned before for the 'normal' cell division in a moss leaf. This formation is strengthened by slight, local cell elongation. The consequence is that each young hyaline cell is surrounded by five chlorophyll cells (Fig. 7.6).

For the direction of this division pattern positional information is necessary, which at first is expressed by the shape of the dividing cell. The first new cell wall is always formed along the longer cell axis, which depends for its determination on the leaf side (Fig. 7.7). In the middle part of the leaf all cells are more rhomboid and no preferential axis can be recognized. The plane of the next division is therefore not predetermined by the division (and elongation) pattern, so other factors must be active. This can be a gradient which goes from cell to cell; all the cells are connected by many plasmodesmata (Sych, 1982) but there is no experimental evidence for any substance that can create such a gradient. A first source of information in the middle part of the leaves could be the sequence of divisions because groups of cells from the same ancestor normally have the same orientation, although sometimes a few cells at the border of a group occur with a division in the 'wrong' direction (Fig. 7.8). For these cells an independent 'gradient' for the orientation of the cell division must be postulated.

All processes such as cell division, elongation, second wave of unequal divisions, and differentiation start at the leaf tip, forming an apical–basal progression. The postulated 'gradient' is, therefore, assumed to have the same direction. The same could be the case for the orientation of the midrib, the cessation of division, and the beginning of elongation, etc. The gradient must be associated with a polar apical–basal structure of the leaf.

Fig. 7.5. Formation of cell triplets by two unequal cell divisions in the leaf of *Sphagnum palustre* (arrows as in Fig. 7.4) (after Sych, 1982).

Fig. 7.6. Unequal cell division in the leaf of *Sphagnum cymbifolium*. The inclined spindles are shown (after Zepf, 1953).

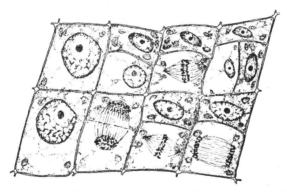

Fig. 7.7. Scheme of the left and right oriented cell walls at the beginning of the unequal cell divisions in the leaf of *Sphagnum palustre* (after Sych, 1982).

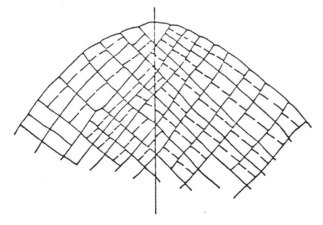

Orientation of the new cell wall

Leading up to the orientation of newly-formed cell walls a series of events takes place all of which have strong positional character. This starts with formation of a 'girdle of microtubules', the so-called preprophase band (Pickett-Heaps & Northcote, 1966; Gunning, Hardham & Hughes, 1978), which also exists in moss leaves, at least in *Sphagnum* (Schnepf, 1973); it is followed by orientation of the spindle, the location of the phragmoplast and finally the direction of the new cell wall. Each of these structures depends on the preceding one, so that only the 'preprophase band' which 'accurately predicts the line of fusion of cell plate and parental walls when it is present' (Busby & Gunning, 1980; Gunning, 1981) has to be determined in its position in the mother cells – which may go back to a specialization of the cell cortex, in particular the plasmalemma, at that site (Hepler, 1981). Disturbance of the microtubular structure of the preprophase band with D_2O, therefore, results in a disturbance of the regular cell pattern in the *Sphagnum* leaf (Schnepf, Deichgräber & Ljubešic, 1976). In the leaf of higher plants detailed examinations have also revealed a close connection between morphological alterations and disorientation of microtubules as well as microfilaments (Hardham, Green & Lang, 1980). In cases where no preprophase band is found, as in the caulonema cells (Schmiedel & Schnepf, 1979), the spindle and the newly-formed phragmoplast can turn during its growth, so that the site of fusion between the new and old cell wall is not at the original position of the phragmoplast but is displaced in such a way that an oblique new cell wall results (cf. Palewitz & Hepler, 1974). However, this is not relevant to the wall positions in moss leaves, because we find no hint of a re-orientation of the spindle and the phragmoplast during cell division.

Two-dimensional patterns in the leaves of liverworts

Basic pattern of division

The leaves of liverworts are much more variable in shape than moss leaves. Even on a single stem, besides two rows of 'normal' leaves, a third row lying in the mid-ventral line of the shoot can consist of smaller and differently formed so-called 'amphigastria' or 'underleaves'. But notwithstanding this variety of types, a basic scheme for a sequence of cell division can be derived from comparison of the early stages of different leaf types. The 'typical' leaf is bifid or bilobate. It always derives from two initials, formed side by side in the outer layer of the shoot from one original cell which has divided anticlinally. These two cells segment to form independent, one-sided (or occasionally two-sided) apical cells which produce a unistratose or uniseriate leaf part of different length (Grill, 1958) with at least two types of cells. To form tri- or quadripartite leaves the elemental pattern is modified by formation of secondary anticlinal divisions of the initials, so that the final number of lobes exists right from the beginning (Crandall-Stotler, 1981). However, the number can be modified subsequently by lateral branching of single cells in the uniseriate parts, which corresponds to polar branching in moss protonema filaments. Very often the apical growth of the initials is arrested so that the greater part of the leaf is formed through basal two-dimensional growth (Fig. 7.9). After only two divisions of the one-sided apical cell in which the spindle position does not change, there starts a very regular two-dimensional division growth which is fully comparable with quadrant formation in mosses, and also shows the typical double Y-shaped wall pattern. Although the division pattern during this phase is strongly determined, a slight displacement of the cell wall by differential elongation makes the cell net apparently irregular.

As in mosses, elongation and differentiation start at the leaf tip. The apical cell differentiates first into either a hair or a gland. At the same time, division activity is translocated to the leaf base where a third type of division pattern appears (Fig. 7.10). Through this type of division, which is also seen at the base of the separate lobes, strips of cells are produced. In most liverworts division activity in the leaves ends with a randomly-distributed and more irregular cell division (Bopp & Feger, 1960).

Fig. 7.8. Cell net of the leaf of *Sphagnum palustre* with cell groups. Three cells with the 'wrong' orientation are labelled with a black dot (redrawn from Sych, 1982).

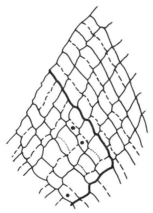

What does this generalized picture of cell division mean in terms of positional effects and leaf shape? It can be concluded that the formation of all the different leaf forms, such as uniseriate lobes, three or four lobes, entire leaf lamina, leaves with two different halves, etc., depends on changing the relative proportions of the four types of division. Whole, round leaves, for example, arise when the activity of the two initials is totally arrested, and the leaf blade is formed by the division types two and three, which give rise to regular growth and strip-growth.

Gradients in leaves of liverworts

Even though the leaves are never formed from one single apical cell, it seems that the gradient of growth and differentiation depends on the division pattern starting at the tips. But characteristic exceptions exist which must be considered in order to understand the problem of positional information in liverwort leaves. The species *Madotheca platyphylla* possesses a two-lobed leaf which starts its development in the usual way (Fig. 7.11) except that the apical cell is differentiated very early to a 'primordial' slime papilla. During further development the upper lobe

Fig. 7.9. *Lophocolea bidentata* (liverwort). Development of the cell net. Stage of apical and regular growth. Arrows in lower diagram show cell walls with a double Y formation (after Bopp & Feger, 1960).

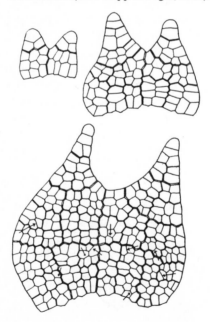

enlarges more than the lower, and furthermore, the two halves of each lobe behave differently, so that each lobe consists of a small and a large part. Through this growth process the papilla at the end of each lobe is dislocated and (especially in the upper lobe) the morphological tip is replaced by a 'geometrical' tip. Together with this dislocation, the axis of the lobe is turned and all further development is then directed to the geometrical (real) tip, where elongation and subsequently differentiation of the cells start (Bopp & Feger, 1960; Fig. 7.12). This example clearly shows that the time period for regulation of gene activity in a cell (responsible for elongation and differentiation) can be uncoupled from the time sequence of the cell cycle.

The example of *Madotheca* also shows that additional information is necessary for the formation of such asymmetrical leaves, because the difference between the smaller and the larger lobes is always relative to the

Fig. 7.10. *Lophocolea bidentata*. Enlargement of the leaf blade by forming cell strips at the base of the leaf and the lobes. Arrows indicate where a change occurs in the pattern of growth (after Bopp & Feger, 1960).

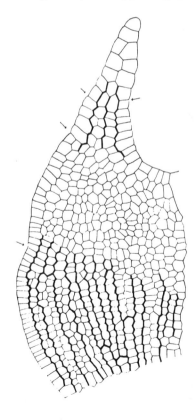

shoot and not to a left-right symmetry located in the leaf itself, as was described for the leaf of *Sphagnum*. The greater division activity, therefore, must be directed through an effect exercised by the shoot, which in most liverworts has a strong dorsiventral-bisymmetrical structure; this will be discussed in a later section. Because the young leaf goes through a symmetrical first stage of growth, the subsequent difference in the development of the two lobes and the two halves of each lobe must be determined by the symmetry of the shoot. The division pattern, therefore, cannot be the source of the postulated positional information, and one has to assume that a special 'growth factor' from the shoot regulates the differences in growth and the sequence of differentiation within the leaves.

Three-dimensional pattern in moss leaves

Cell orientation in the leaf blade

In most mosses a nerve or midrib is differentiated in three dimensions. Many types can be found from completely uniform cords of small cells to

Fig. 7.11. *Madotheca platyphylla* (liverwort). Development of the cell net. The very young symmetrical leaf changes shape to form two different and asymmetric lobes (after Bopp & Feger, 1960).

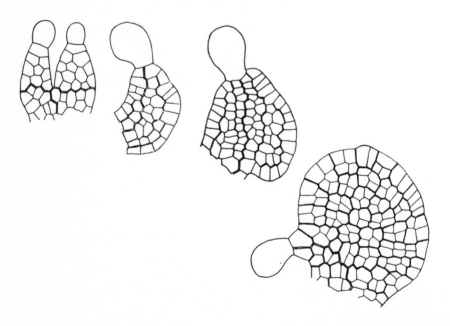

Fig. 7.12. *Madotheca*, older leaf. The direction of cell elongation and differentiation starting at the 'geometrical' leaf tip is seen. Only cell complexes are drawn to characterize the pattern (after Bopp & Feger, 1960).

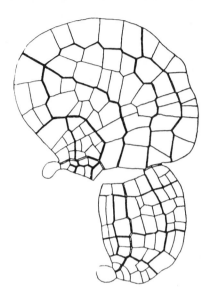

very complex tissues containing cells with different functions. The highest degree of complexity is realized in the Polytrichales and Dawsoniales (Hébant, 1977). Such complex structures are formed in a very regular way (Lorch, 1931), following a strongly determined division pattern with equal and unequal cell divisions being a prerequisite for specific differentiation of the cells. In this case also, formation of the final pattern requires additional information beyond the sequence of divisions, primarily to determine the spatial arrangement of new cell walls during development.

A transverse section through the sequence of leaves in a young *Polytrichum* shoot (Fig. 7.13) shows that formation of the three-dimensional structure begins long before the two-dimensional development, with its anticlinal divisions in the leaf plane, is finished (Bopp, 1961). The earliest stage involves a cell division in the most central cell-file. The new cell wall runs parallel (periclinal) to the leaf surface producing two identical cells. A second wall, parallel to and below the first, outlines a new cell (called the 'central cell') just in the middle of the leaf, which becomes the point of orientation for all further events. Divisions then proceed toward the leaf edges in both directions from this central cell (Fig. 7.14). These new cell

170   *M. Bopp*

walls, however, are not in one line. Each new cell wall that points towards
the edge of the leaf is translocated slightly upwards. The small difference
can be amplified – as we have seen in other cases – through a subsequent
growth process, so that a saw-like structure results. The process is similar
to the formation of double Y-shaped cell walls. With a second wall, below
the first and inclined in the opposite direction, a series of trapezoidal cells
is formed, always with the broader side directed towards the central cell.
In the next step the trapezoidal cells divide by two quickly succeeding,
unequal divisions into a group of three cells whose position is determined,
without doubt, by the geometry of the trapezoidal mother cell. Two
smaller cells lie on the inner, and one larger cell on the outer side. Five or
six such groups are formed on each side of the central cell, and are derived
from the same number of symmetrically-oriented trapezoidal cells. This
means that for all these groups, all the information for cell orientation is
contained in the geometrical structure of the mother cells. Positional
information must be given for orientation of the first oblique cell wall of
the cells adjacent to the central cell. In the central cell itself the walls are
straight. The cell about to divide, therefore, must 'know' whether its

Fig. 7.13. *Polytrichum attenuatum* (moss). Transverse section through a young
shoot tip demonstrating the development of succeeding leaves. The number 1
indicates the shoot apex; the other numbers denote the sequence of leaves; c is
central cell referred to in the text (after Bopp, 1961).

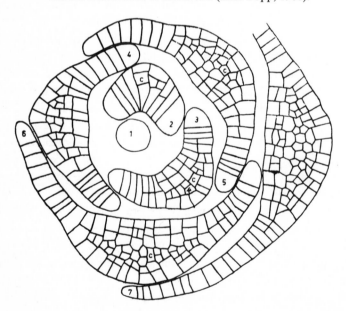

Fig. 7.14. Young leaf of *Polytrichum attenuatum* showing the formation of the central cell (c), the groups of three cells, and the trapezoidal cells (after Bopp, 1961).

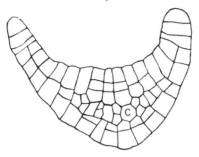

position is to the right or left of the central cell; but because the first wall of the trapezoidal cell inserts above the preceding cell and the second below, only these points of insertion at the old wall must be determined. As soon as this determination has taken place the subsequent steps are fixed. The mechanism of the insertion of the new cell walls could be the same as we have found in the double Y shaped cell pattern in the first dimension of moss or liverwort leaves. However, and this calls for special attention, we really know nothing about the mechanisms which determine the insertion point of the cell wall – or, which amounts to the same thing, the localization of the preprophase band (if present), or special structures of the plasmalemma (Hepler, 1981). Thus, we are unable to say anything about the nature of the determining factors. We can assume, however, that the same gradient which gives information about the mid-line of a leaf (as in *Sphagnum* leaves), allowing the central cells to form along the leaf axis, could also provide orientation for cell-wall insertion. The gradient need not be of a single substance, but could be a 'morphogen'. The 'positional value' provides the cell with its position within the system and this value is used, together with the cell's genome, to specify the molecular or cytological aspects of its differentiation (Wolpert, 1969, 1971).

Equal and unequal cell divisions

The geometrically-determined cell divisions are unequal and consequently further cytodifferentiation leads to the formation of either 'hydroids' (water-conducting cells) or 'leptoids' (Fig. 7.15). It seems that the program to become one or the other is determined either at the unequal cell division or after a sequence of two or three such divisions (Bünning, 1965) which also determine the position of each cell in a coordinate system.

It is important that this type of decision results in a regular sequence of a repeated cell pattern (groups of three cells) rather than in a field of identical cells. When a gradient, therefore, forms the basis of differentiation it cannot be the absolute concentration of the substance forming the gradient, or a unique concentration which decides the fate of a cell or determines cell division, but only the relative amount of the substance between the ends of a dividing cell or, in other words, the direction of the gradient. Just this sequence of identical events makes it necessary that the cells should be able to discriminate between small differences in concentration of a compound if this is to have any influence on the formation of such structures.

After formation of the central group of cells (Frey, 1970) further cells are produced by unequal divisions on the upper side of the leaf, the so-called 'Deuter' (Fig. 7.15). They have the largest area in the transverse section of a fully developed leaf. The upper and lower epidermis are then formed by unequal divisions. After this, in a final wave, occasional irregular but equal cell divisions take place forming the 'stereid bands'; finally, outgrowth of the upper epidermis produces the lamellae on the leaf surface. The leaf is fully developed with this formation, and its division pattern can be recognized very clearly in the adult leaf, even after particularly prominent wall thickening has occurred in the specialized cells.

Positional information in uniform cells

In summary of the whole development of three-dimensional structure in moss leaves, it can be said that the formation of cells with different functions is always a consequence of unequal cell divisions, whereas equal divisions produce identical cells of a uniform tissue such as the two bands of 'stereids'. These equal divisions can be – and very often are – irregular with respect to the orientation of the new cell wall in a field of strong geometric patterns. In the case of *Polytrichum*, and also in the two-dimensional liverwort leaves, such divisions are found in the final wave of division. So all cells derived from such divisions have the same positional information; division only enhances the number of identical members of a 'field'. No new decisions need to be made between different sets of activated genes. Therefore, the positional information already given can be prolonged for one or two divisions. This remark does not include the statement that a cell division, or DNA synthesis, is a pre-requisite for making a decision on the future character of a cell (cf. Holtzer & Rubin-

Fig. 7.15. Adult leaf of *Polytrichum piliferum*. L, lamellae; S, stereid band; D, Deuter; L, leptoid; H, hydroid; Z, central cell (after Frey, 1981).

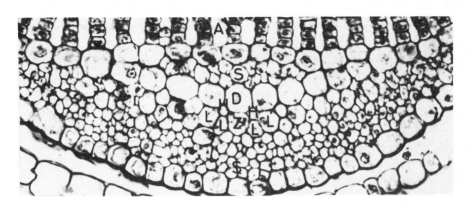

stein, 1977). No experimental data in moss plants are available to support or reject such a hypothesis. In protonema, however, we found that hormonal treatment of a fully differentiated cell could change its differentiated state without any cell division or DNA synthesis (Bopp, unpublished data).

Similar conditions may be found in the three-dimensional, more-or-less complex-structured thalli of liverworts such as Marchantiales or Metzgeriales (cf. Hébant, 1977; Crandall-Stotler, 1981). The history of the development of these plants has not yet been analysed in the same way, so we will refrain from a detailed treatment. However, we will discuss three-dimensional structures which derive directly from one single apical cell in the next section.

The three-dimensional structure of moss and liverwort shoots

Division pattern in the apical cell

The formation of the three-dimensional structure of leaves and the spatial development of three-dimensional shoots are completely different. The two-sided apical cell of a leaf forms only a unistratose organ, which produces in an independent second step the transition to the third dimension. In shoots, however, the three-sided apical cell produces directly all the cells which participate in the formation of the three-dimensional organ. Details of this process, which is a complicated sequence of predetermined cell divisions, are shown in very careful observations by Frey

(1981) for mosses, and by Crandall-Stotler (1981) for liverworts. The discussion that follows is based on these descriptions.

To understand the spatial distribution of the cells one needs to combine observation of transverse and longitudinal sections together and thus to reconstruct the shoot and its lateral organs in three dimensions.

A transverse section of a moss shoot shows the three-sided apical cell delivering derivatives from all three cutting faces (Fig. 7.16). No particular face is preferred; thus, all descendants of the apical cell produced by continuous unequal cell divisions are identical. Between two consecutive unequal cell divisions the spindle in the apical cell is always turned by 120°. Although the time between two divisions can become very long during the development of a moss (Hallet, 1972) the sequence of divisions is not disturbed. If the angle between two divisions is exactly 120° the daughter cells and their decendants – for example the leaves which correspond exactly to the three rows of cells – are arranged exactly in three straight rows (orthostichies) as in *Fontinalis* and *Tetraphis pellucida* (Fig. 7.17) or in the rhizomes of *Polytrichum* (Fig. 7.18a). In young shoots of *Polytrichum*, however, the rows are slightly twisted (Fig. 7.18b). This is due to enlargement of the angle between successive divisions to more than 120° so that the next cell division on the same face of the tetrahedral cell is always rotated by a certain constant amount; this is also the case in most other mosses. In older stems (Fig. 7.18c), the spiral is much more twisted

Fig. 7.16. *Rhacomitrium lanuginosum* (moss). Three-sided apical cell (arrowed) with its descendants, showing the spiral position of the succeeding segments (compare with Fig. 7.13) (after Frey, 1970).

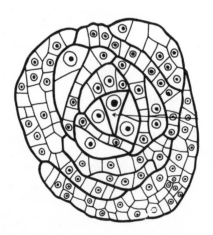

Fig. 7.17. *Tetraphis pellucida* (moss). Transverse section through the apical cell and its descendants oriented in three orthostichies (after Berthier, 1971–72).

Fig. 7.18. Shoot apex of *Polytrichum commune* with different segmentation of the apical cell. (a) Rhizome, (b) young shoot, (c) older shoot. The arrows show the strength of spiralization in succeeding descendants. A, the apical cell; S, its youngest descendant (after Hébant, 1977).

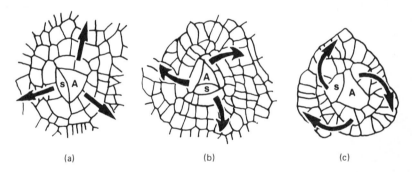

(a)   (b)   (c)

because the angle between two divisions is much greater than 120° (Hébant, 1977; Wigglesworth, 1956).

Nothing is really known about factors which change the direction of division, except that they may be located in the plasmalemma of the apical cell (Hepler, 1981). Nevertheless the lawfulness of this process must be considered. The turning of the division plane may be a time-dependent process because the sequence of cell divisions in the apical cell of older plants is much slower than in younger shoots (Hallet, 1972). The microtubules which must be involved in this orientation must also go through a complicated series of constructive and destructive processes.

On the other hand, as we have seen in the leaves, divisions in which the spindle is not exactly perpendicular to the old cell wall are a very common feature in bryophytes. Therefore, when we regard the oldest cell wall of the triangular apical cell as the plane of reference for the orientation of the spindle (or the insertion of the preprophase band), we can compare this type of division with that in leaves. Because the strong division pattern in the apical cell is responsible for the positioning of leaves, the spiralization of the rows of leaves can be easily explained in this way. Fig. 7.16 illustrates how secondary growth processes need not play a significant role in determining leaf position. This means that the only 'positional information' necessary for formation of the relatively complicated leaf pattern depends on the division of the apical cell.

Dorsiventrality

This idea may be supported by observations on leafy liverworts which show a clear dorsiventral morphology in the adult stage. Dorsiventrality is expressed by the division of the apical cell, whose descendants which form the two lateral flanks are different from that which forms the ventral part where the smaller underleaves develop. In *Trichocolea*, for example, in which the leaves consist of cell-filaments, the descendants of both lateral faces of the apical cell form, in two division steps, four initial cells at the surface of the shoot to make four lobes. The descendant of the ventral face, however, makes only one division which gives rise to two initials for two lobes (Grill, 1958). This stems from the fact that in many Jungermanniales the ventral surface of the apical cell is of much smaller area than either of the lateral surfaces and that the cells segment in perpendicular rather than spiral sequence. This involves a first division on lateral side 'one', a second division on the ventral side, a third on lateral side 'two', then a fourth on the ventral side, and so on, so that a ventral division takes place between every lateral division. The dorsiventrality is, therefore, associated with a fundamental change in the sequence of division orientation. Such rotation of the spindle orientation over the 'back' of the dorsiventral axis is impossible, and thus the formation of a spiral leaf pattern is prevented. In most hepatics this deviation is replaced by the typical spiral pattern with the enlargement of the ventral face of the apical cell on maturation (Crandall, 1969; Crandall-Stotler, 1981). The reason for this may be dependence of these processes on factors like growth speed and nutritional supplements, factors which have been calculated for the developmental regulation of filamentous structures in mosses (Buis, Brière & Larpent, 1976).

Dorsiventrality must be regulated by external factors. Gravity is probably an important source of positional information. This was demon-strated for liverworts as well as for mosses. The gemmae of *Marchantia polymorpha* possess two identical apical cells on each side (Fig. 7.19). Gravity determines which will divide to produce the thallus. Only the lower of the two cells becomes the apical cell for the new dorsiventral thallus (Halbsguth, 1953; Kohlenbach, 1957). Another example is the formation of side branches in the caulonema cells of a moss protonema (*Funaria*). When protonema are experimentally placed vertically, the side branches are always negatively geotropic (Schmiedel & Schnepf, 1979).

Fig. 7.19. *Marchantia polymorpha.* Gemma showing the double dorsiventral struc-ture (X denotes the axes of symmetry) and the two apical cells at each end. Arrows indicate the final levels of the outgrowing thallus (after Halbsguth, 1953).

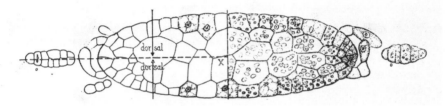

In several dorsiventral liverworts it is possible to induce pheno-variants with three rows of isomorphic leaves, rather then two lateral and one ventral row, when the merophyte-descendants of the apical cell are all identical and the usual reduction of the ventral merophyte is prevented (Basile, 1980). Without going into details, it can be said that such pheno-variants are induced by substances which are reported to be antagonists of normal hydroxyproline-rich protein synthesis. To interpret these re-sults it has been suggested that these special wall proteins regulate mor-phogenesis by suppressing cell enlargement and/or cell division in very localized populations of cells (Basile, 1980) such as the file of ventral cells in the shoots of Jungermanniales. The leaf rows are straight, as they are in untreated shoots, showing that the angle between two divisions is exactly 120°.

The aspect in longitudinal sections

We can summarize by saying that, in the preceding section, we treated the arrangement of cells and leaves around the apical cell as a two-dimen-sional pattern, but, as we have mentioned before, this reveals only one aspect of the shoot structure. The shoot develops by a continuous process and not by independent steps, so to draw the real picture we have to integrate the perspective provided by the longitudinal sections which give us additional information about cell arrangement in shoots.

The three-sided or tripolar apical cell appears in longitudinal section either oval or inverse-egg shaped and sometimes also triangular (Fig. 7.20). Segments are produced with cell walls more or less parallel to the shoot axis. Each segment then divides into an upper half and a lower half through what is, at first, a strongly determined sequence of divisions (Frey, 1970). The lower (inner) cell produces the tissue of the stem by further divisions. The upper part gives rise to the leaves, side branches and 'epidermis'. It is not possible to describe all the details of this development, which is different in acrocarpous as compared to pleurocarpous mosses (Frey, 1970; Berthier, 1971–72). The most important aspects are summarized in the following scheme (*see below*) which shows how, at each of the first four division steps, a decision is made between two courses of differentiation, maybe by turning on or off the specific sets of genes responsible for the further differentiation of a cell or of all its descendants, thus determining the pattern along the shoot axis. Regarding this scheme, one has to bear in mind that each step is accompanied by several divisions transverse to the longitudinal axis, which enlarge the surface of all organs but do not change the future destiny of the cells.

Scheme

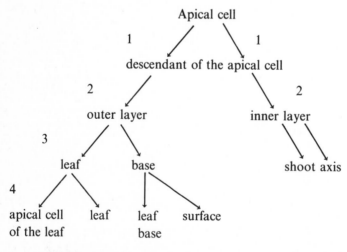

1–4 = division step

Fig. 7.20. *Hypnum lindbergii* (moss). Longitudinal section through the apical region of a shoot. S denotes the apical cell (after Frey, 1970).

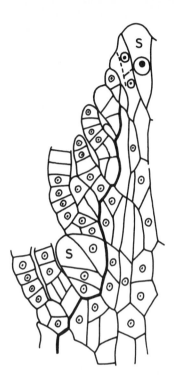

The generalized scheme can vary in details and it is irrelevant whether the tip of a shoot is very small, as in the genus *Neckera*, or flatter as in *Mnium* (Fig. 7.21). In both cases the divisions produce an 'apical dome' and one can regard the product of the outer layer as a plane which is pulled over the inner part like a glove over a finger (compare Fig. 7.20), as was determined for the shoots of higher plants by Green (1980). It may be worth mentioning that as soon as a three-sided apical cell is formed as a young moss bud at the protonema stage, it is vaulted at its surface. This determines the later three-dimensional structures.

As a second important point, it can be seen in longitudinal section that the descendants which lie in the outer layer lose their original orientation and give rise to leaves with a new apical cell and new polarity. This polarity is perpendicular to the original cell orientation in the shoot and is initiated already in the first descendant in which the division can be regarded as H-shaped (new cell wall perpendicular to the previously-formed wall (Green, 1980)). The information for the change in polarity which establishes a new point of orientation, therefore, is based only on the division sequence, and so only this has to be determined. This process is also relevant to the formation of the 'epidermis' or cortical cells.

Fig. 7.21. *Mnium hornum* (moss). Longitudinal section through the apical region of a shoot. A denotes the apical cells (after Berthier, 1971/72).

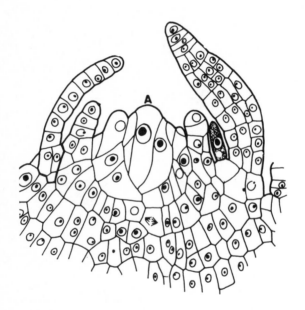

Formation of side branches

Localization of branch initials

The outer cell layers of the stem are produced at the base of each leaf. These cells elongate and produce the cortical structure ('epidermis'). On the abaxial side of the leaves a special tissue is formed, the bud initial; it is a small mound of tissue on which a central, three-faced apical cell can be discerned, occasionally with a few descendants (Fig. 7.22) (Berthier, 1971–72). Such structures represent inhibited buds. They are normal products of the cortex development and can be activated while cells contiguous to the central apical cell simultaneously develop by repeated divisions into the first leaf primordium of a side branch. The division frequency of the apical cell itself is much slower than in the surrounding cells. Already mature cortical cells below the bud, however, are never involved in this development (Nyman & Cutter, 1981).

A very regular distribution of apical cells along the axis is found in many liverworts (Jungermanniales). Only two types (Figs. 7.23, 7.24) out of 11 (Crandall-Stotler, 1972) are selected as examples to demonstrate two main possibilities: either that a cortical cell itself becomes the branch initial, or that a young cortical cell basiscopic to a leaf divides periclinally, forming two cells, of which the innermost enlarges to form the branch initial and the outer cell undergoes many anticlinal divisions, giving rise to initials of the 'collar', a tissue which lies like a collar around the young branch, breaking through this outer cell layer.

The important point is that in each case the positions of the branch initials are fixed by the cell division program and no supplementary information is called for. If one considers, however, that the real initial cell is formed only after a sequence of as many as six or seven cell divisions, a supplementary positional signal might be expected, but the question of whether this is a division-independent factor cannot be answered. The strict regularity, without statistical deviations, of the branch initials in the liverworts supports the assumption that only the division program itself destines the final position.

Regulation of growth activity in side branches

The regulation of the formation of initials is completely different from the regulation of the activity of the newly-formed apical cells. For this, the influence of gradients has to be considered. There must be differences in these gradients since in some cases all initials develop into branches; in other cases only few of them do. For this regulation 'apical dominance' is taken into account, an effect well known in higher plants. As in higher plants, a hormonal regulation for mosses and liverworts has been described by several authors (Sironval, 1952; von Maltzahn, 1959; Maravolo, 1980; Nyman & Cutter, 1981; Koch, 1982). The results, however, are not consistent enough to allow formulation of a general hypothesis valid for all bryophytes. It seems clear that the two hormones auxin and cytokinin have an effect on the outgrowth of side branches and that auxin alone can replace the inhibitory action of the growing apical bud. An apical–basal transport, however, seems not to be essential, because a high concentration of auxin throughout the whole stem produces its effect irrespective of the site of auxin application (Koch, 1982).

Cytokinin can act synergistically (Nyman & Cutter, 1981) as well as antagonistically (MacQuarrie & von Maltzahn, 1959) to auxin. In the latter case, more buds are developed with cytokinin, and the buds grow faster, than without this hormone. It must be noted, however, that cytokinin induces adventitious buds in each case, for example as regeneration products (Bopp, 1981). Apical dominance requires at least one gradient in an basipetal direction. Direct and indirect arguments have been put forward to demonstrate the existence of gradients in bryophytes. A very clear example comes from regeneration experiments with the unistratose liverwort *Riella* (Stange, 1957). Incision in the wing of this plant always induces regeneration at places at which the shortest route of plasmatic transport between the apical meristem and the wound edge cells is interrupted (Fig. 7.25). Cytological and physiological differences between the cells distributed in a regular apical–basal way can be disturbed by auxin antagonists (Stange, 1977), which may support the idea of a hormonal gradient along the wing. Basipetal auxin transport has been found in specialized tissue of

Fig. 7.22. *Mnium hornum*. Side branch initial with three-sided apical cell (the numbers give the sequence of descent) (after Berthier, 1971–72).

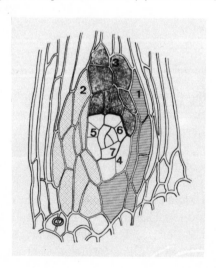

Fig. 7.23. *Bryopteris fruticulosa* (liverwort). Longitudinal section of the stem, showing the formation of branch initials (bi) as the translocated innermost parts of the periclinally-divided cortical cells (Co). X denotes the apical cell. The dorsiventrality of the shoot can be recognized (after Crandall-Stotler, 1972).

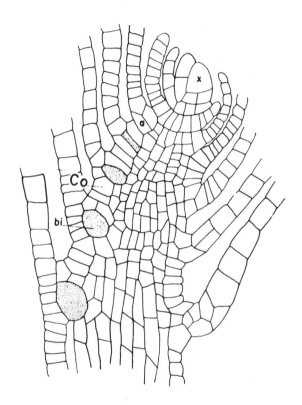

the liverwort *Marchantia* and is responsible for the interrelation between the midrib and the apex in the control of regeneration (always polar in *Marchantia*) and apical dominance (Maravolo, 1980). Polar transport and its mechanism are clearly demonstrated also in the uniseriate protonema filaments of mosses (Rose & Bopp, 1983; Rose, Rubery & Bopp, 1983).

Branching of the filamentous system of the moss protonema

Regular branching

In many mosses two stages of protonema differentiation can be distinguished, the chloronema and the caulonema. Both types are characterized, among other things, by differences in the system of branching. Whereas in chloronema the branches appear to arise in an irregular way (careful analysis is needed to discern the rules of side-filament formation), in caulonema this formation is very regular and can be explained easily by the division pattern alone. In *Funaria*, with the best characterized caulonema, the third cell in a main filament always forms a side branch near the apical-cell end. This place is predetermined by the polar structure of the cells, which is dependent on the cytoskeleton and the position of the nucleus (Sievers & Schnepf, 1981). These branches (at each cell of the main filament, exactly in the same position) grow more slowly than the main filament, but so steadily that the pattern of branching depends on the one hand on the regular cell division, and on the other hand on the constant growth relationship between main filament and side branches.

After five or six cells, a second branch can be formed also at the apical end. The caulonema is also the place of bud formation (the initials of the moss plants). The position of the buds is either the same as that of the branches, or the buds appear at the basal cell of an already-grown side branch (for details see Bopp, 1981).

In summary, we can conclude that the branching system is determined by the heritage of each cell and no external factors are involved (an exception is the experimental translocation of the nucleus, or disturbance of polarity (Sievers & Schnepf, 1981)). For the distribution of buds, how-

Fig. 7.24. *Fontinalis antipyretica* (moss). A type of bud initial (bx) formation which is also found in liverworts. The apical cell for the branch is formed in the second cortical cell below a leaf (after Crandall-Stotler, 1972).

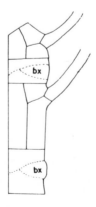

Fig. 7.25. Regeneration pattern in the wing of *Riella helicophylla* (liverwort) induced by different directions (*A, B, C*) of incisions (after Stange, 1957).

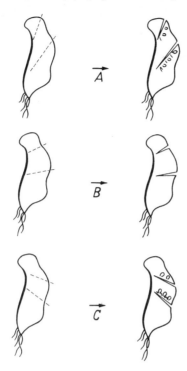

ever, supplementary information of a hormonal nature is necessary; this
has been partly analysed by Bopp (1981).

Models for protonema branching

More important in our context is the information one can gain from the
analysis of mosses with no clear distinction between caulonema and chloronema, as in *Ceratodon purpureus* which has been carefully studied by
Brière (1982).

Spatial organization depends on interconnection of all protonemas in
such a way that a well co-ordinated development results. The single elements are joined together in a 'morphogenetic system' (Bopp, 1965),
which means that all elements influence one another producing the whole
system, and the removal of one component induces reactions in the other
components to reconstitute the original state. An important component in

the system is the form of branching. In an earlier examination, four independent 'factors' involved in this process were analysed (Buis *et al.*, 1976). One was the stimulation of branching as a consequence of apical cell division, a second was related to apical dominance. The two other factors either stimulated branching by cellular elongation or inhibited by a mitosis-elongation antagonism. The first of these factors is a consequence of growth – not necessarily regulated by one single substance – because cell elongation and cell division are closely related processes (Bopp, 1959; Brière, Buis & Larpent, 1979; Brière, 1982).

This statistical research on branching formed the basis of a more detailed analysis which finally resulted in the development of a simulation model (Brière, 1982). Fig. 7.26 shows a comparison of the branching system of *Ceratodon* with a simulated system. Both drawings correspond very well, so the factors used for the calculation of the model can account for the natural branching process. These factors are the apical activity (as an activator), the apical dominance (as an inhibitor), a factor which regulates the density of branches, and finally a factor for the speed of branching. We cannot discuss all the predictions of such models, but we can conclude that the exact description of the branching system needs a complicated interaction of promoting and inhibiting factors, which are not necessarily single or well-defined substances, but which must be involved as long as the branching is not completely determined by a cell-inheritance division program. It may be added, furthermore, that in the caulonema the regular division program can be lost, for example in old protonemata or if the growth is drastically changed.

Importance of Ca^{2+} ions

A very important factor involved in the whole process of branching in protonemata, which are always in close contact with the substrate, is the influx of Ca^{2+} ions. This can be seen not only in a strong apical–basal Ca^{2+} gradient (Reiss & Herth, 1979) or in the distribution of a current influx at the top of growing cells and at the place of branch initiation (Weisenseel & Kicherer, 1981; Bopp & Weisenseel, unpublished), but also in experimental work with Ca^{2+} ionophores. One finds with the latter a strong Ca^{2+} accumulation at the site of side-branch formation (Saunders & Hepler, 1981). But it is important to say that the strong localized Ca^{2+} influx is not the cause of the branching pattern but depends on a given prepattern in peripheral plasma layers. Ca^{2+} gradients are the expression of a preformed plasmatic differentiation.

Fig. 7.26. *Ceratodon purpureus* (moss). Branching of a young protonema. (a), (b), Drawings of naturally-occurring branching systems; (c), (d) simulated system which gives evidence of determinant factors involved in the branching system (after Brière, 1982.)

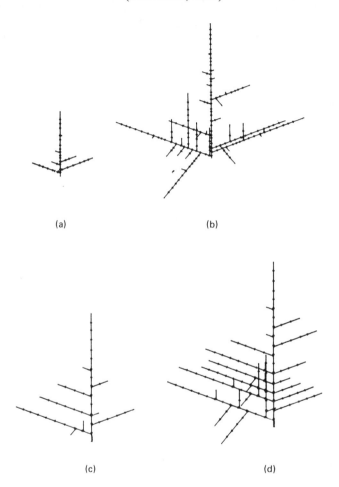

(a)　　　　　　　(b)

(c)　　　　　　　(d)

Conclusion

We have seen that the patterns of cells in mosses are, in most cases, best understood as systems of predetermined sequences of cell division. This means that during consecutive divisions new genetic information is activated, and the next division depends on the set of information in the

M. Bopp

preceding one. In many special patterns, however, the sequence alone is not sufficient to explain the position of cell walls and, with this, the determination of future cell differentiation. We, therefore, have to postulate unidirectional gradients which decide between two (or more) possibilities. Such gradients may determine the position of a cell wall or the direction of unequal cell divisions, which are prerequisite for cytological and molecular differentiation. We know only very little about the real nature of such gradients, though they can be simulated by theoretical models (Brière, 1982). It is impossible, therefore, to distinguish between real gradients of substances such as hormones or ions, and 'factors' such as growth speed and activity, etc., which have to be regarded as the bases of physical gradients. Nevertheless, the formation of cell pattern in bryophytes can present a view of cell differentiation quite distinct from higher plants.

I should like to express my sincere thanks to the following authors who generously granted permission to reproduce the following drawings: Professor Dr Schnepf (2, 5, 7, 8), Professor Dr Hagemann (1), Professor Dr Frey (3, 4, 15, 16, 20), Professor Dr Berthier (17, 21, 22), the late Professor Dr Hébant (18), Professor Dr Stange (25), Professor Dr Kohlenbach (19), Dr Crandall-Stotler (23, 24), and Dr Brière (26).

References

Basile, D. V. (1980). A possible mode of action for morphoregulatory hydroxyproline-proteins. *Bulletin of the Torrey Botanical Club*, **107**, 325–38.

Berthier, I. (1971–72). Recherches sur la structure et le développement de l'apex du gamétophyte feuillé des mousses. *Revue Bryologique et Lichénologique*, **38**, 421–551.

Bopp, M. (1959). Versuche zur Analyse von Wachstum und Differenzierung des Laubmoosprotonemas. *Planta*, **53**, 178–97.

– (1961). Beziehungen zwischen den Teilungs- und Differenzierungsvorgängen während der Blattentwicklung von *Polytrichum attenuatum*. *Revue Bryologique et Lichénologique*, **30**, 253–9.

– (1965). Die Morphogenese der Laubmoose als Beispiel eines morphogenetischen Systems. *Berichte der deutschen botanischen Gesellschaft*, **78**, 44–54.

– (1981). Entwicklungsphysiologie der Moose. In *Advances in Bryology*, vol. 1, ed. W. Schultze-Motel, pp. 11–77. Vaduz: J. Cramer.

Bopp, M. & Feger, F. (1960). Das Grundschema der Blattentwicklung bei Lebermoosen. *Revue Bryologique et Lichénologique*, **29**, 256–73.

Brière, C. (1982). Analyse quantitative et modélisation du développement du protonema des Bryales. Mémorisation de l'activité apicale et ramification. *Thèse d'État de l'Institut National Polytechnique de Toulouse*.

Brière, C., Buis, R. & Larpent, J. P. (1979). Cellular growth and cellular division in relation to age and illumination in the protonema of *Ceratodon purpureus* Brid. *Zeitschrift für Pflanzenphysiologie*, **95**, 315–22.

Buis, R., Brière, C. & Larpent, J. P. (1976). Analyse quantitative du développement du protonéma de *Ceratodon purpureus* Brid. *Physiologie Végétale*, **14**, 817–32.

Bünning, E. (1965). Die Entstehung von Mustern in der Entwicklung von Pflanzen. In *Handbuch der Pflanzenphysiologie* vol. XV/1, ed. W. Ruhland, pp. 383–408. Berlin, Heidelberg & New York: Springer-Verlag.

Busby, C. H. & Gunning, B. E. S. (1980). Observations on preprophase bands of microtubules in uniseriate hairs, stomatal complexes of sugar cane and *Cyperus* root meristems. *European Journal of Cell Biology*, **21**, 214–23.

Crandall, B. J. (1969). Morphology and development of branches in the leafy hepaticae. *Beiheft zur Nova Hedwigia*, **30**, 261 pp.

Crandall-Stotler, B. (1972). Morphogenetic patterns of branch formation in the leafy hepaticae – a résumé. *The Bryologist*, **75**, 381–403.

– (1981). Morphology/anatomy of hepatics and anthocerotes. In *Advances in Bryology*, vol. 1, ed. W. Schultze-Motel, pp. 315–98. Vaduz: J. Cramer.

Frey, W. (1970). Blattentwicklung bei Laubmoosen. *Nova Hedwigia*, **20**, 463–556.

– (1981). Morphologie und Anatomie der Laubmoose. In *Advances in Bryology*, vol. 1, ed, W. Schultze-Motel, pp. 399–477, Vaduz: J. Cramer.

Green, P. B. (1980). Organogenesis – a biophysical view. *Annual Review of Plant Physiology*, **31**, 51–82.

Grill, R. (1958). Blattentwicklung von *Trichocolea tomentella*. *Planta*, **51**, 673–93.

Gunning, B. E. S. (1981). Microtubules and cytomorphogenesis in a developing organ: the root primodium of *Azolla pinnata*. In *Cytomorphogenesis in Plants*, ed. O. Kiermayer, pp. 301–25. Vienna & New York: Springer-Verlag.

Gunning, B. E. S., Hardham, A. R. & Hughes, J. E. (1978). Preprophase bands of microtubules in all categories of formative and proliferative cell division in *Azolla* roots. *Planta*, **143**, 145–60.

Hagemann, W. (1978). Zur Phylogenese der terminalen Sproßmeristeme. *Berichte der deutschen botanischen Gesellschaft*, **91**, 699–716.

Halbsguth, W. (1953). Über die Entwicklung der Dorsiventralität bei *Marchantia polymorpha* L. Ein Wuchsstoffproblem? *Biologisches Zentralblatt*, **72**, 52–104.

Hallet, J.-N. (1972). Morphogénèse du gamétophyte feuillé du *Polytrichum formosum* Hedw. I. Étude histochimique, histoautoadiographique et cytophotometrique du point végétatif. *Annales des Sciences Naturelles, Botanique*, 12ème Série, **13**, 19–118.

Hardham, A. R., Green, P. B. & Lang, J. M. (1980). Reorganization of cortical microtubules and cellulose deposition during leaf formation in *Graptopetalum paraguayense*. *Planta*, **149**, 181–95.

Hébant, C. (1977). *The Conducting Tissues of Bryophytes*. Vaduz: J. Cramer.

Hepler, P. K. (1981). Morphogenesis of tracheary elements and guard cells. In *Cytomorphogenesis in Plants*, ed. O. Kiermayer, pp. 327–47. Vienna & New York: Springer-Verlag.

Holtzer, H. & Rubinstein, N. (1977). Binary decision quantal cell cycles and cell differentiation. In *Cell Differentiation in Microorganisms, Plants and Animals*, ed. L. Nover & K. Mothes, pp. 424–37. Jena: G. Fischer.

Koch, A. (1982). Phytohormoneinfluß auf Wachstum und Verzweigung von Moospflänzchen. *Staatsexamensarbeit Heidelberg.*

Kohlenbach, H. W. (1957). Die Bedeutung des Heteroauxins für die Entwicklung der Dorsiventralität der Brutkörperkeimlinge von *Marchantia polymorpha* L. *Biologisches Zentralblatt,* **76,** 70–125.

Lorch, W. (1931). Anatomie der Laubmoose. In *Handbuch der Pflanzenanatomie,* vol. VII, ed. K. Linsbauer. Berlin: Gebrüder Bornträger.

MacQuarrie, J. G. & von Maltzahn, K. E. (1959). Correlation affecting regeneration and reactivation in *Splachnum ampullaceum* (L.) Hedw. *Canadian Journal of Botany,* **37,** 121–34.

Maltzahn, K. E. von (1959). Interaction between kinetin and indolacetic acid in control of bud reactivation in *Splachnum ampullaceum* (L.) Hedw. *Nature,* **183,** 60–1.

Maravolo, N. C. (1980). Control of development in hepatics. *Bulletin of the Torrey Botanical Club,* **107,** 308–24.

Nyman, L. P. & Cutter, E. G. (1981). Auxin-cytokinin interaction in the inhibition release and morphology of gametophore buds of *Plagiomnium cuspidatum* from apical dominance. *Canadian Journal of Botany,* **59,** 750–62.

Palewitz, B. H. & Hepler, P. K. (1974). The control of the plane of division during stomatal differentiation in *Allium.* I. Spindle reorientation. *Chromosoma,* **46,** 297–326.

Pickett-Heaps, J. D. & Northcote, D. H. (1966). Organisation of microtubules and endoplasmic reticulum during mitosis and cytokinesis in wheat meristems. *Journal of Cell Science,* **1,** 109–20.

Reiss, H. D. & Herth, W. (1979). Calcium gradients in tip growing plant cells visualized by chlorotetracycline fluorescence. *Planta,* **146,** 615–21.

Rose, S. & Bopp, M. (1983). Uptake and polar transport of indoleacetic acid in moss rhizoids. *Physiologia Plantarum* **58,** 57–61.

Rose, S., Rubery, P. H. & Bopp, M. (1983). The mechanism of auxin uptake and accumulation in moss protonema. *Physiologia Plantarum* **58,** 52–6.

Saunders, M. J. & Hepler, P. K. (1981). Localization of membrane-associated calcium following cytokinin treatment in *Funaria* using chlorotetracycline. *Planta,* **152,** 272–81.

Schmiedel, G. & Schnepf, E. (1979). Side branch formation and orientation in the caulonema of the moss *Funaria hygrometrica*: normal development and fine structure. *Protoplasma,* **100,** 367–83.

Schnepf, E. (1973). Microtubulus Anordnung und Umordnung, Wandbildung und Zellmorphogenese in jungen Sphagnumblättchen. *Protoplasma,* **78,** 145–73.

Schnepf, E., Deichgräber, G. & Ljubešic, N. (1976). The effect of colchicine, ethionine and deuterium oxide on microtubules in young *Sphagnum* leaflets. A quantitative study. *Cytobiologie,* **13,** 341–53.

Sievers, A. & Schnepf, E. (1981). Morphogenesis and polarity of tubular cells with tip growth. In *Cytomorphogenesis in Plants,* ed. O. Kiermayer, pp. 265–99. Vienna & New York: Springer-Verlag.

Sironval, C. (1952). Expériences sur la fragmentation des tiges de *Funaria hygrometrica. Bulletin de la Société Royale de Botanique de Belgique,* **84,** 281–9.

Stange, L. (1957). Untersuchungen über Umstimmungs- und Differenzierungsvorgänge in regenerierenden Zellen des Lebermooses *Riella. Zeitschrift für Botanik,* **45,** 197–224.

– (1977). Meristem differentiation in *Riella helicophylla* (Bory et Mont.) Mont. under the influence of auxin or antiauxin. *Planta,* **135,** 289–95.

Sych, A. (1982). Zellteilungsmuster und Plasmodesmenverteilung in jungen Blättchen von *Sphagnum palustre. Staatsexamensarbeit Heidelberg.*

Weisenseel, M. H. & Kicherer, R. M. (1981). Ionic currents as control mechanism in cytomorphogenesis. In *Cytomorphogenesis in Plants,* ed. O. Kiermayer, pp. 379–99. Vienna & New York: Springer-Verlag.

Wigglesworth, G. (1956). Further notes on *Polytrichum commune* L. *Transactions of the British Bryological Society,* **3,** 115–20.

Wolpert, L. (1969). Positional information and spatial pattern of cellular differentiation. *Journal of Theoretical Biology,* **25,** 1–47.

– (1971). Positional information and pattern formation. *Current Topics in Developmental Biology,* **6,** 183–224.

Zepf, E. (1952). Über die Differenzierung des Sphagnumblattes. *Zeitschrift für Botanik,* **40,** 87–118.

8

Axiality and polarity in vascular plants

TSVI SACHS

According to Esau (1977), 'polarity refers to the orientation of activities in space.' As such, it is an essential and relatively simple expression of organization of the various parts of plants behaving according to their position. Both external plant form and the patterns of differentiated cells require differences between axes and directions and their co-ordination in neighbouring cells. It is not surprising, therefore, that the differentiation of polarity is among the first traits to appear, both in ontogeny and phylogeny.

The term polarity, however, has been used to refer to various phenomena (Bloch, 1965a) and these must be separated in any thorough discussion of the topic. For this purpose the following ideas and terms have been found to be most useful. *Axiality* refers to the occurrence of developmental processes along one axis in preference to all others, with no reference to direction. *Polarity*, on the other hand, refers to a directionality expressed by a difference between one side and the rest of a cell, tissue, organ or organism. In a *Polar Gradient* this difference is graded rather than abrupt. In *Axial Polarity* events have both a clear axis and a clear direction. Additional terms are, of course, possible and sometimes useful, but the above-mentioned will suffice for the topics considered in this chapter.

A wide variety of oriented phenomena have been described in plants, the emphasis being on polarity rather than axiality (Bünning, 1948, 1958; Sinnott, 1960; Bloch, 1965a). Experimental work has concentrated on environmental effects that induce or even change polarity, especially in lower plants (Chapters 5, 6, 7). A review of the available fragmentary knowledge concerning all the various expressions of axiality and polarity would be neither possible in the available space nor interesting. In the context of positional controls the major interests of these phenomena are the ways in which orientation is determined, and the co-ordination both of different expressions of axiality and polarity and of their occurrence in neighbouring cells and tissues. The present chapter will concentrate, therefore, on a few central phenomena (Fig. 8.1), aiming at a general hypothesis of their internal control that could serve as a framework for further studies.

The polarity of apical development

A vascular plant can be viewed as a branched axis that grows at its tips. This is a bipolar structure, since there are different apices at the opposite shoot and root tips. Polar development and its stability can depend on the growing apices and on the traits of the connecting axial tissues. Apices are the subject of a number of other chapters so the emphasis here will be on the polarity of the connecting axis: how it is expressed, whether it can be changed, and what its cellular basis could be.

Polarity of regeneration

A clear expression of polarity is found in cuttings from which all apices have been removed. Cuttings of many plants regenerate roots, and it is not unusual for shoots to be formed as well. It is the location of the newly regenerated apices that expresses polarity: roots form on the basal or root side, and shoots on the opposite side of the cutting (Vöchting, 1878; Sinnott, 1960). In other plants, buds and occasionally root primordia are present along the stem and their development, rather than initiation, is localized in the same way. This polar development often occurs even in the absence of any environmental cue that could specify direction, gravity being excluded by means of inverted and horizontal treatments. It is important that the very same parts of the plant axis are able to develop different apices depending on their location relative to the rest of the cutting (Fig. 8.1A; Vöchting, 1884). It is often found that polarity of regeneration is not very clear and it can also be overridden by environmental conditions (see below), but the many examples where it is expressed show that there must be a tissue polarity that is maintained in cut sections.

An innate polarity of regeneration may depend on either or both of two types of controls: a gradient of developmental potentials along the plant axis and a polarity of transport of morphogenetic signals (Fig. 8.2). In the first case, the development of roots and shoots, though possible along the entire axis, would occur most rapidly on the side closest to their original location. The apices that form and grow first would then inhibit similar development in the rest of the cutting by means of correlative signals. The second possible control would involve no difference in the traits of the cells along the axis. A polarity of the transport of signals, however, would lead to their accumulation at specific sides of the cutting. These signals would then induce the appropriate development of apices.

Fig. 8.1. Diagrammatic representation of the problems dealt with in this chapter.
A. Polar regeneration of girdled stem cuttings. Arrows show the original direction to the roots. Roots form on the root side of the tissue and shoots on the shoot side. This can be independent of gravity and the location of cuts and girdles.
B. The axial pattern of vessel members. These specialized cells are arranged in long files, in a matrix of other cells (not shown). An axis is defined by cell arrangement, cell shape and the location of the openings between cells.
C. Polar transport. Auxin is moved from a donor agar block (*D*) to a receiver (*R*) in the same direction as the tissue's polarity (indicated by an arrow) (top diagram) but not in the opposite direction (bottom diagram).
D. The plane of cell divisions. The procambium in the centre of a *Coleus* leaf primordium indicates how this plane can be controlled and co-ordinated in neighbouring cells.
E. Cell polarity defined by unequal divisions. Epidermal cells of a *Lathyrus* leaf dividing unequally in an early stage of stomata formation. These divisions are the first stage in the development of a spacing pattern of the mature stomata.

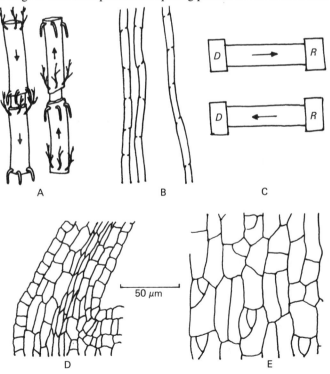

50 µm

Evidence concerning controls of polar regeneration

The available evidence shows that polar transport plays a role in determining the location of regeneration while gradients of developmental traits cannot be very important. Polar transport of one morphogenetic signal, the hormone auxin, is well established (see p. 205). The exogenous application of auxin promotes root initiation in many plants (Went & Thimann, 1937). Known facts thus suggest that the accumulation of auxin transported to a basal cut promotes root initiation (Went & Thimann, 1937; Gautheret, 1944). Auxin also inhibits both bud initiation and development (Thimann, 1977) so that its accumulation could exclude buds from the basal region. It is less clear whether auxin transport could explain the distribution of buds along cuttings. It is possible that most apical buds have an advantage in correlative inhibition and that additional factors are involved. The polar, tissue-dependent transport of signals other than auxin is a possibility but there is no conclusive evidence for it at present (see p. 206).

On the other hand, though gradients of regenerative capacity, especially root formation, can be found along the stems of many plants, correlative inhibition between apices of the same type is not pronounced until they are of noticeable size and are growing rapidly. Small differences in the rate of apical development, therefore, could not lead to the observed localization of the early stages of regeneration. Though the desirable qualitative data are not available, it is therefore unlikely that gradients

Fig. 8.2. Schematic representation of two possible controls of polar root regeneration. A morphogen that determines root formation is indicated by dots. A gradient of this morphogen (top row) could lead to differences in the size of the new roots. The earliest and largest roots would be on the original root side and would suppress other roots by correlative inhibition. In the bottom row the morphogen is not graded, but polar transport concentrates it next to the cut on the root-side. This concentration is followed by localized root formation.

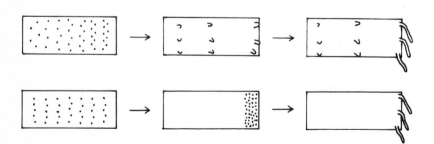

of regenerative capacity play an important role in polar regeneration, especially in cuttings taken from long branches, where the gradients cannot be steep.

In addition to regenerative capacity, there are also a variety of structural and developmental gradients (Bloch, 1965*a*) that have been mentioned as expressions of polarity and considered to depend on underlying gradients of hormones or other factors. These include the capacity to form flowers and roots and other traits connected with juvenility (see Chapters 12, 13). The apical growth of plants, however, results in cells being arranged along the axis according to their age and the state of the meristem at the time they were formed. Cell behaviour may depend on local concentrations of substances; gradients of hormones have been measured, but a causal relation is only a hypothesis for which there is no direct proof. It is quite possible that stable programs of gene expression do not depend on the concentration of any one substance.

Determined aspects of apical development

Finally, the stability of the polarity expressed by apical regeneration and its implications should be considered. The degree of this stability has been the subject of much debate (Bloch, 1965*a*), partly on the basis of work on non-vascular plants (see Chapters 6, 13). In many seed plants the location of apical development, especially root initiation, can readily be determined by local environmental conditions (gravity, low light, low oxygen and high humidity). Although this has been considered to show that polarity can be readily changed, it is also possible that only the expression of polarity is affected, not its underlying cellular basis (Vöchting, 1892). While this appears at first sight to be a matter of semantics, the stability of tissue polarity is demonstrated by its expression in horizontal cuttings in a uniform environment. Further evidence for an underlying tissue polarity comes from plants in which the original shoot-to-shoot orientation of part of the axis was reversed. This is achieved by grafting plant parts together after one of them was inverted and by inducing shoot and root development in an inverse orientation by means of local environmental conditions (Vöchting, 1892; Bloch, 1965*a*). In most cases such inverted plants fail to develop, or develop poorly, producing large swellings at the base of the stems, demonstrating a tissue polarity that is not readily changed (Vöchting, 1892, 1906). There are, however, reports of plants with an inverted stem segment that did develop normally (Castan, 1940; Went, 1941; Sachs, 1968; Sheldrake, 1974). On the basis of the vascular patterns of new roots (Rathfelder, 1955), regeneration of 'reoriented' sections (Libbert, 1956) and auxin transport (Sheldrake, 1974), it has been questioned whether the tissues functioning in the new orientation have been 'truly' inverted. It may be concluded that tissue polarity is stable or determined, though there may be conditions where at least some of its attributes can be changed. Further discussion of this determination requires consideration of additional expressions of polarity at a tissue and cellular level (pp. 203–4).

Tissue determination displayed by cuttings could contribute to the control of the bipolarity of intact plants (Vöchting, 1884). This does not mean that the determined state of the apices themselves is not an important, and even sufficient, factor in this stability (Wareing, 1978). In *Selaginella*, however, a root apex can be converted to a shoot apex in culture, but the same conversion does not occur on the intact plant. A possible basis for this difference is the polar transport of auxin, which maintains the root-inducing conditions on one side of the plant axis (Wochok & Sussex, 1974). The same and similar processes could be a contributing factor for the stability of the bipolar structure of seed plants. In addition, the hormonal differentiation of the apices can make them dependent on one another: root initiation requiring auxin from the growing shoots and being essential for these shoots as sources of cytokinins (Sachs, 1975*a*; Gersani, Lips & Sachs, 1980). This positive feedback relation could stabilize the bipolar structure quite independently of the determination of the apices and the intervening axis.

Vascular differentiation

The cellular expression of polarity and its controls

The conclusion that the regeneration of shoot and root apices expresses an innate polarity suggests searching for its expressions at an anatomical level. This was realized by Vöchting (1892) who found that grafts in which one of the members was inverted, so that tissues of opposite polarities were in close contact, formed complex whirlpools of xylem elements. The vascular tissues are composed of cells whose axis is clearly defined; this is especially true of the vessels and sieve tubes, where it is expressed by cell arrangement, shape and the location of specialized connecting wall structures (Fig. 8.1B). Controls of vascular differentiation, therefore, must specify oriented events at a cellular level. The study of these controls, reviewed recently by Sachs (1981*a*), shows that a positive feedback between the flow of correlative signals and axial differentiation accounts for major aspects not only of polarity but also of the patterned differentiation of cells in organized strands. Only a few major results and conclusions can be included here:

(a) The vascular system continues to develop throughout the life of the plant, connecting new leaves with new root tissues (Fahn, 1974; Esau, 1977). In most plants with a vascular cambium, continued vascular tissue formation along the axis adjusts the functional vascular system to the combined size of the organs it connects (Benayoun, Aloni & Sachs, 1975). The vascular system in these plants is also able to regenerate readily around wounds (Fig. 8.3A), expressing the effects of constant developmental controls. The three-dimensional course of vessels, furthermore, can be readily observed in cleared herbaceous tissues. These are among the traits that make vascular differentiation remarkably suitable for the study of controls of oriented development.

(b) The formation of vascular tissue is correlated with the development and presence of the leaves to which it connects. Removal of these leaves, or cutting the direct contacts with them (Fig. 8.3B), reduces or even stops differentiation. The growth of buds, or their graftage (Fig. 8.3F), on the other hand, induces vascular differentiation. This correlation of leaves and differentiation is very general though there are exceptions that might indicate that it is not the only control of vascular development (Sachs, 1981*a*). It is possible, and by no means contradictory, that vascular differentiation promotes leaf formation (Larson, 1975) though conclusive evidence for this is as yet not available. In short-term experiments the removal of the roots, or cutting the direct contact with them (Fig. 8.3C), does not stop differentiation leading in their direction (Sachs, 1968). These and other facts show that a major control of vascular differentiation involves the polar flow of signals from the leaves towards the roots.

(c) Many aspects of the influence of leaves can be replaced by a localized source of auxin (Fig. 8.3D,E). These include the differentiation of both vessels and sieve tubes from procambium, cambium and young parenchyma, the formation and induction of cell divisions of the cambium, and the control of the orientation of cell divisions and vascular differentiation. As it is known that auxin is produced by leaves, its replacement of their effect means that auxin is part of the signal by which leaves control vascular differentiation and orient it so that it connects them with the rest of the plant (Jacobs, 1952; for review see Sachs, 1981*a*). This does not imply that auxin is the only inductive signal originating in the leaves.

(d) Differentiation induced by both the leaves and exogenous auxin fol-

lows the shoot-to-root polarity of the tissue whenever possible (Fig.
8.3C–E). This polarity can be quite strict (Aloni & Jacobs, 1977) and is
found even in plants cut so that the polar direction does not lead to the
formation of functional contacts, as differentiation in another direction
would (compare Fig. 8.3A,C). The influence of original tissue polarity on
vessel formation is also evident in grafts where the lower member has been
inverted (Fig. 18.3G; Vöchting, 1892). It may be concluded that vessel
differentiation is dependent on auxin transport and expresses its direction
at a cellular level.

(e) When the polar direction is blocked by cuts, differentiation proceeds
readily towards a contact with the basal or root direction at any angle
with the original tissue polarity (Fig. 8.3J,K; Janse, 1914). Thus, by the
criterion of vessel differentiation, polarity can be changed readily when no
polar alternative is available. This re-orientation occurs next to lateral

Fig. 8.3. Controls of vessel differentiation. Schematic representation of new vessels (wavy lines) formed in bean hypocotyls cut as shown. Original vessels not
indicated. Absence of line above or below indicates contacts in the direction of the
shoot or the root, respectively. Darkened regions are sources of auxin (1% indolyl-
acetic acid in lanolin), and arrows show the original direction to the roots. The
results shown can be obtained in plants kept horizontal, indicating that gravity is
not an important factor. As discussed in the text, these treatments show that vessel
differentiation is controlled by a feedback relation between auxin flow and tissue
polarity.

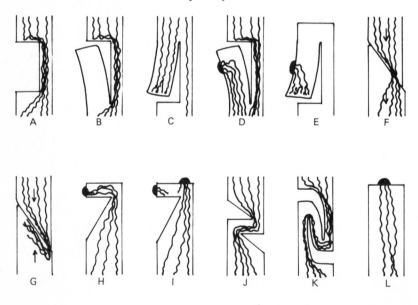

buds released from dominance (Neeff, 1914), in regeneration around wounds (Fig. 8.3A), and it can even involve a complete reversal of the shoot-to-root polarity over short distances (Fig. 8.3K). The axis of the first vessels formed during such re-orientation does not correspond to the elongation and cell wall thickenings of the cells that compose them (Fig. 8.4A); these are normally at right angles to the axis of the vessel. Thus not all parameters are changed with the first re-orientation. The new polarity occurs along the route of the expected diffusion of auxin from where it is accumulated by polar transport to the channels that transport it away in the basal direction. Once the new polarity is induced it presumably becomes the preferred direction of auxin flow, and this leads to further induction of the new polarity. Thus it may be suggested that tissue polarity and auxin are linked by a positive feedback. This would account for all known facts and bring differentiation involving re-orientation of polarity into line with the differentiation of undisturbed tissues (Sachs, 1981*a,b*). While a variety of factors influence the differentiation of new vessels (Roberts, 1976), localized auxin is the only one that is known to determine their orientation.

(f) The induction of re-orientation and differentiation is gradual and occurs only after a long-term flow of auxin through the cells. Conditions which prevent an auxin gradient or flow also prevent differentiation (Fig. 8.3I). When conditions are changed during differentiation, as by the application of an additional auxin source (compare Fig. 8.3H and I), the effect of original conditions is not expressed. When one source is applied above another, so that polar transport can be expected but no long-term gradient established, differentiation does occur. When polar transport leads to a cut and diffusion is possible only in the direction opposite to polarity, circular vessels are formed (Fig. 8.4B; Sachs & Cohen, 1982). These and other facts can be readily understood if vessel differentiation is a response to a continued flow of auxin through the cells (Sachs, 1981*a,b*).

(g) A gradual polarization of auxin transport by the response of the cells to auxin flow could lead to the canalization of differentiation to the defined strands seen in vessels and sieve tubes (Figs. 8.1B; 8.3L). Thus, as polarization continues, limited files of cells suffice to transport the available signals and finally specialize in various ways (Sachs, 1969). This suggestion relates a major cellular pattern to the induction of differentiation and to the expressions of polarity considered above. Evidence for this hypothesis comes from observations that very young leaf primordia influence the entire subtending stem tissues and only later is their effect canalized to the vascular system (Sachs, 1981*a*). It is also supported by studies of the transport of radioactive auxin (pp. 205–8) showing that auxin is transported best in the vascular system (Sheldrake, 1973), that auxin can maintain its own transport capacity in cut tissues, and indicating that auxin treatments can induce transport in a direction new to the tissue (Sachs, 1975*b*; Gersani and Sachs, in preparation). Finally, the patterns of

Fig. 8.4. Polarity and vessel differentiation. A. New vessels formed in a bean hypocotyl above a wound. The axis of the vessel does not correspond to cell shape nor to the orientation of cell wall thickenings, normally at right angles to this axis. B. Circular vessels formed where inductive flow has no outlet (as in Fig. 8.3, C and E). C. Veins of *Cyrtomium* leaf. Though a network, it is possible to assign a polarity or direction towards the roots to all cells of this system. D. Veins of *Pisum* leaf. An example of a complex network where some of the veins have no definable polarity. E. Early stage of the vascular bridge in the first node above the cotyledons of beans. Connections to leaves are on the two sides and to the roots, downwards. The individual vessels have a clear polarity, connecting the leaves with the roots, but opposite polarities occur in neighbouring vessels of the vascular bridge.

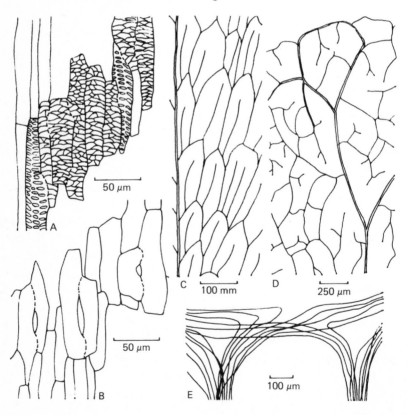

vascular contacts in both experimental and intact systems indicate that the vascular strands are the preferred channels of auxin transport (Sachs, 1969).

(h) If differentiation is a response to signal flow it can be expected to have a direction and vascular tissues would have a shoot-to-root polarity. The general form of the vascular system is in fact that of the expected drainage system, connecting the tissues of the shoot with the root tips. The vascular networks of the leaves of most angiosperms and some ferns, however, include regions to which no polarity can be assigned (Fig. 8.4D; Sachs, 1975c). The structure of large transverse vascular strands, however, suggests an answer which could be an important modification of the hypothesis. In these strands (Fig. 8.4E) there is no overall direction or polarity, yet the individual vessels do have a clear shoot-to-root direction. Vessels with opposite polarities, therefore, are found next to one another within the same strand. This suggests that veins with a clear axis but no defined polarity are formed by a signal flow whose direction changes during the process of vascular differentiation and this change may occur a number of times. The relation of differentiation to the axis rather than direction of flow is supported by some experimental evidence (Sachs, 1975c) and by the occurrence of veins with no defined polarity only where growth is unsynchronized and could cause repeated changes in the direction of signal flow.

(i) A source of auxin alone can orient and induce the differentiation of sieve tubes and a cambium in addition to the vessels that were the basis of the results summarized here. How one signal can have such complex effects, and whether additional signals are involved in the plant, are among the major unsolved problems. Information concerning sieve tube differentiation is meagre, because of technical difficulties, but it is known to be influenced by tissue polarity and can be oriented by auxin flow in the same way as vessel differentiation. For a discussion of the possibility that transverse gradients contribute to the control of the differentiation of different cell types within one longitudinal system see Chapter 10.

A positive feedback hypothesis

A general hypothesis emerges: polar differentiation is both a response and a cause of a polar flow of signals through the plants, and one of these signals is auxin. This positive feedback between flow and differentiation accounts for the determined nature of polarity and suggests that it depends on the distribution of carrier molecules rather than on controls of gene expression. The hypothesis also accounts for the lability of polarity, expressed by new differentiation, which occurs when flow cannot follow the original polarity (Sachs, 1981b). The response of the cells to flow is gradual, and continued differentiation increases the transport capacity of the cells. This results in the flow being canalized to rows of cells which differentiate in a special way, as do vessels and sieve tubes. The hypothesis thus accounts for the major axial arrangement of specialized cells in the plant. It is based on some concrete evidence and has been analysed mathematically (Mitchison, 1980a, 1981). It differs from Meinhardt's model (Chapter 1) in requiring only one signal, whose drainage into the preferred channels of transport both promotes the differentiation of the cells along the channel and inhibits the differentiation of neighbouring cells. A further difference is that the response of the tissue is gradual. Using the vascular tissue as an indication of orientation, it is found that the earliest step in differentiation is not a polarization of the cells but rather their axialization: the cells transport signals preferentially along an axis but not necessarily in one direction. This is important because, as considered in the next section, polar transport is known only for auxin while the transport of other hormones might have a preferred axis.

Transport of hormones and other substances

It was seen above that both the regeneration of apices and vascular differentiation point to a major role of hormone transport, especially that of auxin, in axial and polar organization. This section will therefore attempt a short review of oriented transport. It will be limited to the passage of materials between cells and across tissues and will not deal with the large topic of polar transport in or out of a single cell or organelles, whose implications to tissue organization are less direct.

Transport of materials other than hormones

Almost the entire transport between organs is through the highly differentiated phloem and xylem. As may be expected from the structure of the sieve tubes and vessels, this transport through mature tissues has a clear axis. The direction of transport, on the other hand, appears to depend on source–sink relations and not on any traits of the transporting tissues, which are not polar. Hormones have been considered to direct transport (Patrick, 1976), and since at least auxin can induce and orient vascular differentiation this must be true on a long term basis (Gersani et al.,

1980). The evidence for a short-term, direct effect of hormones on transport, however, is conflicting, and in many cases the measurements may reflect modifications of source and sink activities by hormones.

For understanding controls of development, however, the significant transport would be in meristems and tissues capable of regenerative development, rather than the highly differentiated vascular system. Here the available evidence is limited. Transport of both inorganic ions and organic substances occurs through parenchyma at rates that are considerably greater than expected on the basis of diffusion alone (Arisz, 1952; Lüttge & Higinbotham, 1979). Evidence indicating transport capacity also comes from measurements of electrical coupling between cells (Spanswick, 1974).

The direction of such transport of radioactive materials is often referred to as polar, but it appears to depend on the location of sources and sinks rather than on any polar traits of the transporting cells. These sources and sinks often involve differentiated cellular activities, including transport across membranes, as in various secretory glands. A preferred axis of transport through parenchyma cells could be important for developmental controls (see previous section) but the subject appears not to have been studied. Some such axiality might be expected on the basis of the distribution of the plasmodesmata (Juniper, 1977). Most important from a developmental point of view would be studies of transport in meristems, especially apical meristems, but here information is virtually unavailable. The main problem is the small size which results not only in extreme technical difficulties but also in diffusion being a rapid and presumably important process (Crick, 1970). Information about substances whose diffusion is limited by their being confined to the symplast could be all the more important (Goodwin, 1983).

Polarity of hormone transport

Hormones, like other organic substances, move most rapidly in the phloem and can also be transported through parenchyma and sometimes in the xylem sap. For phloem transport of auxin and other hormones there is no evidence that direction depends on factors other than source and sink locations of the hormones and the sugars with which they are transported. There is thus no known transporting tissue polarity of the type indicated in the previous two sections. For the specific transport of auxin outside the sieve tubes, on the other hand, the following main types of evidence point to a determined tissue polarity (Goldsmith, 1969, 1977).

T. Sachs

(a) The direction of transport for any type of tissue is fixed, independently of the presence of growing apices or any other metabolic centre. Thus transport towards the roots continues not only in their absence but also when their sink activity is prevented by cold (Eliezer & Morris, 1978). (b) The polar effects of auxin are clearly expressed in the differentiation events considered in the previous two sections and its existence at a tissue level can be seen clearly in grafts where one of the two members was inverted (Fig. 8.3G). (c) Polarity is maintained even when both the source and the sink are agar blocks, whose location can readily be changed (Fig. 8.1C). Finally, (d) polar auxin transport continues against an overall concentration gradient, as shown both by measurement (van der Weij, 1934) and differentiation events (previous two sections), though this does not mean that such gradients have no effect on transport.

There are many reports of polar transport of non-auxin hormones through sections and at rates that show sieve tubes cannot be the dominant factor (table in Goldsmith, 1977; Jacobs, 1979). The question of whether the direction of transport is due to a determined trait of the transporting tissue or to source and sink relations appears not to have been considered. The observed directions of transport are variable, even for the same hormone in the same tissue, and they are influenced by the presence of auxin, whose polarity could be expected to determine the location of metabolic sinks. This and the lack of developmental evidence for polar effects would indicate that tissue traits are not involved. On the other hand, many studies do show transport into agar blocks that occurs only in one direction. It may be concluded that directional transport of various hormones occurs, and that it could be an important control of development. However, its relation to the transporting tissue is not clear.

Polar auxin transport has been the subject of much research (reviewed by Goldsmith, 1977; Thimann, 1977) and only the following points, of major significance to the general subject of polarity, can be mentioned here:

(a) The direction of transport is towards the root apices. This generalization, simpler than references to basal or apical parts of individual organs, applies to numerous reports concerned with a wide variety of organs and plant species. The one exception is conflicting evidence concerning the direction of transport within the root apex (Goldsmith, 1977). There appear to be no studies of the direction of transport within the lamina of developing broad leaves where there is no one polarity of growth and differentiation.

(b) The rate of transport has been measured by extrapolation of the shortest time required for auxin to pass through a given distance and by following pulses of radioactive auxin (Goldsmith, 1969). Rates vary, but they are generally between 5 and 15 mm h^{-1}. This is about an order of magnitude slower than phloem transport and much too fast for diffusion alone. Transport in the direction opposite to polarity, towards shoot apices, does not appear to be anything more than diffusion, possibly aided by cyclosis.

(c) Polar transport can occur through a variety of living cells. It has been found in sections consisting only of parenchyma and epidermal cells, though the most rapid transport appears within the vascular system, especially in the general region of the cambium (Sheldrake, 1973). Polarity has been demonstrated in the embryonic axis of pines and beans but it has not been studied within the apical meristems of shoots.

(d) Vectorial influences of the environment (light, gravity) can divert part of the transported auxin (Goldsmith & Wilkins, 1964). An extreme case of reversal of polarity by gravity has been reported recently (Wright, 1981). This is unusual and it is not clear whether the effect on the tissue is permanent. Reversal of transport in inverted cuttings (see p. 197) was reported by Went (1941) but could not be confirmed (Sheldrake, 1974). A re-orientation of transport polarity where vascular differentiation was re-oriented by wounds (Fig. 8.3J) has been found recently (Gersani & Sachs, in preparation).

The question of the mechanism of auxin transport will be considered here only briefly, primarily from the point of view of the nature of the structural polarity that could account for the measured transport. The following appear to be major facts that must be taken into consideration; for references to the original evidence see Goldsmith (1969, 1977). Polar transport is specific to auxin and molecules with similar biological activity (Hertel, Evans, Leopold & Sell, 1969) and it can be specifically inhibited. Considerable advance has been made in the isolation of auxin receptors (Jacobs & Hertel, 1978). Polar transport continues when the cells have been plasmolysed or centrifuged in ways that have been observed to tear the plasmodesmata. Transport in and out of cells, through the apoplast, must therefore be involved. Transport also continues when cyclosis has been inhibited. The entrance of auxin into tissues is not polar and continues under conditions that inhibit polar transport. It is the exit of auxin from the cells, therefore, that must be polarized in some way. The ratio between transport in opposite directions, or the degree of polarity, is dependent on the length of the tissue, supporting the idea that each cell acts as a polar unit and the arrangement of these results in the observed polarity (Leopold & Hall, 1966; Mitchison, 1980*b*; Goldsmith, Goldsmith & Martin, 1981).

The simplest available hypothesis accounting for all known facts is the chemiosmotic theory (Rubery & Sheldrake, 1973; Raven, 1975; Goldsmith, 1977). Auxin is expected to enter the cells as an uncharged, small molecule. The internal pH is higher, and the auxin anion formed cannot pass through membranes without special channels. It is these channels that are localized at the root side of each cell, and the anions that leave through them become uncharged molecules in the lower pH of the cell walls; some of these enter the next cell along the file. The movement of auxin within each cell would be down a concentration gradient caused by its leaving only, or primarily, on one side. Its movement out of the cells would also be due to concentration differences of the anion and to the effect of the negative electrical potential of the cells. Though auxin can be transported against a gradient of its total concentration, the movement of the undissociated acid and the anion are along their gradients and do not require energy. The one stage in the process where this is required is the maintenance of the non-specific pH difference across the cell membranes; specificity of transport, on the other hand, resides in the channels for the anion and their localization. This picture might be an oversimplification, but calculations and computer models based on known parameters show that it could account for observed polar transport (Mitchison, 1980*b*; Goldsmith *et al.*, 1981).

The plane of cell divisions

A polar flow of signals, one of them auxin, was shown above to be a major control of the polarity of the regeneration of plant structure and the axiality of the vascular system. The present and the following sections are much more speculative and deal with oriented properties expressed at a cellular level and their possible control by a flow of signals. The possibility of tissue-joining in grafts has often been considered to depend on individual cell polarity, even by major authorities (Bünning, 1948; Sinnott, 1960), but cells with opposite polarities join readily (Fig. 8.3G). There are problems of continued development in such grafts but these are presumably due to the polarity of vascular differentiation (p. 198 *et seq.*). For reasons of space this section will deal only with orientation as it is expressed in cell divisions. This orientation is of major importance in the determination of the shape of organs and individual cells as well as the form of entire organs. For example, it is the orientation of divisions, rather than their overall rate, that results in the formation of the leaf primordium (Lyndon, 1972).

Longitudinal divisions in relation to signal flow

A most common orientation of cell divisions is at right angles to the long axes of organs. It follows that, as expected, divisions are also at right angles to the major axes of growth. Where divisions parallel to the plant axis are found the question is what is known of the controls of this orientation? Three major locations of such parallel divisions will be mentioned. (a) Much of the cellular width of the axis is found at its very tips, in the promeristems. The tunica is characterized by divisions only at right angles to the surface and similar divisions are found also in the corpus and the root apex. (b) Divisions at various angles to the axis form not only the bulges that are leaf primordia but also the stem buttress below them. These combined buttresses together form most of the stem and the divisions appear as the 'Residual Meristem' in cross-sections (Esau, 1977). (c) Divisions parallel to the axis are a basis for the formation of the elongated cells in the vascular tissues. Such oriented divisions are among the first characteristics of the procambium as well as the secondary thickening of the cambium.

This picture is, of course, a great oversimplification, as it ignores not only scattered parallel divisions (R. M. Sachs, 1965) but also the phellogen and the primary thickening meristem of some monocotyledons. It can still serve, however, as a basis for asking whether there is any relation of parallel divisions to the polarizing flow of signals considered above. The question being asked is not whether the plane of all divisions is controlled by an inductive flow but rather if and where such a control is expressed. The clearest evidence concerns the cambium: organ removal and hormonal treatments all point to its divisions being controlled by the same orienting signal flow as other processes of vascular differentiation (Sachs, 1981a). Though there are conditions that increase divisions relative to cell maturation and *vice versa*, there is no good evidence for the common intuitive assumption that different processes indicate separate controls rather than different cellular responses. Cell division can also be re-oriented by the same conditions of changed auxin flow that orient vascular differentiation (Fig. 8.3J; Neeff, 1914; Kirschner, Sachs & Fahn, 1971). The re-orientation of the cambium occurs by oriented divisions and intrusive cell growth with no movement of the cells relative to one another (Fig. 8.5). It may be concluded that the hormone flow considered above can be a control of the orientation of cell divisions.

Primary vascular differentiation also responds in the same way as the cambium to the presence of leaves and roots, to polarity and to re-orientation by wounds. This includes the cell divisions of primary differentiation, but within meristematic apices the procambial divisions are not induced by auxin (Young, 1954). It is not known whether other signals are required and how important are the problems of applying a local source

Fig. 8.5. Reorientation in the secondary xylem of *Robinia*. Branches cut as in Fig. 8.3J, and the drawings show the wood formed at different times. Top left is the original state, and bottom right is a re-oriented system that is almost normal. New, abnormal vessels form immediately, as seen in the second drawing. Transverse divisions followed by oriented cell growth gradually re-orient the system without any movement of the cells relative to one another. (Drawn from Kirschner *et al.*, 1971.)

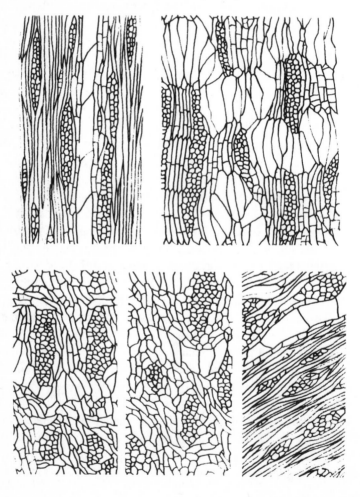

of auxin to apices that are small enough for diffusion to be a major factor. In maturing primary tissues regenerative differentiation induced by auxin includes, in its later stages, cell divisions that are parallel to auxin flow. The available evidence thus suggests that the orientation of divisions during vascular differentiation is a response to the flow of signals considered in previous sections, but signals other than auxin may well be involved and even dominant.

The procambium of the shoot apex first appears in a region of parallel divisions that form the cortex as well. These divisions depend on the presence of leaf primordia (Sachs, 1972). Close to the apex, furthermore, the cortex is capable of regeneration after wounds (Sachs, unpublished). Though auxin does not cause oriented divisions at this stage, it may be suggested (p. 201) that the formation of the cortex is a response to the same orienting signal flux that becomes canalized to the procambium and induces vascular differentiation. If this be true, the formation of most of the cells of the plant axis is controlled by a polar flow, the exceptions that actually support this suggestion being the unpolarized promeristems and young leaf primordia at the very tips of the axis, through which no signal flow can be expected.

Departures from axial development

A thorough review of the exceptions to the polarized development of plant tissues would be beyond the scope of this chapter, but three general cases may be considered in relation to the suggestion of a general polarization by signal flow. The first is the absence of a clear orientation in the cells of broad leaves, most clearly seen in the epidermis. This, however, is associated with similar orientations and changes of orientation (p. 203) in the adjoining veins and can be interpreted as a response to a changing flow of signals during the unsynchronized development of the leaf surface. If this be true, orientation would still depend on developmental centres that are a source of orienting signals, but in leaves these centres change their location with time. The rules of these changes are not known, but they are not random: they are restricted to one plane and result in the formation of a flat surface, which means that all the different parts grow equally.

The second departure from orientation along the axis is the emergence of hairs and various other specialized structures. As with leaf development, these appear to be associated with new centres of inductive influence and, possibly, an oriented flow of signals. This topic, about which little is known, will be mentioned briefly below in connection with the polarity of unequal divisions and the activity of meristemoids. Finally, a third large group of departures from integrated, oriented development are the callus and tumourous growths on damaged plants and of tissues isolated in culture (Gautheret, 1959). Initial callus development, however, is oriented by the determined traits of the damaged tissue and by the surface of the wound. Only in the absence of shoot and root apices and during tumour development does growth continue and include the formation of nodules of meristematic activity and product cells with no consistent axes of growth and cell division (Fig. 8.6). These centres lack definite interrelations and developmental polarity. The vascular contacts of continued callus development, however, indicate that from a hormonal point of view it may represent either partially differentiated shoot or root apices (Kirschner *et al.*, 1971; Sachs & Cohen, 1982). Hormonal flows could therefore be involved in determining the structure of the callus, and it might be

Fig. 8.6. Unoriented cell divisions in a tumour. Section of a crown gall on *Helianthus annuus*, showing a variety of cell types, including tracheary elements. The features distinguishing the tumour from normal tissue are the absence of consistent orientations of cell division and cell growth.

100 μm

much more organized or consistent in three dimensions than appears from sections. The continued growth of tumours and their ability to dominate the plants are associated with hormonal autonomy which makes each developing region of a tumour a source of orienting flow that is not dependent on the hormonal response of the plant (Sachs, 1975*a*).

This discussion does not account for all oriented growth and division, but it does suggest that they are meaningful and in many cases may be amenable to experimental study. The possibility that they are controlled by the same hormone flow as other polar phenomena offers a working hypothesis that appears correct, at least, for the cells that form part of the vascular system.

Polar, unequal divisions and the formation of spacing patterns

Polarity of meristemoid formation

Unequal divisions define one side or corner of a cell relative to all others (Fig. 8.1E) and are thus an expression of polarity at a cellular level. They are complementary to the plane of equal divisions as they define a direction within the cells but not necessarily an axis, though this is often apparent from the shape of the cells. These divisions form two cells different not only in size but also in the structure of their cytoplasm, seen in both light and electron microscopes (Palevitz & Hepler, 1974), and in their fate in the mature tissue. Unequal divisions are, therefore, an early expression of a polar differentiation. Such divisions are found in all stages of plant development (Bloch, 1965*b*); for example, in apical cells of ferns, where they are an early stage in the formation of precise cellular patterns (Gunning, 1982); in the cambium during the formation of new ray initials; in primary meristems, where the smaller product cell is the ultimate source of many different specialized cells or idioblasts; and in the separation of sieve elements from their adjoining companion cells (Fahn, 1974; Esau, 1977). Because of space limitation and relevance to other topics considered here, only the divisions leading to various idioblasts will be stressed in the following discussion. The main object considered will be the formation of the stomata; these have been relatively well studied because of their importance and location on the surface.

The polarity of the first unequal division during the formation of any given structure is the same as that defined for the same tissues by the other criteria considered above, and the smaller cell is generally found towards the shoot or root apex from which the primary meristem originated. An exception, probably not the only one, are the trichoscereids of *Monstera*, that form in the opposite direction (Bloch, 1965*b*). In broad leaves where there is no one orientation of veins and other tissues there is also no consistent polarity of unequal divisions. This polarity does not always conform even to the axis of very close veins, but such conformation was observed in divisions that occurred during vein formation (Smith, 1935). The general suggestion emerges, therefore, that the signal flow involved in other developmental events also orients unequal divisions. As this flow may involve intracellular gradients (p. 208), it could also be a determining factor in cytoplasmic inequality before it is expressed by division; this, however, is completely hypothetical.

Once the division has occurred, however, the smaller cell, which is the

Fig. 8.7. The developmental basis of stomata pattern in *Sedum sediforme*. A. An example of stomata distribution. Measurements of such patterns show that the presence of a minimal distance between neighbouring stomata is the only deviation from random distribution. B. Stages in the development of stomata. Unequal divisions arranged in a spiral, form the stoma together with the neighbouring cells. C. Detail of neighbouring stomata, showing that the cells formed together with the stomata can be the only separation between them.

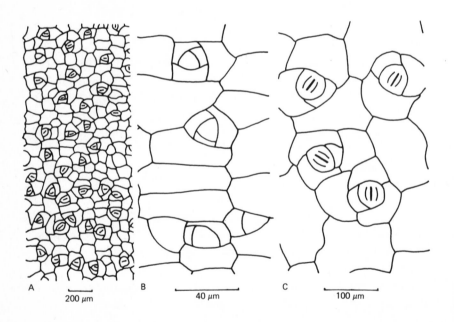

A   200 μm   B   40 μm   C   100 μm

one with a dense cytoplasm, becomes a centre of specialized activity and often also of new axes of polarity in its products as well as in neighbouring cells. This is expressed by the axis of further equal and unequal divisions of the small cells, at various angles to the original polarity (Fig. 8.7). In monocotyledons there are unequal divisions in neighbouring cells that are oriented towards a stoma cell (Fig. 8.8; Stebbins & Jain, 1960). There is some very incomplete evidence that the nuclei of neighbouring cells are displaced towards the small cell (Bünning, 1965). Such indications, together with the size and dense cytoplasm of this cell, have led to the suggestion that it is a relatively independent centre of meristematic activity in a maturing tissue and the term 'meristemoid' was suggested (Bünning & Sagromsky, 1948). This term is useful and appropriate, suggesting a cellular analogy to an apical meristem. It should therefore be retained even though the original definition, of a cell that continues to divide when the neighbouring tissue has ceased to do so, is certainly wrong.

Stomata spacing patterns

Directly, or following further divisions, the meristemoids form specialized cells or structures, such as stomata, whose distribution appears intuitively to be non-random (Bünning, 1965). The most common distribution is a simple spacing pattern (Wolpert, 1971) where distances between the special structures are more uniform than they would be if these structures were placed by chance. Statistical measurements have confirmed Bünning's claim for stomata (Korn, 1972; Sachs, 1978a; see also Chapter 2 by R. W. Korn). Thus the cellular polarity of unequal divisions is the basis of a two-dimensional pattern of specialized cells and, in cases other than stomata, three dimensions as well. The basis for this has been studied in any detail only for stomata, where the pattern does not correspond closely to any other distribution of structures in the plant and must therefore depend on events within the epidermis (Korn, 1972; Sachs, 1978a).

The possible controls of stomata spacing may be classified into two broad groups: (a) Interactions between a developing stoma and its surroundings. This could be an inhibitory effect of the formation of new stomata nearby (Bünning & Sagromsky, 1948; Bünning, 1965; Korn, 1981; see also Chapter 1) but it could also involve an induction by a future stoma of growth and division in neighbouring cells. (b) The divisions leading to the formation of a stoma necessarily form neighbouring cells also. The products of the original unequal divisions would touch one

216   T. Sachs

another but the individual stomata could always be separated. The developmental rules, dependent on events within cells, could result in minimal distances between stomata, so that 'cell lineage' could be the basis of the spacing pattern (Bünning, 1965; Sachs, 1978a).

These two general possibilities are not mutually exclusive, and it is their relative contribution in different cases that should be sought. Interactions are indicated by the inductive effects of a stoma mentioned above and the simplicity of the control they suggest has led it to be widely accepted as correct with no further proof. Two lines of evidence, however, prove that cell lineage controls are the major basis of stomata distribution. The first is that the first signs of differentiation, unequal divisions, occur next to one another and next to mature and maturing stomata (Figs. 8.7 and 8.8). This should not occur if stomatal differentiation inhibits similar events in neighbouring cells, but is expected if the distances are determined by cell lineage. The second type of evidence is that the minimal distances produced by cell lineage, predicted by following development (Figs. 8.7 and 8.8), correspond very well to the minimal distance between

Fig. 8.8. The developmental basis of stomata pattern in *Zebrina pendula*. As in Fig. 8.7, A shows the general pattern, B developmental stages, and C neighbouring stomata developed from developmental processes initiated in neighbouring cells. This is an example of orderly development where the same polarity in initial unequal divisions prevents the occurrence of stomata that touch one another (Sachs, 1974, 1978a). It is also an example of stomata inducing the formation of the cells that ultimately surround them.

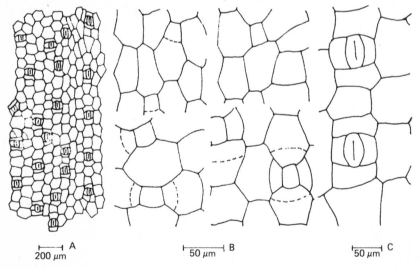

A 200 μm   B 50 μm   C 50 μm

neighbouring stomata that were measured and shown to be the only non-random aspect of their distribution (Sachs, 1974, 1978*a*; Marx & Sachs, 1977). In most cases, however, stomata development does not involve the formation of epidermal cells on all sides, and the polarity in neighbouring stomata is essential for the formation of the minimal distance between them (Fig. 8.8; Sachs, 1974, 1978*a*). The changing polarity of leaf development can result in neighbouring stomata complexes with opposite polarities, and it is in such cases that direct observations of development of specific stomata (by epi-illumination microscopy) are essential. The limited work that has been done indicates that interactions between developing stomata are important (Sachs, 1978*a*, 1979). Further evidence on this point is needed.

Determination or stability of unequal divisions

A related question to both polarity and pattern formation is the degree of determination of the developmental processes that start with unequal divisions. Again, evidence is available only for stomata, where the mature state indicates that development can be very variable within the same leaf, a variability expressed by both the number of divisions that lead to a mature stoma and their relative orientation (Paliwal, 1965). This was confirmed by following the changes of individual cells by epi-illumination microscopy (Sachs, 1978*a*, 1979) which also showed that the smaller product of unequal divisions sometimes grows to be an epidermal cell and need not produce a stoma. Experimental disruption of development, by colchicine and centrifugation, prevents the maturation of normal stomata but the underlying polarity is stable enough to be expressed in the mature structures (Bünning, 1958).

It may be concluded that unequal divisions are the result of a stable polarity that is related to or even identical with the one expressed by other aspects of development. These divisions lead to developmental events that are not strictly determined and yet are orderly enough to form a spacing pattern of specialized cells based primarily on polarity of cell lineage. Further work is needed to check these generalizations for other idioblasts.

Conclusion and discussion

A general hypothesis

The central conclusion to emerge is an hypothesis: various expressions of axiality and polarity depend on the same feedback-control involving a flow of auxin and other, unknown, signals. This flow both induces and depends on the transport polarity of the tissues to which oriented cell division and differentiation are coupled. A further feedback aspect is that sinks for auxin (roots) are induced where it accumulates and the development of the major sources (buds) depends on the availability of channels that draw it away. It is thus suggested that one system controls both the special differentiation of the tips and the orientation of the cells along the plant axis. A growing apex is therefore a polarizing centre for the cells it forms and for adjacent tissues, as is readily seen in vascular differentiation, and its differentiation and development depend on the possibility of these tissues becoming polarized in its direction. These feedback relations, in both transport and source-sink differentiation, result in the stability of organized structure and, at the same time, assure its regeneration when organization is severely damaged.

Major open problems

This suggestion would appear to require too much from a control whose signals are to act both between the cells of a developing vessel and between shoot and root apices (though not necessarily the promeristems). It is meant, however, as a working hypothesis, useful for the definition of problems and additional controls. A central problem it raises is the nature of the other signals involved in the controlling flow. Auxin, stressed because it has been so useful for experimental work, is certainly not the only signal. This is shown, for example, by its failure to orient events within apices. As discussed above, the additional signals need not be associated with any tissue polarity but only with preferred transport along one axis, or axiality, and this has not been studied. These signals should act in embryonic tissues, within the apices, and most physiological work has been performed on tissues that are either mature or at least highly determined. Gibberellins might be candidates for such a role, as is also indicated by their effect on vascular differentiation (Hess & Sachs, 1972), but this and other possibilities require further work.

Another general problem is how, in molecular terms, flow could control orientation and differentiation. The answer is not known, of course, but it may be useful to indicate some possibilities. As was mentioned in the section on transport of hormones on p. 207, the polarity of auxin transport probably depends on the localization of channels for the exit of the auxin anion. The first question, therefore, is how flow could be associated with this localization. A similar localization has been considered by Jaffe

(1981), who has suggested that potential differences resulting from ionic currents polarize cells by self electrophoresis. Hormone flow could not, by itself, generate potential differences, but it could do so indirectly, as a result of local responses to intracellular gradients associated with polar transport (Mitchison, 1980*b*; see also p. 208). Other possible mechanisms would involve, for example, small molecules associated with intracellular flow that would be deposited, and change the membrane, where the hormone leaves the cell. Control of the orientation of organelles such as microtubules, and changes of gene expression resulting in observable differentiation, would be later stages, possibly related to internal gradients set up by polarization. The need for continued flow would then reflect the requirements for these gradients during a relatively long period.

Generalizations about pattern control

Axiality and polarity are related to positional controls not only because they are expressions of orientation and localized differentiation but also because, as shown above, they can be the basis of important cellular patterns, strands of cells in vessels and spacing of stomata. A final point requiring discussion, therefore, is the comparison of the controls suggested here with other models and hypotheses of pattern formation in biological systems. Some general traits of all these suggestions must be considered first. Two components of pattern formation can be distinguished: spatial controls, correlating events in neighbouring regions, and temporal controls, causing a dependence on past developmental history of the cell, tissue or organ (Sachs, 1978*b*). The operation of such controls in non-biological patterns may be seen, for example, in clouds, waves and streams. Processes that reinforce themselves serve as temporal controls which counteract the levelling influence of diffusion, while the drainage of water from neighbouring regions, in various forms, is a spatial control. Similar systems, known as reaction-diffusion mechanisms, have been mathematically modelled and shown to be a possible basis for a wide variety of biological patterns (chapter 1 by Meinhardt; see also Brenner, Murray & Wolpert, 1981). In biological systems, however, there may also be relatively complex stepwise determination processes and a separation between pattern specification, at early stages of development, and its expression by elaborate differentiation that can produce complex patterns not readily apparent in the early specification of positional information (Wolpert, 1971, 1981). These possibilities allow for a very wide array of controls, and since biological simplicity is not identical with chemical simplicity and is not understood, they cannot be evaluated on the basis of models only. A discussion of all possible controls is impossible here, but some comments on the basic properties of the suggestions made above relative to various published ones may be in place.

Distinguishing characteristics of control by signal flow

The controls of pattern suggested here depend on the distribution and movement of signals. This contrasts with physical forces suggested in some cases (Chapters 3, 4). The two possibilities are not mutually exclusive, but the evidence considered above does suggest that complex responses to signals that carry information only in their quantity and direction of flow are of major importance. These signals could influence physical phenomena by orienting microfibrils and the resulting growth.

The distribution of the signals considered here becomes patterned by a gradual process, and this patterning is not separate from the differentiation of the cells. Thus the differentiation of transport channels canalizes the signal flow and this flow induces differentiation. This 'differentiation dependent' pattern formation (Sachs, 1978*b*) contrasts with positional information (Wolpert, 1971) where a pre-pattern of signals precedes differentiation and is not influenced by it. Pre-pattern, or early differences that are not observable and yet determine later differentiation, are found in plants as well as animals (Mohr, 1978). The separation between observable and pre-pattern differences, however, is only a matter of available techniques. It might be, therefore, that 'differentiation dependent' processes control the establishment of pre-patterns that are at present difficult to study.

The controls suggested here emphasize the role of flow and of orientation, the topic of this chapter. To the extent that orientation is considered in other models, except that of Jaffe (1981), it is ascribed to gradients of substances whose concentration is very precisely interpreted by the cells. This might reflect a difference between plants and animals, for in plants gradients of cell behaviour, expressed most clearly when cells normally distant are brought together, are as yet unknown.

Finally, the controls suggested here are remarkably economical, both in the number of signals involved and in the basic cellular processes. Surprisingly, the same positive feedback results in both temporal determination and spatial interactions. The same signals act both between organs and between neighbouring cells. Specialized cells may first be localized by chance and yet the gradual determination, involving constant feedback controls, results in patterns that, though not precise, are free of mistakes of functional significance. In this they resemble the patterns that are actually found in both plants and animals.

The general characteristics of the positional controls suggested above, therefore, are the gradual feedback determination that involves the processes of differentiation and oriented signals. It has been shown that it can account for major features of patterning and determination at a level that is removed from immediate controls of gene expression. In all these traits it resembles the control by self-electrophoresis (Jaffe, 1981) and differs from other hypotheses found in the literature.

References

Aloni, R. & Jacobs, W. P. (1977). Polarity of tracheary regeneration in young internodes of *Coleus* (Labiatae). *American Journal of Botany,* **64,** 395–403.

Arisz, W. H. (1952). Transport of organic compounds. *Annual Review of Plant Physiology,* **3,** 109–30.

Benayoun, J., Aloni, R. & Sachs, T. (1975). Regeneration around wounds and the control of vascular differentiation. *Annals of Botany,* **39,** 447–54.

Bloch, R. (1965*a*). Polarity and gradients in plants: a survey. In *Encyclopedia of Plant Physiology,* vol. 15(1), ed. W. Ruhland, pp. 234–74. Berlin: Springer.

– (1965*b*). Histological foundations of differentiation and development in plants. In *Encyclopedia of Plant Physiology,* vol. 15(1), ed. W. Ruhland, pp. 146–88. Berlin: Springer.

Brenner, S., Murray, J. D. & Wolpert, L., editors (1981). Theories of biological pattern formation. *Philosophical Transactions of the Royal Society of London,* B, **295,** 427–617.

Bünning, E. (1948). *Entwicklungs- und Bewegungsphysiologie der Pflanze.* Berlin: Springer.

– (1958). Polarität und inäquale Teilung des pflanzlichen Protoplasten. *Protoplasmatologia,* 8/9a, 1–86.

– (1965). Die Entstehung von Mustern in der Entwicklung von Pflanzen. In *Encyclopedia of Plant Physiology,* vol. 15(1), ed. W. Ruhland, pp. 383–403. Berlin: Springer.

Bünning, E. & Sagromsky, H. (1948). Die Bildung des Spaltöffnungsmusters in der Blattepidermis. *Zeitschrift für Naturforschung,* **3b,** 203–16.

Castan, M. R. (1940). Sur le rôle des hormones animales et végétales dans le développement et l'organogénèse des plantes vasculaires. *Revue Générale de Botanique,***52,** 285–304.

Crick, F. H. C. (1970). Diffusion and embryogenesis. *Nature,* **225,** 420–2.

Eliezer, J. & Morris, D. A. (1978). Effects of temperature and sink activity on the transport of ^{14}C-labelled indol-3yl-acetic acid in the intact pea plant (*Pisum sativum* L.). *Planta,* **147,** 216–24.

Esau, K. (1977). *Anatomy of Seed Plants,* 2nd edn. New York: Holt, Rinehart & Winston.

Fahn, A. (1974). *Plant Anatomy,* 2nd edn. Oxford: Pergamon Press.

Gautheret, R. J. (1944). Recherches sur la polarité de tissue végétaux. *Revue de Cytologie et de Cytophysiologie Végétales,* **7,** 45–217.

– (1959). *La Culture des Tissus Végétaux.* Paris: Masson & Cie.

Gersani, M., Lips, S. H. & Sachs, T. (1980). The influence of shoots, roots and hormones on sucrose distribution. *Journal of Experimental Botany*, **31**, 177–84.

Goldsmith, M. H. M. (1969). Transport of plant growth regulators. In *Physiology of Plant Growth and Development*, ed. M. B. Wilkins, pp. 125–62. New York: McGraw-Hill.

– (1977). The polar transport of auxin. *Annual Review of Plant Physiology*, **28**, 439–78.

Goldsmith, M. H. M., Goldsmith, T. & Martin, M. H. (1981). Mathematical analysis of the chemosmotic polar diffusion of auxin through plant tissues. *Proceedings of the National Academy of Sciences of the USA*, **78**, 976–80.

Goldsmith, M. H. M. & Wilkins, M. B. (1964). Movement of auxin in coleoptiles of *Zea mays* L. during geotropic stimulation. *Plant Physiology*, **39**, 151–62.

Goodwin, P. B. (1983). Molecular size limit for movement in the symplast of the *Elodea* leaf. *Planta*, **157**, 124–30.

Gunning, B. (1982). The root of the water fern *Azolla:* cellular basis of development and multiple roles for cortical microtubules. In *Developmental Order: Its Origin and Regulation*, ed. S. Subtelny & P. B. Green, pp. 379–421. New York: A. R. Liss.

Hertel, R., Evans, M. L., Leopold, A. C. & Sell, H. M. (1969). The specificity of the auxin transport system. *Planta*, **85**, 238–49.

Hess, T. & Sachs, T. (1972). The influence of a mature leaf on xylem differentiation. *The New Phytologist*, **71**, 903–14.

Jacobs, M. & Hertel, R. (1978). Auxin binding to subcellular fractions from *Cucurbita* hypocotyls: *in vitro* evidence for an auxin transport carrier. *Planta*, **142**, 1–10.

Jacobs, W. P. (1952). The role of auxin in the differentiation of xylem around a wound. *American Journal of Botany*, **39**, 301–9.

– (1979). *Plant Hormones and Plant Development*. Cambridge: Cambridge University Press.

Jaffe, L. F. (1981). The role of ionic currents in establishing developmental pattern. *Philosophical Transactions of the Royal Society of London*, B, **295**, 553–66.

Janse, J. M. (1914). Les sections annulaires de l'écorce et le suc descendant. *Annales du Jardin Botanique de Buitenzorg*, **28**, 1–90.

Juniper, B. E. (1977). Some speculations on the possible roles of plasmodesmata in the control of differentiation. *Journal of Theoretical Biology*, **66**, 583–92.

Kirschner, H., Sachs, T. & Fahn, A. (1971). Secondary xylem reorientation as a special case of vascular tissue differentiation. *Israel Journal of Botany*, **20**, 184–98.

Korn, R. W. (1972). Arrangement of stomate on leaves of *Pelargonium zonale* and *Sedum stahlii*. *Annals of Botany*, **36**, 325–33.

– (1981). A neighboring-inhibition model for stomata patterning. *Developmental Biology*, **88**, 115–20.

Larson, P. R. (1975). Development and organization of the primary vascular system in *Populus deltoides* according to phyllotaxy. *American Journal of Botany*, **62**, 1082–99.

Leopold, A. C. & Hall, O. F. (1966). Mathematical model of polar auxin transport. *Plant Physiology*, **41**, 1476–80.

Libbert, E. (1956). Untersuchungen über die Physiologie der Adventivwurzelbildung. IV. Physiologische Untersuchungen über Castans 'Polaritatsumkehr'. *Berichte der deutschen botanischen Gesellschaft*, **69**, 429–34.

Lüttge, U. & Higinbotham, N. (1979). *Transport in Plants*. New York: Springer.

Lyndon, R. F. (1972). Leaf formation and growth at the shoot apical meristem. *Physiologie Végétale*, **10**, 209–22.

Marx, A. & Sachs, T. (1977). The determination of stomata pattern and frequency in *Anagallis. Botanical Gazette,* **138,** 385–92.

Mitchison, G. J. (1980*a*). A model for vein formation in higher plants. *Proceedings of the Royal Society of London* B, **207,** 79–109.

– (1980*b*). The dynamics of auxin transport. *Proceedings of the Royal Society of London,* B, **209,** 489–511.

– (1981). The polar transport of auxin and vein patterns in plants. *Philosophical Transaction of the Royal Society of London,* B, **295,** 461–71.

Mohr, H. (1978). Pattern specification and realization in photomorphogenesis. *The Botanical Magazine, (Tokyo),* Special Issue No. 1, 199–217.

Neeff, F. (1914). Über Zellumlagerung: ein Beitrag zur experimentellen Anatomie. *Zeitschrift für Botanik,* **6,** 465–547.

Palevitz, B. A. & Hepler, P. K. (1974). The control of the plane of division during stomatal differentiation in *Allium.* I. Spindle reorientation. *Chromosoma,* **46,** 297–326.

Paliwal, G. S. (1965). The development of stomata in *Basella rubra* L. *Phytomorphology,* **15,** 50–3.

Patrick, J. W. (1976). Hormone-directed transport of metabolites. In *Transport and Transfer Processes in Plants,* ed. I. F. Wardlaw & J. B. Passioura, pp. 433–46. New York: Academic Press.

Rathfelder, O. (1955). Anatomische Untersuchungen zu Castans 'Polarisationsumkehr' bei *Pisum sativum. Berichte der deutschen botanischen Gesellschaft,* **68,** 227–32.

Raven, J. A. (1975). Transport of indoeacetic acid in plant cells in relation to pH and electrical potential gradients, and its significance for polar IAA transport. *The New Phytologist,* **74,** 163–72.

Roberts, I. W. (1976). *Cytodifferentiation in Plants: Xylogenesis as a Model System.* Cambridge: Cambridge University Press.

Rubery, P. H. & Sheldrake, A. R. (1973). Effect of pH and surface charge on cell uptake of auxin. *Nature New Biology,* **244,** 285–8.

Sachs, R. M. (1965). Stem elongation. *Annual Review of Plant Physiology,* **16,** 73–96.

Sachs, T. (1968). The role of the root in the induction of xylem differentiation in peas. *Annals of Botany,* **32,** 391–9.

– (1969). Polarity and the induction of organized vascular tissues. *Annals of Botany,* **33,** 263–75.

– (1972). The induction of fibre differentiation in peas. *Annals of Botany,* **36,** 189–97.

– (1974). The developmental origin of stomata pattern in *Crinum. Botanical Gazette,* **135,** 314–8.

– (1975*a*). Plant tumors resulting from unregulated hormone synthesis. *Journal of Theoretical Biology,* **55,** 445–53.

– (1975*b*). The induction of transport channels by auxin. *Planta,* **127,** 201–6.

– (1975*c*). The control of the differentiation of vascular networks. *Annals of Botany,* **39,** 197–204.

– (1978*a*). The development of spacing patterns in the leaf epidermis. In *The Clonal Basis of Development,* ed. S. Subtelny and I. M. Sussex, pp. 161–83. New York: Academic Press.

– (1978*b*). Patterned differentiation in plants. *Differentiation,* **11,** 65–73.

– (1979). Cellular interactions in the development of stomatal patterns in *Vinca major* L. *Annals of Botany,* **43,** 695–700.

- *(1981b)*. Polarity changes and tissue organization in plants. In *Cell Biology 1980–81*, ed. H. G. Schweiger, pp. 489–96. Berlin: Springer.

Sachs, T. & Cohen, D. (1982). Circular vessels and the control of vascular differentiation in plants. *Differentiation*, **21**, 22–6.

Sheldrake, A. R. (1973). Auxin transport in secondary tissues. *Journal of Experimental Botany*, **24**, 87–96.

- (1974). The polarity of auxin transport in inverted cuttings. *The New Phytologist*, **73**, 637–42.

Sinnott, E. W. (1960). *Plant Morphogenesis*. New York: McGraw-Hill.

Smith, G. E. (1935). On the orientation of stomata. *Annals of Botany*, **49**, 451–77.

Spanswick, R. M. (1974). Symplastic transport in plants. *Symposia of the Society for Experimental Biology*, **28**, 127–37.

Stebbins, G. L. & Jain, S. K. (1960). Developmental studies of cell differentiation in the epidermis of monocotyledons. *Developmental Biology*, **2**, 409–26.

Thimann, K. V. (1977). *Hormone Action in the Life of Whole Plants*. Amherst: University of Massachusetts Press.

van der Weij, H. G. (1934). Der Mechanismus der Wuchsstoff -transport II. *Recueil des Travaux Botaniques Neérlandais*, **29**, 380–496.

Vöchting, H. (1878). *Über Organbildung im Pflanzenreich*, vol. 1. Bonn: Max Cohen.

- (1884). *Über Organbildung im Pflanzenreich*, vol. 2. Bonn: Emil Strauss.

- (1892). *Über Transplantation am Pflanzenkorper*. Tubingen: Verlag H. Laupp'schen Buchhandlung.

- (1906). Über Regeneration und Polarität bei höhre Pflanzen. *Botanisches Zeitung*, **64**, 101–48.

Wareing, P. F. (1978). Determination in plant development. *The Botanical Magazine, Tokyo*, Special Issue No. 1, 3–17.

Went, F. W. (1941). Polarity of auxin transport in inverted *Tagetes* cuttings. *Botanical Gazette*, **103**, 386–90.

Went, F. W. & Thimann, K. V. (1937). *Phytohormones*. New York: Macmillan.

Wochak, Z. S. & Sussex, I. M. (1974). Morphogenesis in *Selaginella*. II. Auxin transport in the root (rhizophore). *Plant Physiology*, **53**, 738–41.

Wolpert, L. (1971). Positional information and pattern formation. *Current Topics in Developmental Biology*, **6**, 183–224.

- (1981). Positional information and pattern formation. *Philosophical Transactions of the Royal Society, London*. B, **295**, 441–50.

Wright, M. (1981). Reversal of the polarity of IAA transport in the leaf sheath base of *Echinochloa colonum*. *Journal of Experimental Botany*, **32**, 159–69.

Young, B. S. (1954). The effect of leaf primordia on differentiation in the stem. *The New Phytologist*, **53**, 445–60.

9

Control of tissue patterns in normal development and in regeneration

J. and P. M. WARREN WILSON

Within a higher plant, each individual tissue occurs in a characteristic spatial pattern that is more or less obviously related to its function. For example, the endodermis forms a hollow cylinder, the primary xylem is in longitudinal strands, and abscission zones form transverse layers. The various tissues are derived from meristems consisting of masses of relatively uniform undifferentiated cells, and the tissue patterns become manifest as the derivative cells differentiate. This chapter examines the way in which meristematic derivatives receive positional information which controls their pathway of differentiation.

'Tissue' is here used broadly to cover any group of cells with common structure and function. A tissue may be either simple, consisting of only one type of cell, or complex, consisting of two or more cell types as in phloem or epidermis. The arrangement of different cell types within a complex tissue, although relevant to the theme of this book, is not considered here. Further, the examples of tissue patterning mentioned in this chapter are drawn primarily from dicotyledons, which, in this regard, have been more studied than monocotyledons or lower vascular plants.

Many studies of positional control have employed surgical operations on plants. The pattern of tissues formed after wounding, often developed in the dedifferentiated tissue of a wound meristem, can give useful insight into positional control systems. Other studies have used tissue culture, in which the explant or callus can be regarded as a more extreme type of wound or wound meristem. Such experimental systems provide opportunities for controlling the physical environment of the differentiating tissues and for examining the effects of supplying chemical substances that may modify tissue patterning, but the cells in such systems are, of course, in an unnatural environment. However, we present the view that some of the same basic processes of positional control operate in tissue in aseptic culture, as well as in wound healing, in much the same manner as they do in the normal plant.

The first part of our chapter describes the tissue patterns that arise during regeneration after wounding and identifies some major features of patterns in both regenerated and normal organs. The second part attempts to identify positional control mechanisms that may be responsible for this patterning.

DEVELOPMENT OF TISSUE PATTERNS

Tissue regeneration after various types of wounding

Many workers have examined the effects of wounding, usually of stems, occasionally of roots and rarely of other organs. The response to wounding often results in more or less complete regeneration of tissue patterns, including especially the continuity of the vascular system. However, the course of regeneration is affected by the character of the wound. For example: (a) the size of the wound in an internode affects the way in which continuity is restored; (b) for wounds in an older internode where differentiation has occurred, dedifferentiation of cells near the wound may be a prerequisite for differentiation of regenerated tissues, but for wounds in very young internodes at the apex where the cells are relatively undifferentiated, dedifferentiation is less important; and (c) for material explanted in tissue culture, the long-distance transport needs of the whole plant are no longer relevant, and although vascular differentiation may occur the pathways have no clear relationship to long distance transport and the word 'regeneration' is less commonly applied.

For these reasons the discussion below is based on four categories which, though not entirely distinct, are convenient: 'large wounds' which are typically 10 mm or more in length; 'small wounds' not more than a few millimetres long; 'wounds in apices' that affect the more or less meristematic regions; and 'explants and callus cultures' which involve excision of small parts, perhaps up to 10 mm in size, for aseptic culture.

Large wounds

This category is described at some length because it displays clearly a wide range of regeneration phenomena but has been less adequately described in the literature. Various types of large wound can be made, but a basic form is that in which one side of an active young internode is cut away over a length of some 10–30 mm and to a depth of about one-third of the stem diameter, interrupting the vascular cylinder (Fig. 9.1a). The following generalized description of regeneration is based mainly on our own observations of species from many families.

Although cells near the cut surface (except for those of the cambial zone) have ceased dividing, wounding stimulates them to divide after a short 'lag' phase. Mitotic figures may be seen one day after wounding, and divided cells on the second day. This response is limited to the region next to the wound surface; in some species only the two cell layers closest to the surface divide. The new cell walls tend to be oriented parallel to the cut surface.

Cell division continues throughout a 'division' phase; initially cells throughout the proliferating callus divide and the increase in cell number may be logarithmic (Fig. 9.2a). Cell enlargement lags behind cell division for a time, with the result that the mean cell size decreases, but by about day 5 a constant, smaller mean size has been attained throughout the proliferating callus (Fig. 9.2b); this callus may approach 0.5 mm in thickness and has a uniform appearance.

Soon afterwards the mitotic figures become increasingly confined to two zones that are roughly parallel to the surface, one about 0.2 mm below it and the other at first about 0.4 mm beneath it (Fig. 9.2c); by

Fig. 9.1. Diagrams of young dicotyledon stems in transverse section (a) immediately after wounding, (b) 10 days and (c) 20 days after wounding. Xylem is hatched, cork and vascular cambia are indicated respectively by broken and continuous heavy lines, and all regenerated tissues are stippled.

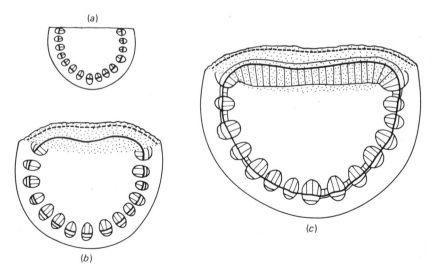

(a)

(b)

(c)

about day 10 (Fig. 9.1b) these two layers are identifiable as cambial zones, with cells having a radial diameter about half that of the other callus cells. The inner regenerated cambium unites with the cut ends of the original vascular cambium. Cell division ceases except in the cambial zones; this marks the transition from the 'division' phase to the 'differentiating' phase.

Cells external to the outer (cork) cambium differentiate as cork cells and those internal to the inner (vascular) cambium differentiate as lignified cells which represent the first-formed regenerated xylem. Not all these differentiated cells are derived from the cambia – some formed

Fig. 9.2. Typical progress of regeneration of tissues with time after wounding a young stem. (a) Mean number of cells along radial transects in the regenerating region derived from the original pith. (b) Mean radial diameter of cells along the same transects. (c) Distribution, with distance from the external surface, of various tissues; the cross-hatched region represents the wound meristematic tissue where divisions commence 1–2 days after wounding and within which the cork and vascular cambia subsequently form. (Based on data of the authors and of J. R. Evans, R. I. Grange and M. J. Hurley, using *Capsicum annuum, Datura stramonium, Helianthus annuus* and *Lycopersicon esculentum*.)

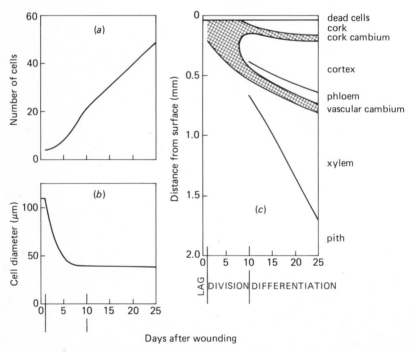

Days after wounding

before they were clearly defined – but subsequent cambial derivatives add to the cork (for a limited time) and to the secondary phloem and xylem on the outer and inner sides respectively of the regenerated vascular cambium, restoring the normal vascular structure (Fig. 9.1c).

Callus parenchyma lies both internal to the regenerated xylem, where it may be regarded as regenerated pith, and also between the phloem and periderm, as regenerated cortex. Thus the radial sequence, cork – cork cambium – cortex – phloem – vascular cambium – xylem – pith is restored.

Other tissues also may be regenerated in some species. In about half the several woody genera examined, the sclerenchyma cylinder that lies outside the vascular cylinder is regenerated within the wound callus between the phloem and periderm. In some cases it is less uniformly positioned than in the normal stem, and it always consists exclusively of sclereids even when fibres alone form the original sclerenchyma cylinder. The regenerated sclerenchyma may form a continuous ring, but more often it is arranged in clumps which roughly correspond in tangential dimensions with the groups of fibres in the original stem (Fig. 9.3a). Whereas the original sclerenchyma normally forms longitudinally continuous strands, the regenerated sclerenchyma is interrupted longitudinally, and in this it differs from other regenerated tissues (Fig. 9.3b).

Another example of a regenerated tissue is the occurrence of oil glands in the outer cortex of *Eucalyptus* species, just beneath the periderm (Fig. 9.3c). Still another occurs in roots where, following wounding, the endodermis may regenerate and join the cut ends of the original endodermis thereby restoring a continuous cylinder (Fig. 9.3d). The regenerated endodermis lacks the precision and uniformity of cell structure of the normal endodermis; however, it is clearly identifiable as one or two layers of cells of which the walls, especially the radial walls, are thickened and stain like those of normal endodermis. Another aspect of pattern manifestation in wounded roots is the regeneration of rays between vascular bundles reflecting a periodic spacing of ray initials in the regenerated vascular cambium.

Of all the tissues that regenerate, the vascular tissues are probably the ones to do so most consistently; they are also functionally the most important. The positions of the phloem and xylem derived from the regenerated vascular cambium depend on the position and orientation of this cambium, and it is therefore proper to give special weight to studies of its position. This can be modified by some operations which are variants of the simple wound described above.

For example, Kny (1877) showed that when a young internode was split longitudinally into two half-internodes, the half-cylinder of vascular tissue in each part regenerated to form a complete cylinder (Fig. 9.4a). This was confirmed by Snow (1942) and by J. & P. M. Warren Wilson (1961), who compared the pattern of regeneration when split internodes were held apart with that when they were held together. In the latter case (Fig. 9.4b) the regenerated cambium reunited the cut ends of the original cambium into a single cylinder. This response is similar to that occurring

Fig. 9.3. Diagrammatic sections of (a) *Syringa vulgaris* stem 24 weeks after wounding, in transverse section showing regeneration of the sclerenchyma ring (fibre groups outlined, sclereid groups in black); (b) *Syringa vulgaris* stem 7 weeks after wounding, in radial longitudinal section (continuous fibre strand on left, regenerated sclereid groups on right); (c) *Eucalyptus viminalis* stem 10 weeks after wounding, showing distribution of oil glands (asterisks); in this older stem, dead xylem has resulted in separate calluses proliferating from the pith and from outer tissues; (d) *Vicia faba* root 4 weeks after wounding, in transverse section showing position of the endodermis (beaded line).

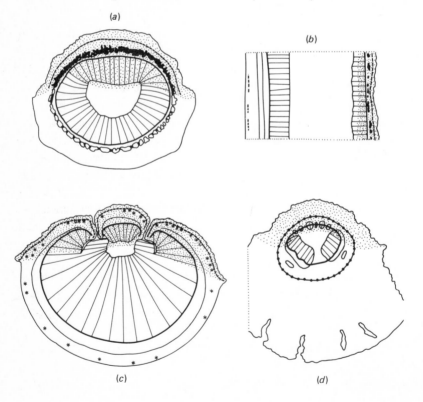

(a)

(b)

(c)

(d)

in an approach graft between two stems. Janse (1921) and Rzimann (1932) showed that if the two halves of a split vine stem or carrot root were held together but displaced so that cambia of opposite orientations were brought together, direct union of these cambia never occurred: the regenerating cambia formed along pathways that ensured unions with compatible orientations (Fig. 9.4c).

Snow (1942) noted that in some of his half-internodes the regenerated cambia did not join together the cut ends, but instead turned out and butted against the surface, though he did not understand the reason for this. J. & P. M. Warren Wilson (1961) showed that this response could be obtained consistently by covering the wound surface with Vaseline and maintaining this covering during subsequent growth (Fig. 9.4d). They also showed that if a wounded stem was grafted to another stem which had been wounded only superficially, and not deeply enough to interrupt the cambial cylinder, the regenerating cambia took a pathway such as to form a second complete cylinder (Fig. 9.4e).

Fig. 9.4. Diagrams of regeneration patterns of vascular cambia in transverse sections of (a) split stem held apart, (b) split stem held together, (c) split stem held together laterally displaced, (d) wounded stem with wound surface covered with Vaseline, and (e) cut stem : scraped stem approach graft.

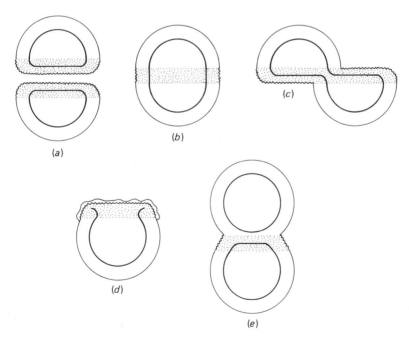

Small wounds

In this type of operation a transverse wound in the stem, interrupting one or more vascular bundles, is made by means of either a lateral incision which removes no tissue or a V-cut which removes only a small wedge; in both cases the wound is commonly not more than three millimetres deep. Many such operations have used *Coleus blumei* (e.g. Simon, 1908; Sinnott & Bloch, 1945; Jacobs, 1952; Aloni & Jacobs, 1977; Behnke & Schulz, 1980) but other species have shown similar regeneration, e.g., *Impatiens* (Kaan Albest, 1934), *Nicotiana tabacum* (Sussex, Clutter & Goldsmith, 1972), and *Pisum sativum* stems (Benayoun, Aloni & Sachs, 1975) and roots (Robbertse & McCully, 1979; Hardham & McCully, 1982); severing of leaf veins leads to comparable regeneration processes (Freundlich, 1908).

In response to such wounding, tracheary strands are clearly apparent within a few days, characteristically bridging the wound (Fig. 9.5a). The strands, each of which is usually only one or a few cells wide, are composed of cells with reticulately thickened and lignified walls, joined end to end and sometimes with perforated end walls. The individual cells are thus tracheary elements that in some cases are no more elongated than the parenchymatous cells around them. The first such strands can be seen

Fig. 9.5. Diagrams showing typical patterns of (a) tracheary strands and (b) sieve tube strands formed around a small wound, as seen at different depths in a cleared tangential slice including four vascular bundles.

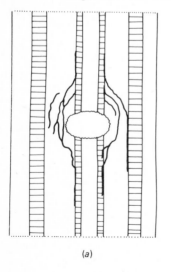 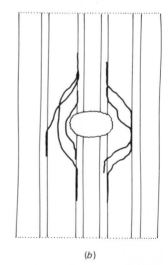

(a) (b)

three or four days after wounding (e.g., Thompson, 1967); the number of strands increases over a week in *Coleus* (Aloni & Jacobs, 1977; Jacobs, 1979) but there is no increase after two weeks, at least in *Nicotiana* (Sussex *et al.*, 1972).

In so far as these wounded tracheary strands appear before the regenerated cambium is formed, they cannot be derived from it. It appears that they may form either by differentiation of existing parenchyma cells without change in cell shape (Jacobs, 1952; Sinnott, 1960), or from cells of the dividing wound callus (Sussex *et al.*, 1972), or from derivatives of the original vascular cambium around the wound (Benayoun *et al.*, 1975).

Sieve tube strands (Fig. 9.5b) comparable to these tracheary strands, also form in response to small wounds, though they may appear a little earlier and take longer to differentiate to maturity. The sieve tube elements may be associated with companion cells and phloem parenchyma cells, all formed from parenchyma cells which divide in response to wounding (Behnke & Schulz, 1980), or may be differentiated from products of the original vascular cambium around the wound.

Both tracheary strands and sieve tube strands tend to differentiate basipetally. In both cases they are associated with new vessels or sieve tubes above and below the wound, rather than joining the ends of wounded pre-existing strands.

The wound strands arise at a distance below the surface, and the tracheary strands are internal to the sieve tube strands. Where they form from the existing cambium around the wound, they are on opposite sides of the cambium, and in this case the pathways of tracheary strands may obviously correspond to those of sieve tube strands on the other side of the cambium (Aloni & Jacobs, 1977).

Most studies of tracheary and sieve tube strand formation have examined changes over only the first week after wounding, though Sussex *et al.*, (1972) continued observations for three weeks and observed that the regenerating vascular cambium arose between the tracheary and sieve tube strands (which in this case did not correspond in position with one another).

Wounds in apices

Four main categories of wounds on stem apices are most relevant to tissue patterning:

(1) Split apex. Karzel (1924), Pilkington (1929) and Dircks (1981) described the effects of a median longitudinal split on the patterns subse-

quently developed. Each half-apex, which immediately after wounding has one flat face without leaf primordia and one curved face with primordia, gradually becomes restored as a normal dome-shaped apex with leaf primordia all around. Transverse sections at successive levels some weeks after the operation (Fig. 9.6a–e) show the following sequence. (a) At the lowest level is the normal stem below the split. (b) Above this is a region where callus from the two wounded surfaces has joined, as in a graft, and the vascular cambia have become united by regenerating cambia (cf. Fig. 9.4b). (c) and (d) In a region where the two half-stems remain separate the regenerated region contains no primary vascular bundles but only secondary tissues from the regenerated vascular cambium – as at level (b). In the woody *Populus deltoides* × *P. fremontii*, screids (in place of the normal fibres) regenerated at least in the lower part of this region, but in the herbaceous species examined this level lacks any regenerated sclerenchyma ring. (e) The two almost fully regenerated stems have complete vascular rings with normal fibre strands, all derived from the fully restored apices.

(2) Isolated apex. Wardlaw (e.g. 1950) and Ball (e.g. 1952) isolated shoot apices from leaf primordia by making three or four longitudinal cuts. Such isolated apices continued growth and gave rise to leafy shoots. Two aspects of the regenerated vascular pattern are of particular interest. Firstly, a tube of procambium differentiated within the prism of isolated tissue, and conformed in outline with the triangular or rectangular surfaces of the isolated plug as seen in transverse section (Fig. 9.7a, b). Secondly, in the lower region where no leaf primordia were present the cylinder was uninterrupted; higher up, where leaves regenerated, the procambial cylinder became interrupted by leaf gaps which alternated with strands of prevascular tissue.

(3) Isolated leaf primordia. Sussex (1955) has shown that when a longitudinal incision is made in such a position as to isolate a very young leaf primordium from the apex the leaf may develop with a radial vascular structure instead of the usual dorsiventral structure.

(4) Removal of leaf primordia. Helm (1932) and Young (1954) surgically removed young leaf primordia and found that the procambial trace associated with the absent leaf did not develop: parenchyma cells differentiated in the appropriate part of the procambial ring (Fig. 9.7c).

Explants and callus culture

When explants are grown on appropriate media in aseptic culture, vascular tissues often differentiate. Their character and spatial arrangement

Fig. 9.6. Diagrammatic transverse sections at successive levels in a poplar stem that has regenerated after an apical split. (Symbols as in Figs. 9.1 and 9.3. Based on material of S. J. Dircks.)

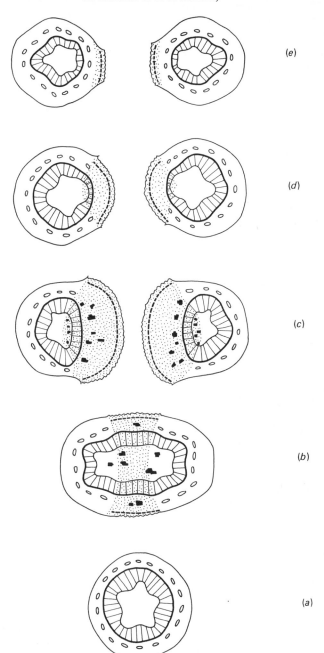

(e)

(d)

(c)

(b)

(a)

vary greatly with the material and the cultural conditions, but two distinct types of tissue development can be recognized.

Tracheary elements in clusters and strands

There are many reports of lignified tracheary elements differentiating in pith explants. Dalessandro (1973) observed them after only three days culture of *Helianthus tuberosus* pith explants, Dalessandro & Roberts (1971) observed them after four days culture of *Lactuca sativa*, and Comer (1978) after six days culture of *Coleus blumei* pith explants; most workers have found them well developed after 7–14 days, for example in lettuce (Dalessandro & Roberts, 1971) and *C. blumei* (Earle, 1968). The

Fig. 9.7. *Above*, diagrams showing regenerated procambial ring (cross-hatched) within central plug of apex isolated by (a) four or (b) three cuts, seen in transverse section. *Below*, diagrammatic transverse sections of apices after removal of a leaf primordium, showing (c) resultant gap arising in procambial ring, and (d) closure of gap with meristematic tissue after auxin application to the primordial stump.

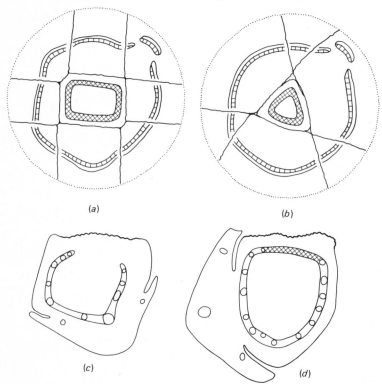

(a)   (b)

(c)   (d)

tracheary elements occur as single cells, as clusters, and most obviously as strands which are usually one or two cells wide, are often branched, and follow somewhat erratic courses. They are largely or entirely confined to the callus that arises on the outside of the pith explant, but may be absent from the outermost 0.5 mm or so beneath the surface (e.g. Dalessandro & Roberts, 1971; Warren Wilson, Roberts, Gresshoff & Dircks, 1982). Sieve tube strands have not usually been recorded in association with these tracheary elements, either because they are absent or because they are more difficult to observe.

Vascular nodules and bundles

Several reports exist of vascular nodules forming in callus cultures or in the callus proliferating from pith explants. These spherical structures develop from pockets of meristematic cells within which xylem cells first appear (Wetmore & Rier, 1963; Jeffs & Northcote, 1966). The fully-developed nodule, which may be 0.3–1.0 mm in diameter, normally has a core of xylem surrounded by phloem with a vascular cambium between these two tissues (Fig. 9.8a). The occurrence of files of cells, particularly in the xylem, shows their derivation from the cambium. These nodules are relatively slow to develop. Wetmore & Rier (1963) and Jeffs & Northcote (1966) record them in callus blocks after 30–60 days on inductive media. In pith explants they have been found after nine weeks in *Nicotiana tabacum* (Forest & McCully, 1971).

Fig. 9.8. Diagrams of sections of vascular structures in callus: (a) nodules in proliferated callus, (b) vascular groups around a pith explant, (c) a single nodule in the centre of a callus fragment. Cambium is indicated by a heavy line, xylem (hatched) is internal and phloem external to the cambium. (All diagrams are drawn to the same scale.)

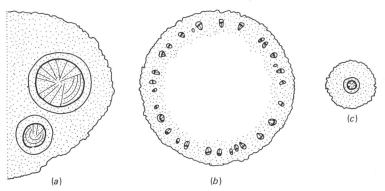

In sections through the callus the nodules are often seen arranged in an irregular discontinuous ring about 0.5 mm beneath its surface. Sometimes the symmetry of individual nodules tends to be bilateral rather than radial; in this case xylem lies towards the centre of the callus or explant and the phloem towards the surface, with a tangential cambium between (e.g. Wetmore & Rier, 1963). Such structures (Fig. 9.8b) grade into vascular groups or bundles (e.g. Sevenster & Karstens, 1955; Gautheret, 1966) reminiscent of the structure of normal stems but without longitudinal continuity.

In the simplest situation a small body of callus may contain a spherical vascular structure like a nodule (Gautheret, 1959) with central xylem surrounded by cambium and phloem, the latter at a distance beneath the surface of the callus body (Fig. 9.8c).

It is noteworthy that vascular nodules occur not only in tissue culture but also in wound callus, for example in graft unions of apple (Mosse & Labern, 1960). In the latter case amphivasal nodules (xylem peripheral) were more common than amphicribral nodules (phloem peripheral): amphivasal nodules occurred in tissue derived from the xylem. In tissue cultures, amphicribral nodules are more common, but amphivasal nodules can again occur in xylem explants (Gautheret, 1966).

General concepts in the development of tissue patterns

Dedifferentiation, and the loss and restoration of pattern

The initiation of spatial patterns of tissues involves the creation of differences where these were not present previously. There is a change, in some regions at least, from a less-differentiated to a more-differentiated state.

In normal development, relatively undifferentiated derivatives are produced by the meristem. In regeneration after wounding of stems or roots, on the other hand, most or all of the tissues around the wound are initially differentiated, not meristematic. If regeneration of normal patterns is to occur, tissues must first dedifferentiate and become competent for subsequent differentiation into appropriate new patterns. If such dedifferentiation does not occur, restoration of pattern may fail. For example, in an approach graft between a petiole and stem (Fig. 9.9a), restoration of a continuous ring of vascular tissue fails (cf. Fig. 9.4b); here (Fig. 9.9a), gaps in the vascular ring correspond to petiolar ground parenchyma that is not competent to form a regenerated cambium although this tissue is traversed by the cambial pathway (P. M. & J. Warren Wilson, 1963).

Dedifferentiation is characteristically associated with cell division. Wounding often stimulates division in cells close to the wound surface, but the degree of activity depends on the type of tissue (Gautheret, 1966), the internodal age (Sussex *et al.*, 1972) and (in tissue culture) the composition of the medium. Thus the lag period, before division commences, may be negligible in a very young internode or in the cambial region of an

Fig. 9.9. (a) Approach graft between petiole and stem of *Atropa belladonna* in transverse section: *left*, diagram of arrangement at grafting; *right*, diagrammatic section 4 weeks after grafting, with vascular cambium failing to regenerate across groups of undivided cells of the petiolar parenchyma. (b)–(e) Diagrammatic transverse sections of *Datura stramonium* stems wounded and then covered with Vaseline (b) immediately, (c) after 2 days, (d) after 3 days, (e) after 4 days, and (f) control, not covered; all were harvested 2 weeks after wounding.

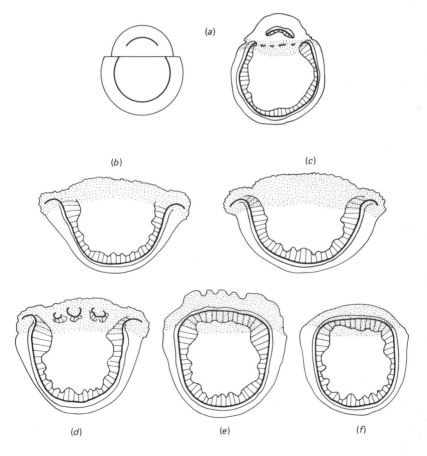

older internode, but will be longer in the pith of an older internode. Division results in a fall in cell size (Fig. 9.2b); anatomically the cells tend to become isodiametric, thin-walled, with reduced intercellular spaces and relatively dense cytoplasm, as in the callus of wounds or cultured explants. There may be reduced 'coupling' of the symplast, through loss of plasmodesmatal connections (Goodwin, 1976). The tissue appears meristematic and unspecialized, and is regarded as undifferentiated. It is likely to be competent to respond to stimuli inducing differentiation, but such competence may have been acquired before anatomical dedifferentiation has been completed.

Competence and determination, i.e. the commitment to a particular differentiation pathway, are essentially biochemical, and may occur before changes are apparent under the microscope. This is illustrated by an experiment in which young internodes were wounded as in Fig. 9.1a and the wound was covered with Vaseline either immediately, or after one, two, three or four days, or not at all in the control (Fig. 9.9b–f). Regenerating vascular cambial patterns in Fig. 9.9b (covered immediately) and Fig. 9.9f (control, not covered) accord with the generalized diagrams in Figs. 9.4d and 9.4a, but the wounds with delayed covering show that the tissues had become competent and that determination of the regenerated pattern of vascular cambia was occurring on day 3: the discontinuous committed regions subsequently rounded off as vascular nodules (Fig. 9.9d). Day 2 is when the first cell divisions occur.

However, even well developed callus may not be completely undifferentiated; *Hedera helix* callus retains differences associated with the 'juvenile' or 'adult' character of the shoot from which it was explanted (Stoutemyer & Britt, 1965). As Wareing (1978) points out, stem and root apices can still less be regarded as fully undifferentiated, though they are of course meristematic.

Thus differentiation and pattern formation can be seen as progressive phenomena, some of the successive stages for a shoot axis being the organization of an apex with histogenetic layers, the formation of the procambial ring, its division into strands, and the differentiation of protoxylem and protophloem and of the vascular cambium.

Categories of regeneration

Both in wounded stems and in explants, two types of regeneration can be distinguished: they are here termed 'short-term' and 'long-term' regeneration. The difference is most obvious in the vascular tissues.

The short-term structures consist of tracheary or sieve elements which most characteristically are arranged in strands usually not more than a few millimetres long. Their differentiation is often preceded by cell division, but this is localized and no callus is formed. They appear within a few days of wounding or explanting and they may show little further development after two or three weeks.

The 'long-term' structures consist of xylem and phloem derived from a thin sheet of regenerated vascular cambium that may complete a cylinder (especially in regenerating stems where it may be a few centimetres in length) or may be a hollow sphere (especially in tissue cultures). These structures develop within a callus that forms at the wound surface. They appear two to six weeks after wounding or explanting and continue activity for some weeks (in explants) or indefinitely (in wounded stems).

Few studies describe both types of regeneration. The 'short-term' structures have been observed mainly by harvesting one to two weeks after wounding or explanting, and examining cleared material. The 'long-term' structures have been observed by harvesting later (often one to two months after wounding or explanting) and usually by examining transverse sections in which the strands are difficult to identify. The 'long-term' structures may not be obvious where only small wounds have been made, and the 'short-term' structures may not occur around large wounds. However, both types of structure are recorded by Sussex *et al.* (1972) in *Coleus blumei* stems, wounded with a 3-mm V-notch, harvested one, two and three weeks later, and sectioned longitudinally.

The distinction between 'short-term' and 'long-term' regeneration can be extended to some non-vascular tissues that occur in wounded stems but do not develop in pith explants or callus cultures. As regards mechanical tissues, the sclerenchyma cylinder may be restored by a more or less interrupted ring of sclereid groups that, in favourable conditions, can appear within two weeks of wounding but do not develop further. This short-term tissue develops within the wound callus at an early stage, sometimes before cambium formation. The long-term regeneration of mechanical tissue is by differentiation of vascular cambial derivatives as sclerenchyma fibres in the xylem and, in some species, in the phloem; they first appear some two or three weeks after wounding and continue to be formed indefinitely.

As regards protective tissues, the short-term response to wounding is the formation of waxy, relatively impermeable coats of suberin and cutin within a few days on the walls of cells exposed to the outside atmosphere as a result of wounding, while the long-term regeneration is by formation of cork, some two weeks later, from a cork cambium that develops within the wound callus.

The functional significance of these two types of regenerated structure is that a measure of short-term wound repair is achieved within days (surface protection, vascular strands, sclerenchyma cylinder), and may serve during the period of two weeks or more required for continuing longer-term restoration (cork, phloem and xylem formation from regenerated cambia).

These relations are summarized in Fig. 9.10, which also introduces two terms that have been used to describe animal regeneration (e.g. Garrod, 1973; Grant, 1978) but which seem appropriate also for these plant processes of pattern restoration after wounding:

(1) 'Morphallaxis' occurs when the part of the organism that remains is remodelled to restore the whole form without the addition of new material. In animals it may involve migration of cells; in plants the cells do not move but some become changed in structure and function, i.e. redifferentiated, either after very localized division or without division.

(2) 'Epimorphosis' occurs when cells near the wound proliferate and dedifferentiate to form new and initially undifferentiated material (blastema in animals, callus in plants) within which differentiation occurs in such a pattern as to restore the whole form.

The discussion in this section has referred only to wounded stems and explants. With operations on apices, less drastic changes are involved. The wounded apical region consists of relatively undifferentiated tissue:

Fig. 9.10. Derivation of tissues regenerated after wounding.

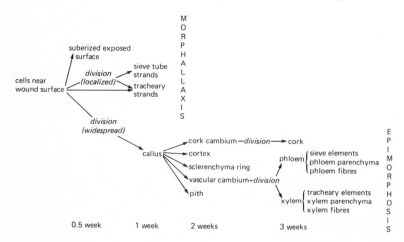

within a week of splitting, new apices have regenerated. Both dedifferentiation and redifferentiation are at a more modest level than in the stem
wound or explant.

Common features of tissue patterns in various types of regeneration

Certain basic features of the spatial arrangement of tissues are common to
the various types of wound and explant described above. Some of these
are summarized diagrammatically, for vascular tissue, in Fig. 9.11a–e.
Three major characteristics discussed in turn below are sequences, conti-
nuities and periodicities.

Sequences

By far the most important sequence is that which tends to develop
beneath exposed surfaces. The 'free surface theory' (Bertrand, 1884;

Fig. 9.11. Diagrams of operations on stems (not to scale). *Above*, surface view;
below, longitudinal sections (regenerated tissues stippled, vascular tissues in
black): (a) small wound, (b) large wound, (c) split stem, (d) split apex, and (e) pith
explant.

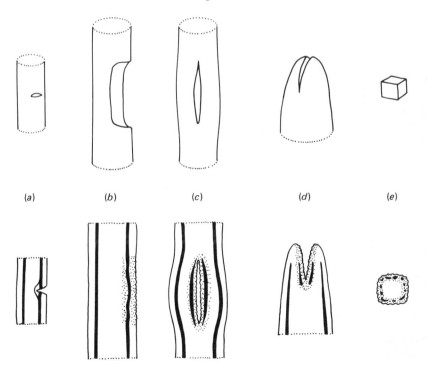

(a)   (b)   (c)   (d)   (e)

Vöchting, 1892; J. & P. M. Warren Wilson, 1961) states that a cambium forms beneath, and generally parallel to, each natural or artificial free surface; it produces phloem on the side towards the free surface, and xylem on the opposite side.

Thus there arises a sequence of three tissues – phloem and vascular cambium and xylem – within the ground tissue; this latter can be regarded as composed essentially of cortex external to the phloem, and pith internal to the xylem. This type of sequence is found in the various examples of regeneration described above and illustrated in Fig. 9.11: in small wounds where tracheary strands and sieve tube strands differentiate on the inner and outer sides of either the existing vascular cambium on each side of the wound or the site of the subsequently regenerating cambium; in large wounds where a vascular cambium regenerates within the wound callus and forms xylem inwards and phloem outwards; in wounded apices where a procambium regenerates and subsequently forms bundles with phloem, cambium and xylem; and in small explants or callus bodies which, after some weeks, develop nodules with a spherical cambium which produces phloem peripherally and xylem internally. In some cases the cambium forms first, so that its position seems especially critical; but in other cases tracheary and sieve tube strands appear in appropriate positions within a week, before any cambium arises.

Where the cambium or procambium forms first, its pathway typically follows closely the form of the surface; striking examples come from the triangular and rectangular shapes beneath the three-sided and four-sided apical plugs in Fig. 9.7a, b. Its distance beneath the surface, at initiation, is about 0.3 to 0.5 mm.

The importance of the free surface in influencing the position of the cambium is reinforced by experiments in which the surface is covered, leading to an absence of the cambium beneath the covered surface (Fig. 9.4d). Again, the creation of an internal free surface (Fig. 9.12a) can result in a cambium forming beneath the new surface but in inverted orientation so that the phloem is still towards the surface.

A cambium may form without relation to a free surface in some special situations, e.g. the additional internal cambium in wounded *Campanula pyramidalis* stems (Fig. 9.12b), the regenerating cambium in a scraped stem : cut stem graft (Fig. 9.4e), and the amphivasal nodules formed within the wood in grafts and tissue cultures.

In regeneration after wounding – but not in explants – other tissues may differentiate within the spatial sequence. The most obvious is perhaps the phelloderm; this, too, is related to the free surface both in distance beneath it (about 0.2 mm, consistently less than for the vascular cambium) and in orientation (cork towards the surface). Other tissues have been mentioned above, so that a sequence of tissues can be listed, not all of which will occur in any one case:

Cork
Cork cambium
Cortex
Endodermis
Sclerenchyma
Phloem
Vascular cambium
Xylem
Pith

Such radial sequences in axes are the most obvious, but other sequences can be recognized. The sequence of tissues from abaxial to adaxial side of a dorsiventral leaf is related to that in an axis. A longitudinal sequence can be recognized in the apical meristems (see below) and in abscission zones, where a transverse layer may include:

Separation layer
Cork
Cork cambium

Fig. 9.12. Diagrammatic transverse sections of (a) *Atropa belladonna* stem 8 weeks after boring out an oblique hole, seen in the centre of this section with a vascular cambium regenerated around part of it, and (b) *Campanula pyramidalis* stem 8 weeks after wounding, with the cambium regenerated beneath the surface of the pith callus extending internally at both ends (arrows).

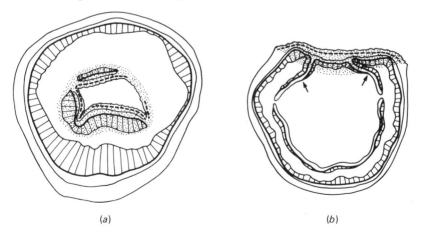

(a)   (b)

between, usually, normal petiolar tissues distally and stem tissues proximally.

Continuity

At the lowest level, this can be seen in the tendency for homeogenetic induction of differentiation, as exemplified by the tendency for clusters or groups of similarly-differentiated cells to occur more often than would be expected on a random basis. Phillips (1980) refers to tracheary element differentiation in single cells or very small groups spreading to surrounding cells so that larger groups are formed, and Warren Wilson, Dircks & Grange (1983) describe the formation of clusters of sclereids.

At the next level is the formation of more or less well defined strands of tracheary or sieve elements, usually directed longitudinally. The continuity is not merely within the strand, but extends to junctions or anastomoses of differentiating strands with strands formed simultaneously or previously, including strands existing prior to wounding and severed by the operation.

Thirdly, continuity is present in sheets, as in the regenerating vascular and cork cambia. Within these sheets the continuity is pronounced: the ability of cambia to unite with like cambia, including ends of cambia severed at wounding, is particularly striking, subject always to the absolute requirement that the orientation (phloem/xylem or cork/secondary cortex) corresponds.

Periodicity

Some tissues show a degree of regularity of spacing, especially in the tangential direction. Examples are the spacing between tracheary element strands or the occurrence of rays at intervals in the regenerated vascular cambium and its derivatives, or the fairly regular spacing of clumps of sclereids in the regenerated sclerenchyma cylinder in both tangential and longitudinal directions. This is perhaps to be regarded as involving heterogenetic induction.

Relation of tissue patterning in regeneration and in normal development

The view has been presented above that the vascular differentiation in tissue culture reflects the differentiation that occurs in wound repair, though clearly this is on a limited basis since only vascular tissues normally regenerate in tissue culture whereas the cork cambium, cork, sclerenchyma cylinder, endodermis and other tissues may regenerate in wounded stems and roots.

Highly-developed positional control systems must be required to achieve the very complete restoration of pattern that can occur in wound healing. This might be seen as a very successful adaptation that helps to counter physical damage to plants by grazing, trampling and climatic or soil disturbance. However, the view proposed here is that, although de-differentiation and callus formation are special responses to wounding, the subsequent regeneration of appropriate patterns of differentiated tissues involves essentially the same control systems as operate in the normal plant, where meristematic derivatives are differentiating to produce tissue patterns.

Thus the formation in the apical meristem, a certain distance beneath the surface, of procambial strands differentiating into a branching system of vascular strands may be comparable to wound strand formation, though more regularly organized and subject to the influence of leaf primordia. The formation of interfascicular cambium may be comparable to the development of the regenerating vascular cambium in the wound callus, and cork cambium formation in normal and wounded structures may similarly be compared. If this comparison is justified and the same physico-chemical processes of pattern control participate in both regeneration and normal differentiation, observations on wounded plants and explants are the more likely to be relevant to interpretation of the normal processes of tissue pattern control.

Meristem structure, dimensions and asymmetry

Tissue patterns arise among cells derived from the meristems – either wound meristems, normal apical meristems (shoot and root), or lateral meristems (vascular and cork cambia). Cells of the meristems and their immediate derivatives are usually regarded as undifferentiated, as indicated by their fairly uniform unspecialized character, small size, dense cytoplasm, thin walls and lack of intercellular spaces. Nevertheless, meristems have characteristic internal organization:

(1) Planes of cell division. For example, a wound meristem has new walls predominantly parallel to the wound surface; a cambial meristem has predominantly tangential walls; and within the shoot apical meristem the tunica has mainly anticlinal divisions, whereas the internal corpus has both anticlinal and periclinal divisions.

(2) Rates of cell division. For example, in the vascular cambial zone, the maximum rate of division is within the xylem mother cells, resulting in perhaps 4 to 10 times as many xylem derivatives as phloem derivatives

(Bannan, 1955); and in the root meristem the rates of division in the quiesscent centre are only about one-tenth of those in the surrounding cells of the central cylinder and root-cap initials (Clowes, 1969).

(3) Rates of cell enlargement. Differences occur both in the total enlargement and in the relative enlargement in different directions, for various derivatives of the meristem; a striking difference occurs between ray and fusiform cells of the vascular cambium.

Thus a measure of differentiation exists within meristems and their immediate derivatives. However, differences subsequently increase greatly as these derivatives undergo more or less enlargement, wall thickening, organelle development and other differentiation.

The meristems are the site of the first stages of tissue pattern formation and two general features of their structure are relevant in this respect. First, meristems are always small in at least one dimension: cork, vascular and wound cambial zones may form extensive sheets but these are only 0.05 to 0.2 mm in thickness; apical meristems are small in all dimensions, typically 0.2 to 0.5 mm for shoots and 0.1 to 0.4 mm for roots depending on how the limits are defined. Secondly, meristems are sited between contrasting regions, and are asymmetric in their activity: the cork cambium lies between the cork and the secondary cortex, the vascular cambium between the phloem and xylem, the root meristem lies between the root cap and the body of the root tissue, and the shoot meristem (being superficial) between its external surface and the body of the shoot itself. This asymmetry is a stable state, relating both to the physiological environments on the two sides of the meristem, and to the difference in the rate of production and the subsequent differentiation of the derivatives on the two sides, as discussed later. If a cambium is removed from its normal environment by explanting into tissue culture, it loses its identity and forms a callus mass which, as it increases in size, has divisions increasingly limited to a layer near the periphery, where a new, stable cambium may form (Steeves & Sussex, 1972; Lancaster & Rowan, 1974).

POSITIONAL CONTROL OF TISSUE PATTERNS

Components of positional information in the plant

The development of tissue patterns requires that new cells, which are derived from meristems and which initially appear undifferentiated, should possess information on their positions within the organism as a whole. On the basis of this positional information, individual cells would differentiate in such a way as to produce the pattern of various tissue types.

The view proposed here is that this information is specified in terms of three basic components of position in the plant, at least for axial organs (Fig. 9.13):

(1) Radial. As indicated in the figure, a sequence of different tissues lies along the radius from axis to surface of a stem or root; this is true both for the normal organ and where regeneration has occurred. In the very simple situation of a spherical tissue mass in sterile culture, the radial component dominates, for no longitudinal component is present. In the case of a dorsiventral leaf, positions may be specified in much the same way as in a half-stem (P. M. & J. Warren Wilson, 1963).

(2) Longitudinal. In stems and roots there is a clear relationship between the tissue pattern at one level and that above or below: continuity predominates, but complications occur, for instance in the primary vascular network at a node. Longitudinal sequences are found occasionally, as in the abscission zone already discussed (p. 245) and at apices.

(3) Tangential. A third component of position is required to specify tangential patterning, which often involves periodicity as in the alternating arrangement of primary vascular bundles and rays in a stem, or of protoxylem and protophloem in a root.

Fig. 9.13. Three directional components of positional information in a stem.

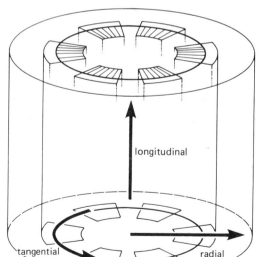

The positional control mechanisms discussed below operate in terms of these three major components of position.

Fields, gradients and morphogens in positional control

Concepts

It has long been suggested that the development of tissue patterns, whether normal or regenerated, depends on the existence of morphogenetic fields. A field is conceived as a region within which some agency co-ordinates the development of parts – such as cells – in relation to their position in the whole. This concept has been applied mainly to animals (e.g. Child, 1941; Waddington, 1966; Garrod, 1973; Ede, 1978) but also to plants (e.g. Prat, 1948; Sinnott, 1960; Wardlaw, 1965).

The agency for organization within a field may be a gradient in some physiological factor, and two or more gradients in different directions might provide a co-ordinate system. Increasingly it has been thought that gradients in diffusible and physiologically active substances may be involved, with the concentrations in different positions specifying the types of tissues that are to differentiate. This concept is compatible with the existence of one or more 'organizing centres' for a field; these would be the sites where the substances ('organizers' or 'inducers') are produced.

Turing (1952) introduced the term 'morphogen' for substances which can diffuse through tissue masses, react together chemically, and produce changes in form or differentiation. He showed mathematically how two such diffusible substances, if the concentration of one affects the rate of production, destruction, or interconversion of the other, could generate stable spatial patterns in concentrations even though the initial system was virtually homogeneous. Such 'diffusion-reaction' systems have been widely adopted and developed in models of, for example, development of vascular patterning (Ans, 1979), and initiation and maintenance of polarity (Warren Wilson, 1980), and switching between alternative pathways of differentiation (Thornley, 1981). The term 'morphogen' is a broad one, and can embrace hormones and evocators or inducers produced at organizing centres. Wolpert (1971) stressed the importance of the location of the sources and sinks (the boundary regions) in morphogenetic fields where pattern is specified in terms of morphogen distribution.

The concepts discussed in this section were developed for animals but have been extended to plants and provide a plausible formal framework. However, a basic limitation which has militated against their acceptance has been the failure to establish the chemical identity of the morphogenetic substances on which the positional information is based. Despite many attempts, none has been definitely identified for animals (see Ede, 1978; Grant, 1978); for plants only recently have one or more known substances been established as morphogens mediating pattern-forming fields by means of gradients and associated phenomena. One of these substances is auxin (indole acetic acid); others may be sucrose and cytokinin (e.g. Warren Wilson, 1980; Fosket, 1980; Mitchison, 1980; Sachs, 1981). Not infrequently it seems that the balance or ratio of concentrations of two morphogens is important rather than absolute concentration of a single morphogen.

Evidence for the role of such substances as morphogens comes from knowledge of (a) the location of their sources, sinks and transport pathways; the factors affecting their rates of production, loss and transport; and their resulting concentration gradients; and (b) their morphogenetic effects on (where tissue patterns are concerned) cytodifferentiation processes both *in vitro* and in the plant.

Morphogen sources and sinks

Gains and losses of a morphogen within a morphogenetic field may result from three processes: (a) synthesis or destruction, (b) release from or removal to a storage form, and (c) import or export by specialized long-distance transport to or from the morphogenetic field, as distinct from local movement within the field.

The complexities of these relations even for a single morphogen are great, and it is not practicable in the present chapter to do much more than outline some of the relations for the most clearly established morphogen, indole acetic acid (IAA), the principal natural auxin.

Synthesis of this auxin occurs in young expanding leaves, where concentrations are higher than in the apex itself; as the leaves expand the concentration falls several-fold over each node (Jacobs, 1979; Allen & Baker, 1980). Auxin is also thought to be synthesized in the young xylem during the autolysis that accompanies differentiation (Sheldrake & Northcote, 1968), and concentrations just inside the cambium may be of the same order as in young leaves (Sheldrake, 1971) or even greater (Savidge & Wareing, 1981), with a steep gradient to progressively lower concentrations in the cambium and phloem (Warren Wilson, 1978). In tissue culture, accordingly, autotrophy for auxin occurs with excised apices of *Coleus blumei* provided they are large enough to include three pairs of primordial leaves (Smith & Murashige, 1982), and with internodal explants of *Nicotiana tabacum* where secondary xylem differentiation is occurring (Sheldrake & Northcote, 1968); in contrast, pith explants do not normally produce auxin (Cheng, 1972).

Destruction of auxin by a peroxidase acting as an IAA-oxidase (Thimann, 1969) is widespread but is not uniformly distributed in the plant. The enzyme is especially active in the epidermis, in meristems including the sites of leaf primordia, and in vascular tissue and particularly throughout the primary and secondary phloem system (Van Fleet, 1959; Andreae & Ysselstein, 1960; Zenk & Müller, 1964; Jacobsen & Caplin, 1967). However, the most striking site of auxin destruction is at a wound surface, where peroxidase is formed in advance of cell division but in a pattern corresponding to the site of subsequent cell division (Briggs, Steeves, Sussex & Wetmore, 1955; Ray, 1958; Van Fleet, 1959); it results in especially rapid destruction of auxin.

Storage of auxin, in 'bound' as opposed to 'free' form, occurs by conjugation with a range of compounds, resulting in its physiological inactivation. These auxin conjugates protect the IAA against peroxidative attack and, since the conjugation is reversible, they provide a homeostatic system from which free auxin can be released in response to a stimulus (Jacobs, 1979; Bandurski, 1980). The storage pool is substantial; the amount of bound auxin exceeds the amount of free auxin (Bandurski & Schulze, 1977).

Long-distance transport of auxin occurs as basipetally-polarized active movement. Although this transport can occur in ground tissues, the main pathway is within the vascular tissues. Auxin applied experimentally to leaves is transported in the sieve tubes, although the normal pathway for long-distance transport of endogenous auxin is not in the phloem but rather is in the cambium and differentiating xylem (Bonnemain, 1971; Morris & Kadir, 1972; Morris, 1977; Jacobs, 1979).

The concentrations of auxin within a morphogenetic field may depend on interaction between these various sources and sinks, and on other transport phenomena discussed below. It has not yet proved possible to construct a realistic auxin balance sheet for any particular instance of development; nevertheless, where substantial spatial variations are observed in gain or loss of auxin, or in its concentration, this can be considered in relation to patterns of histogenesis.

For sucrose, like auxin, source-sink relations are complex, but the prime source must be the phloem, where the concentration of sucrose (or related transport sugars) in the sieve tubes is about 10–25% which is far higher than in other tissues (e.g. Peel, 1974; Baker, 1978). Sucrose tends to be taken up selectively into phloem tissue *in vitro* (Bieleski, 1966), and within the phloem it is translocated about the plant as the main transport form of carbohydrate, protected by virtue of this tissue's lack of sucrose-metabolizing enzymes (Arnold, 1968; Ziegler, 1975). However, sucrose is off-loaded from the phloem in meristematic regions, where it provides the carbohydrate for metabolic activity and growth.

The mechanism of off-loading is not fully understood, but it has been suggested that the local presence of high activities of invertase in sink tissues may reduce sucrose concentration through hydrolysis and, by steepening the gradient, encourage sucrose import (Walker & Thornley, 1977; Walker, Ho & Baker, 1978; Russell & Morris, 1982; Morris, 1982). Moreover, hormones including auxin can stimulate invertase synthesis (Sacher, Hatch & Glasziou, 1963; Morris, 1982). It is therefore of interest that one response to wounding or explanting is a great increase in the invertase content in the wounded tissue within the first 2–4 days, followed by a gradual fall (Edelman & Hall, 1965; Bacon, MacDonald & Knight, 1965; Thorpe & Meier, 1973; Passam, Read & Rickard, 1976). Perhaps this encourages sucrose transport to the wound, where the regeneration processes provide an active sink for carbohydrate.

Morphogen transport

It is generally supposed that the transport of morphogens within a morphogenetic field is predominantly by diffusion. This assumption was made by Turing (1952) and has been adopted in most subsequent treatments (e.g. Cohen, 1971; Wolpert, 1971; Meinhardt, 1978; Thornley, Warren Wilson & Colley, 1980). Such diffusive movement, down a concentration gradient from a higher concentration at the source to a lower concentration at the sink, is passive (i.e., not involving respiratory energy); the flux of morphogen depends directly on the difference in concentration and on the diffusion coefficient of the morphogen in the tissue. Jeffs & Northcote (1967) applied radioactive phenylalanine, sucrose and auxin to *Phaseolus vulgaris* callus blocks and found that the concentration gradients set up over 2–4 days were consistent with transport by simple diffusion.

Crick (1970), discussing diffusion in embryogenesis, pointed out that the diffusion transport process is such that the time taken for a concentration gradient to approach its final form is proportional to the square of the distance between the source and sink. The process is slow except over short distances. Making reasonable assumptions about tissue characteristics and the coefficients of diffusion and facilitated diffusion, Crick calculated that for a linear gradient to be set up within three hours, the distance from source to sink has to be not more than about 0.7–0.9 mm. Wolpert (1971) tabulated the sizes of morphogenetic fields, mainly in animal embryos, and showed them mostly to range between 0.2 and 1.0 mm. It has been noted above that apical and cambial meristems in plants are usually less than 0.5 mm in thickness, and so are of a size within which diffusion could provide a sufficiently rapid transport mechanism for production of morphogen gradients.

Transport of morphogens can also occur by mass flow with the solution in which they are dissolved. For example, both auxin and sucrose are present in the phloem sap (Baker, 1978) and are transported relatively rapidly in it; and they can move radially from the phloem into the xylem sap (Peel, 1974; Zamski & Wareing, 1974) where, again, mass flow occurs. However, such long-distance transport is less likely to mediate morphogen transport within morphogenetic fields of undifferentiated cells than to bring morphogens to an off-loading point for the field. Nevertheless, pressure-driven mass-flow transport could presumably occur within a field of undifferentiated tissue, provided that driving forces for solution flow are present.

For auxin, two more specialized forms of transport are known:
(1) Active polar transport. Polarized basipetal transport has been found consistently in shoots, with a velocity of 5–15 mm h^{-1}, faster than diffusive transport but slower by an order of magnitude than mass-flow transport in the phloem (Jacobs, 1979). Similar polarized acropetal transport is found in roots (Wareing & Phillips, 1981). This type of transport is active and, unlike diffusive transport, can transport auxin up a concentration gradient (e.g. Goldsmith, 1966). It has been suggested that the polarity resides in individual cells, and for a tissue mass increases exponentially with the number of cells or length of tissue traversed in the direction of flow (Leopold & Hall, 1966; de la Fuente & Leopold, 1966). Polar transport of auxin occurs in internodal explants (Jacobs & McCready, 1967; Thompson, 1970; Sheldrake, 1973), and has been shown also in pith explants, but may be weaker than in internodal explants containing vascular tissue (Nash & Bornman, 1972; Warren Wilson *et al.*, 1982). Callus tissue of various species has shown no polarized transport (Wetmore & Rier, 1963; Jeffs & Northcote, 1967). There are indications that polarized transport may not be well developed initially in embryos and in stem tips where meristematic tissue predominates (Fry & Wangermann, 1976; Jacobs, 1979).
(2) Flux-dependent facilitated transport. Auxin is known to enhance its

own transport and, as discussed later in relation to strand formation, this phenomenon can be expected to lead to canalization of auxin flow, both for diffusive and polar transport (Sachs, 1978; Mitchison, 1980).

A further local type of transport is represented by active uptake: auxin accumulates in cells in stem explants, pith explants, and suspension culture (Nash & Bornman, 1972; Rubery & Sheldrake, 1974; Rubery, 1980), and sucrose is readily accumulated (e.g. Giaquinta, 1980), with phloem tissue being especially active (Bieleski, 1966).

For sucrose in particular, 'hormone-directed transport' is an important further type of movement, though its mechanism is not well understood. An example is provided by the experiments of Gersani, Lips & Sachs (1980) on cylindrical 15-mm explants of *Phaseolus vulgaris* hypocotyls. When roots and buds were allowed to regenerate at the ends of the cylinder, radioactive sucrose applied to the middle of the cylinder was found to accumulate in the active sinks at both ends of the cylinder within five hours. When the ends (with roots and buds) were removed and replaced with agar blocks containing auxin or a cytokinin, movement of sucrose to the ends was increased above that to control agar blocks.

The discussion in a later section suggests that most or all of the transport processes outlined above are involved in one or other of the mechanisms for vascular patterning: diffusive movement is involved in radial positioning, mass flow participates when vascular bundles have been severed by wounding, canalized flow in the formation of veins and strands, polar transport in promoting longitudinal extension of vascular differentiation, and hormone-directed transport contributes to the development of the vasculature linking hormone sources and sinks.

Possible other morphogens or agents forming gradients

Morphogens resemble hormones in being synthesized by the plant and in having an effect on development at a distance – admittedly often only a small distance – from their site of synthesis. Auxin and cytokinin, mentioned in the earlier discussion on morphogens, are hormones, and sucrose has been said to have 'an almost hormone-like function' in xylem differentiation (Torrey, Fosket & Hepler, 1971). However, the role of a morphogen has been tentatively ascribed to some other substances which are not hormones: these include carbon dioxide, oxygen and ammonia (e.g. Sorokin, 1962; Kessell & Carr, 1972), pH and calcium (e.g. Priestley, 1928; Jaffe, 1979). These variables interact with one another and may affect auxin and other hormones. They tend to occur in gradients wher-

ever cells metabolize under crowded conditions, with the character of the
gradient being affected by surface/volume ratio, aeration and nutrient
supply (Loomis, 1959). This may relate to the tendency for tracheary
elements to differentiate when clusters of cells attain a certain size by
growth or by experimental aggregation (Steward, Mapes & Mears, 1958;
Wilbur & Riopel, 1971). Such effects may also be influenced by physical
pressure, which is believed to be necessary for maintaining orderly cambial activity and differentiation of the derivatives (Brown & Sax, 1962);
however, as Steeves & Sussex (1972) noted, there is nothing to indicate
that pressure is responsible for the initiation of the spatial pattern.

The substances discussed in this section (other than auxin, sucrose and
cytokinin) are not treated as morphogens in what follows; this is because
there is insufficient information, so far, on the location of their sources
and sinks in regions where cytodifferentiation is taking place, and on their
roles in inducing specific pathways of cell differentiation.

Major mechanisms for positional control

Sequences associated with gradients of morphogens

Surface effect

The 'free surface theory' of Bertrand and Vöchting recognized the role of
the surface in controlling the position of cambia and the orientation of the
phloem – vascular cambium – xylem sequences in tissue regenerating after
wounding. Wardlaw (1950) noted that in wounded apices the regenerating
procambial ring also conformed with the surface outline, and Sinnott
(1960, p. 238) commented that this suggested 'that its position was
dependent on a gradient of some sort'. J. & P. M. Warren Wilson (1961)
discussed these and other theories about the positional control of vascular
cambia regenerating in wounds and grafts, and suggested that 'there
arises within the callus a quantitative trend or gradient, from the exposed
surface inwards, in some factor as yet unknown; and that a cambium can
arise at only one position on this gradient (where the factor is at an
appropriate level), while its orientation, as regards phloem and xylem
formation, is determined by the direction of the gradient'. It was pointed
out that this 'gradient induction hypothesis' might also be applicable to
the positional control of cambia in some tissue cultures and of procambia
in apical meristems.

The 'free surface theory' and 'gradient induction hypothesis' were concerned primarily with the control of the cambium, and only secondarily with the positioning of the derivative phloem and xylem. However, it has become clear that when vascular strands develop in response to wounding, their depth below the surface may be co-ordinated so that the cambium which subsequently forms passes between the sieve tube strands and tracheary strands. Thus the positional control system should probably be regarded as co-ordinating sieve-tube strands or phloem, procambium or cambium, and tracheary strands or xylem, both in 'short-term' (or primary) and in 'long-term' (or secondary) differentiation.

One consequence of vascular cambia forming at a certain distance beneath the surface is that if a tissue mass is too small (with a diameter less than twice this distance) no cambium formation is to be expected. Thus Wardlaw (1952) showed that if the size of a stem apex is progressively reduced during its growth, by removal of leaf primordia or by growing in feeble light, there is a corresponding change from a solenostelic vascularization (hollow cylinder) to a protostele (narrower solid cylinder) to almost complete disappearance. The sequence was reversed when the shoot was allowed to resume normal growth and apical size increased again. Similarly, in tissue culture, if the tissue mass is below a certain size, no vascular differentiation occurs; a sufficient size results typically in a central body of xylem with cambium and phloem around it; and at a larger size there may be a sphere of vascularization around a parenchyma core.

In referring to differentiation at a certain distance beneath the surface, a 'physiological distance' rather than 'measured distance' is intended: J. & P. M. Warren Wilson (1961) noted that cambia tend to regenerate at a shorter measured distance from the surface in small-celled callus than in large-celled callus.

The 'free surface' and initial 'gradient induction' proposals were deficient, like other field and gradient proposals, in not specifying the agent or factor defining the gradient or distance beneath the surface. However, later work revealed, as argued by Warren Wilson (1978), several characteristics of auxin that make it a contender for this role. (a) Auxin is formed in the differentiating secondary xylem, and high concentrations occur there. It is transported basipetally in the differentiating xylem and cambium, and this region provides an auxin source. (b) Auxin is transported radially outwards from the xylem to the phloem (Peel, 1974; Zamski & Wareing, 1974). It is also transported within callus by diffusion (Jeffs & Northcote, 1967). (c) Auxin is degraded by IAA oxidases which, although widely distributed, are most active at epidermal and especially wound surfaces. (d) Presumably as a consequence of this spatial distribution of auxin sources and sinks, a gradient of auxin can be observed with higher values in the outer xylem and lower values in the inner phloem. The gradient is steep, with a difference in concentration between two- and ten-fold over a distance of less than a millimetre (Sheldrake, 1971; Savidge & Wareing, 1981). (e) Auxin is necessary for vascular differentiation in both tissue culture and wound strand formation (e.g. Jacobs & Morrow, 1957; LaMotte & Jacobs, 1963; Wetmore & Rier, 1963). In wounds, Thompson & Jacobs (1966) found that tracheary differentiation required higher concentrations of auxin than sieve tube differentiation; Aloni (1980) observed the same to be true for some tissue cultures. (f) The oxidative destruction of auxin that normally occurs at exposed surfaces can be reduced by covering them with water or gelatin, or to a greater extent with lanolin or celloidin (Fiedler, 1936; Shoji & Addicott, 1954; Zenk & Müller, 1964). Correspondingly, covering a wound surface with Vaseline or (less consistently) lanolin alters the position of regenerating vascular tissue (J. & P. M. Warren Wilson, 1961).

Such considerations led Warren Wilson (1978) to suggest that auxin was involved in the morphogenetic gradient proposed in the 'gradient induction' hypothesis to account for the positioning of vascular tissues beneath free surfaces.

Homeogenetic induction

Experiments described above (Fig. 9.4a–c) show the tendency of regenerating vascular cambia to join the cut ends of original cambia; this results in restoration of closed cambial rings as seen in transverse section, even where there is no surface effect (Fig. 9.4e). Often the regenerating cambia that appear in the wound callus extend from the cut ends of the original cambia, just as interfascicular cambia may extend from the fascicular cambia; occasionally, isolated patches of cambium regenerate in the callus and later unite into a continuous sheet. The ability of these cambia to extend and reunite, always in corresponding orientation, is striking: cambium joins cambium, phloem joins phloem and xylem joins xylem. The extension appears to be a form of homeogenetic induction.

Janse (1921) and Snow (1942) considered the possibility that this induction might be brought about through the influence of an active substance or hormone diffusing from the cells of the original cambium into the callus. Camus (1949) showed in somewhat comparable grafts in tissue culture that the influence could cross a permeable film of gelatin or Cellophane: he concluded that the induction was caused by a diffusible hormone, probably auxin. J. & P. M. Warren Wilson (1961) felt that the diffusion of such a hormone could not alone explain the restriction of the extending cambium to a particular narrow pathway; they supposed that the level of some factor becomes fixed in differentiated xylem, cambium and phloem and that such established levels tend to induce similar levels in adjacent undifferentiated callus.

Warren Wilson (1978) suggested that a cambial pathway could be defined by a combination of two hormones or morphogens forming opposed gradients, with the cambium being initiated at a particular ratio of concentrations (Fig. 9.14). It was suggested that one morphogen was auxin, with its source in the differentiating xylem and sinks in the phloem

Fig. 9.14. Idealized representations of (*above*) gradients in auxin and sucrose concentrations with position across the vascular cambium, and (*below*) the gradient in auxin:sucrose concentration ratio, showing its proposed relationship to cambial activity and to xylem and phloem differentiation according to the gradient induction hypothesis.

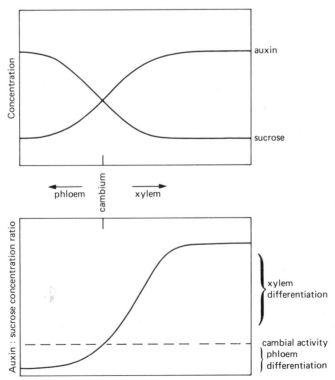

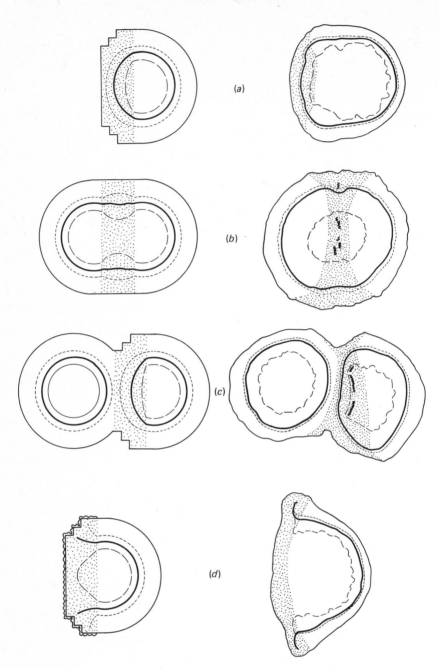

(a)

(b)

(c)

(d)

and at the surface, and the other morphogen might be sucrose, with its source in the phloem and sinks in the cambium and differentiating xylem.

Sucrose (or in some species the oligosaccharide raffinose sugars, with one or more galactose residues attached to the sucrose) is the prime substance for long-distance transport of carbohydrate in the phloem, and its concentration in the sieve tube sap is commonly 10–25%. It is presumably actively metabolized in the cambium and differentiating derivatives, and this sink results in a steep concentration gradient across the cambium, to concentrations of less than 1% – often scarcely detectable – in the vessels of the xylem (see Warren Wilson, 1978). Tissue culture studies show that sucrose promotes formation of both tracheary and sieve tube elements: although tracheary elements can develop in the absence of sucrose in some material, sucrose greatly increases the number formed (other sugars are less effective) up to a certain concentration – perhaps about 2% but varying greatly with species – above which sucrose inhibits tracheary element formation (e.g. Beslow & Rier, 1969; Warren Wilson *et al.*, 1982). Sieve tubes may not appear at sucrose concentrations of 1% or less but increase at concentrations up to about 4% (Wetmore & Rier, 1963; Fadia & Mehta, 1973). Thus the optimum sucrose concentration for induction of sieve tubes tends to be higher than that for tracheary elements.

Gradient-induction hypothesis

This hypothesis integrates the surface effect and the homeogenetic induction effect in terms of opposed gradients of the two morphogens, which have sources and sinks sufficiently defined to permit a simple simulation

---

Fig. 9.15. *Left*, simulations based on the gradient induction hypothesis, for vascular regeneration in (a) wounded stem, (b) approach grafted cut stems, (c) approach grafted cut stem : scraped stem, and (d) wounded stem with wound surface covered with Vaseline. The simulation procedure (Warren Wilson, 1978) allows auxin and sucrose to diffuse from differentiated tissues in the original stem, where concentrations are maintained, into the callus, with auxin breakdown at exposed callus surfaces; cambium then regenerates where the auxin : sucrose concentration ratio equals that at the original cambium. Dotted and broken lines indicate auxin : sucrose ratios half and twice that at the cambium, i.e. phloem and xylem sides respectively. *Right*, diagrammatic transverse sections of the same four types of operation; outer limit of phloem is shown by a dotted line and inner limit of xylem by a broken line. Small zig-zag lines in (b) and (c) indicate crushed cells at the original interface between opposed calluses.

of the processes of positional control. The simulation assumes that levels of the morphogens are maintained in the differentiated phloem and xylem, but that both morphogens can diffuse into adjacent callus, where auxin is destroyed at the surface. In Fig. 9.15a–d the results of some simulations are compared with observed vascular patterns regenerating after various types of wounding and grafting operation. The qualitative correspondence confirms that the hypothetical positional control system based on specified morphogens could indeed account for these vascular patterns.

The hypothesis has been extended to simulations of regeneration in grafts between stems and petioles. These simulations, in which dorsiventral petioles were represented essentially as half-stems, satisfactorily represented observed patterns in various such grafts (J. & P. M. Warren Wilson, 1981).

The gradient induction hypothesis has been applied only to vascular tissues. It makes no reference to the radial sequence of other specialized tissues such as cork cambium, sclerenchyma cylinder and endodermis. The entire spatial sequence is consistently reproduced in normal growth and in regeneration. Thus, if the gradient induction system controls the vascular positioning, it seems that control of the other tissues cannot be independent of this; their positional control systems must relate to the vascular positioning mechanism but in ways as yet unknown.

Axial sequences

The above account of sequences associated with morphogenetic gradients has concerned the radial component of positional information. In the longitudinal direction, continuity predominates over sequences. However, gradients in the longitudinal direction may occur where there are discontinuities in the axis, for example, at shoot and root tips where meristematic regions occur, and at nodes where abscission zones may develop.

An abscission zone consists of a transverse zone which is continuous across a petiole, pedicel, stem or other axial structure. Within the zone is a distal separation layer and a proximal protective layer; thus the abscission zone is asymmetric. The zone is characteristically sited just above a junction where a stem joins a leaf, pedicel, or other axial structure. It serves to cut off a region that is less active, for example through senescence or damage, from a more major pathway of transport.

The mechanism controlling the position of the abscission zone is not known, but it is tempting to speculate that auxin is involved. Perhaps the zone arises where the basipetal auxin flow in a less active part becomes very small compared with the flow in the major pathway, so that the zone is positioned in a band where there is a sharp fall in concentration of auxin diffusing acropetally from the main basipetal pathway. The positioning of the abscission zone just above the junctions accords with a control mechanism that includes a polar component; it is noteworthy that application of triiodobenzoic acid, which inhibits polar auxin transport, can induce abscission in the middle of internodes (Whiting & Murray, 1948). Further, Pierik (1980) has shown that application of auxin to the base of pear pedicels in sterile culture causes abscission to take place at a position higher up the pedicel than in control pedicels without auxin.

At the shoot and root tips, where a boundary is imposed by the end surface of the plant body, longitudinal sequences also are present. Thus at the root tip the most distal tissue is the oldest part of the root cap; proximal to this there occur younger root cap cells derived from 'founder' cells of the next zone, which is the quiescent centre; proximal to this are the actively dividing cells derived from the proximal face of the quiescent centre and forming the meristem; these cells divide a certain number of times as they become progressively displaced further behind the root tip, and the zone of division grades into a zone of enlargement. Barlow (1976) has proposed that the maintenance of this longitudinal sequence of zones in the root meristem can be interpreted in terms of gradients in two morphogens (Fig. 9.16). One, perhaps a cytokinin, is synthesized in the quiescent centre and moves down concentration gradients towards the cap and the root proper. The other morphogen, perhaps auxin, is synthesized in maturing cells of the cap and of the root proper and moves towards the quiescent centre, down a gradient maintained by dilution and destruction in the meristem. Barlow suggests, giving supporting evidence, that the cytokinin/auxin ratio, which will be maximal in the quiescent centre and decrease away from it, controls the mitotic cycle: above a value P all events of the mitotic cycle occur; below P, only endomitosis can occur; and below a lower ratio P', the endomitotic cycle cannot occur. (This system is also discussed in chapter 10, pp. 284–5.)

Opposed gradients

The positional control system proposed by Barlow for the root meristem and by Warren Wilson for the vascular cambium both invoke two morphogens with opposed gradients, the ratio of concentrations providing the positional information that controls the tissue patterning. Crick (1970) and Wolpert (1971) have briefly noted the possibility of such a mechanism, though without comment. It seems worth pointing out that such a system has advantages over a positional control system based on a concentration gradient in a single morphogen, in that variations in the absolute concentrations of the two morphogens, provided they both vary similarly, will have little effect on the positional information residing in the ratio. Thus, variation associated with temperature or with stage of development may be less disturbing to the regulatory processes. Further, for most of the cases simulated by Warren Wilson (1978), where morphogens were diffusing into a region of callus, the pattern based on the ratio took up essentially its final, steady-state form at an early stage when absolute concentrations were still changing rapidly.

In addition, the occurrence of opposed gradients across a meristem seems to accord with the characteristic meristematic organization summa-

Fig. 9.16. Diagram, based on proposals of Barlow (1976), of *above*, concentration gradients in two morphogens, perhaps auxin and cytokinin, with longitudinal distance through a root tip, and *below*, the gradient in cytokinin/auxin concentration ratio showing its proposed relationship to meristematic activities.

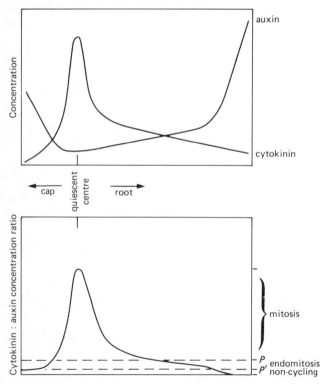

rized in an earlier section, with stable asymmetry either longitudinally as in apical meristems or radially as in lateral meristems.

It may be noted also that MacWilliams & Papageorgiou (1978) have developed a model for translating the ratio of concentrations of two morphogens into a biochemical signal by assuming that an allosteric protein, present in uniform concentration throughout the morphogenetic field, can bind up to two molecules of each morphogen, and is active when one molecule of each morphogen is bound.

Strand continuity associated with flux-dependent facilitation of transport

Tracheary and xylem strands

The tracheary elements that are the most obvious sign of differentiation in tissue culture may occur as isolated cells or in small groups, but characteristically form strands of tracheary elements joined end to end, only one or few cells wide but extending over a length of many cells. Tracheary strands of similar appearance, but tending to follow more clearly defined pathways, have long been recognized around wounds, usually bridging the gap between severed xylem bundles. In tissue culture their pathways may have no such clear function, but they may nevertheless show some radial or, more commonly, longitudinal pattern, for example in the pith explants of Dalessandro & Roberts (1971) and Warren Wilson *et al.* (1982).

In pith explants, tracheary element differentiation depends on provision of auxin, and the number of elements increases with auxin concentration. Around wounds, similarly, the formation of tracheary strands depends on the presence of young leaves, presumably because they provide auxin; if the leaves are excised their effect can be replaced by providing auxin (e.g. Jacobs, 1979). It is thought that where vascular strands are interrupted by wounding, the auxin normally transported in them is pumped out, accumulates locally, and stimulates strand formation (Jacobs & Morrow, 1957). This accords with the observations that (a) the strands differentiate basipetally, and (b) the amount of strand formation is closely related to the number of pre-existing bundles severed (Aloni & Jacobs, 1977). Moreover, when auxin is supplied locally to explants or callus, the pattern of tracheary strands is obviously related to the sites of application (e.g. Sachs, 1981; Warren Wilson *et al.*, 1982). This range of evidence leaves no doubt that auxin is closely concerned in controlling the pattern of tracheary strands in wounds and in tissue culture.

The same is probably true in normal stems and roots for the control of the pathway of the primary xylem strands, or rather the procambial strands in which differentiation subsequently occurs. In the case of these primary strands, the auxin sources seem likely to lie mainly in the leaf primordia into which the strands eventually pass as leaf traces. If a leaf primordium is surgically removed, a parenchyma gap develops in the corresponding part of the procambial ring which would have become a leaf trace; but if auxin is applied to the primordial stump, this region remains meristematic (Young, 1954), as in Fig. 9.7d.

The pathways of wound tracheary strands are defined some time before tracheary differentiation is apparent. Kirschner & Sachs (1978) found oriented cytoplasmic strands appearing in cells of pea stems within two hours of wounding in the locations and orientations predicted for vascular strands. Plasmolysis can reveal cytoplasmic reorientation preceding vascular differentiation around wounds, 16 h after wounding (Sachs, 1981); and not long after this, oriented cell divisions may occur.

Canalization by flux-dependent facilitation of transport

The precision with which tracheary and xylem strands are differentiated, as narrow channels sharply distinct from surrounding tissues, suggests some autocatalytic or self-reinforcing process. For example, where cell division or enlargement is so oriented as to cause elongation of cells along the path of the future strand, or reduction of cell dimensions perpendicular to the strand, this may promote symplastic and apoplastic transport along the pathway and cause it to contain higher concentrations of auxin transported from the source. Also, as soon as differentiation of tracheary or xylem strands occurs, the accompanying autolysis of cells may produce auxin, as in differentiating secondary xylem (Sheldrake, 1971).

However, enhanced auxin transport around wounds has been detected within 16 h of auxin application, some two days before wound vascular strands differentiated (Sachs, 1975). This accords with the known enhancement by auxin of its own transport (Leopold & Lam, 1962; Hertel & Flory, 1968; Rayle, Ouitrakul & Hertel, 1969). The concept of flux-dependent differentiation (e.g. Park, 1975) has been invoked to account for the autocatalytic early delineation of wound tracheary strands; it was proposed by Sachs (1978) that auxin flow facilitates the auxin transport and this brings about canalization of discrete strands. Sachs (1981) has presented evidence that the definition of longitudinal vascular strands is dependent on the flux of auxin, in contrast to the gradient of auxin which has been presented above as the basis for the radial sequence of xylem, cambium and phloem.

Mitchison (1980) has developed the concept of flux-dependent facilitation in a mathematical model which shows that destabilization and canalization can occur as a result of the diffusion constant or polar transport rate for auxin increasing with the auxin flux. This model illustrates how autocatalytic flow facilitation can create preferred pathways of auxin flow, which represent the sites for subsequent differentiation of strands or veins. The particular pattern of strands that develops will depend on the extent to which the auxin flow is due to polar transport or to diffusive transport.

In the case of polar transport in normal tissues in which basipetal, polar auxin flow is well established in a field in which there is a diffuse supply of auxin, canalization will lead to basipetally-extending longitudinal strands, spaced at intervals from one another. If the source of auxin is localized, as by experimental application, strands will extend from this site in a mainly basipetal direction. On the other hand, if the polarity in the tissue is not longitudinal but is confused in orientation, as for example in callus of grafted or explanted tissue, strand direction will be inconsistent; in such calluses there may occasionally occur groups of cells with polarity tending to give a circulatory flow. (For further discussion see the preceding chapter by Sachs.) Mitchison (1980) points out that autocatalytic reinforcement of such a flow could account for the closed loops of strand that have been reported in such tissues for tracheary elements (e.g. Neeff, 1922; Earle, 1968; Kirschner, Sachs & Fahn, 1971; Sachs & Cohen, 1982) and for sieve elements (Aloni, 1980). Closed loops cannot arise from diffusive, as opposed to polar, transport.

In the case of diffusive transport, tracheary strands will tend to become canalized between spatially separate auxin sources and sinks, perpendicular to the concentration gradient. There may be limits to this spatial separation: if the source and sink are close (less than a few tenths of a millimetre apart) there may be a tendency instead for cambial differentiation between them; if the source and sink are far apart (more than a few millimetres) strands may be slow to form. However, once a pathway of auxin flow has canalized, it can be expected to extend progressively so long as the concentration gradient continues.

Tracheary strands may branch, as for example in pith explants of Dalessandro & Roberts (1971). Branching can result from the presence of spatially separate sources, or perhaps an extensive diffuse source, or can be due to change in source position; this is thought to be the basis for formation of vein networks in leaves (Mitchison, 1980). In the same way, where the primary vascular system of the stem consists of a branching system of vascular bundles, with leaf traces making their exit at the nodes, the pattern can perhaps be ascribed to the changes with time in distribution of auxin sources at the stem apex, associated particularly with leaf primordia. In this case, the longitudinal course of the bundles may reflect in part the polarization of the stem tissue. Indeed, in many situations the pattern of xylem strands, veins and bundles seems to reflect both diffusive and polar influences.

Sieve tube and phloem strands: phloem–xylem interrelations

The spatial arrangement and the elongated discrete character of wound sieve strands are so closely comparable with those of tracheary strands that it may be conjectured that similar facilitated-flow processes control their canalization. There appears to be no direct evidence that this auxin-based mechanism is involved, but it is known that auxin can be transported through phloem and that auxin is needed for sieve tube differentiation, though in lower concentrations than for tracheary strands.

Moreover, the differentiation of sieve tube strands is preceded by a meristematic phase: Behnke & Schulz (1980) found that three consecutive divisions were required for producing the phloem mother-cells in *Coleus blumei*. Thus the definition of the pathway has taken place at a meristematic stage (comparable to the procambial strand of the normal stem apex) some time before tracheary or sieve tube differentiation. In some material the sieve tube strands and tracheary strands seem unrelated to one another in positioning (Sussex *et al.*, 1972), but in many cases the path of wound tracheary strands corresponds to that of wound sieve tube strands (Thompson, 1967; Benayoun *et al.*, 1975; Aloni & Jacobs, 1977). In such cases perhaps the canalization of a single meristematic strand may subsequently yield both tracheary and sieve tube strands, though radially separated by formation of a cambium between them. This is the pattern of development in the normal stem apex, where each procambial strand differentiates, in due course, into primary phloem and primary xylem separated by the vascular cambium.

If both sieve tube or phloem strands, and also tracheary or xylem strands, have a similar early development as flow-facilitated meristematic strands, it is necessary to understand how subsequent differentiation into one or other type of vascular tissue is controlled. For reasons discussed earlier it seems likely that the auxin/sucrose ratio is responsible, causing phloem to differentiate on the outer side nearer the surface, and xylem on the inner side.

The processes of formation of vascular strands, in explants and in

wounded and normal stems, can somewhat speculatively be envisaged as
follows (Fig. 9.17). Where a source of auxin exists, a flow away from it
will occur because of polar transport and/or diffusive transport, the direction of flow tending to be basipetal where due to polarity, or down the
concentration gradient where due to diffusion. This flow will tend to
become canalized by flux-dependent facilitation. Those channels will be
preferred which have access to sucrose, because the meristematic activity

Fig. 9.17. Representation of suggested morphogenetic processes of vascular
strand formation in (a) an explant initially lacking vascularization but with an
applied auxin source (e.g. Sachs, 1981), (b) a stem with a small wound interrupting
two of the four vascular bundles seen in a tangential slice, (c) a stem apex in
longitudinal section. *Above*, location of main auxin sources (A +) and sinks (A −)
and sucrose sources (S +) and sinks (S −), and pathways of main longitudinal
fluxes of auxin (continuous arrows) and sucrose (broken arrows). *Below*, positions
of differentiating tracheary or xylem strands (black), sieve tube or phloem strands
(open) and procambium (cross hatched).

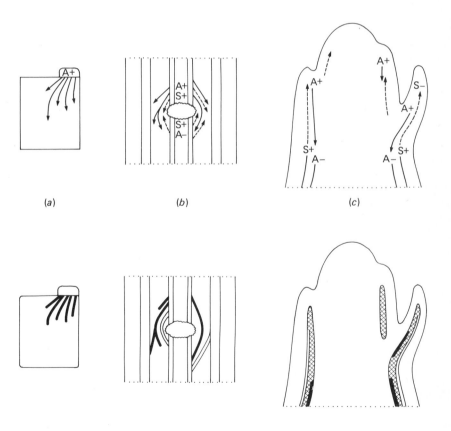

(a)   (b)   (c)

which is an early stage of strand development depends on the availability of both auxin and sugar (especially sucrose). Hormone-directed transport may promote movement of the sucrose towards the auxin source, perhaps by increased invertase activity there. Thus along a pathway joining an auxin source and a sucrose source, there may occur enhanced flow of auxin and sucrose, with consequent increase in concentrations of both substances, canalization, and the promotion of meristematic activity. The canalization may lead to lower auxin levels in the surrounding region which is, or is destined to become, ground tissue. In the stem tip these meristematic or procambial strands will tend to link the leaf primordia (auxin sources) with the acropetally extending ends of differentiating primary phloem a short distance behind the apex (sucrose sources).

The elevated concentrations of auxin and sucrose in meristematic or procambial strands can be expected to show gradients away from the external surface because of the destruction of the auxin at this surface and also – if pre-existing vascularization is present, as behind an apex or a wound – because of homeogenetic induction reflecting concentration gradients in the pre-existing differentiated vascular tissues. When vascular differentiation occurs in the meristematic strand, therefore, sieve tubes or phloem will tend to occur where the auxin/sucrose concentration ratio is lower, on the side of the strand towards the surface, while tracheary elements or xylem develop where the auxin/sucrose ratio is higher, on the side away from the surface. Similarly, it might be expected that phloem would extend from the sucrose source and xylem from the auxin source. This is reflected in the normal apex where the protophloem differentiates acropetally while the protoxylem differentiates in part basipetally from the bases of the leaf primordia.

Vascular differentiation will reinforce the self-organizing properties of the strand. Differentiation of xylem cells will enhance the auxin level in a tracheary strand or in the inner, xylem part of a collateral vascular bundle, while the tendency for phloem to accumulate sucrose will enhance the sucrose level in the sieve tube strand or the outer, phloem part of a vascular bundle. The increased transport efficiency of differentiated tissue will enhance long-distance import of sucrose from sources elsewhere in the plant. In the vascular bundle the region between the phloem and xylem, where concentrations of both auxin and sucrose are fairly high and there is an intermediate auxin/sucrose ratio, will remain meristematic and form the cambium.

Also of relevance to strand formation is the pattern in the sclerenchyma ring which occurs in many species on the outside of the phloem as a bundle cap, or in the pericycle or inner cortex. In the normal stem, in many species, it consists of longitudinal strands more or less closely associated with the phloem, and composed of fibres. Sachs (1972) has shown that removal of a leaf primordium in *Pisum sativum* prevents formation of the fibre strand associated with that leaf. Splitting of an apex has been shown to result in absence of sclerenchyma strands beneath the split surface – though a ring of vascular tissue regenerates – until the apex has restored its radial structure and leaf initiation has recommenced. (Fig. 9.6). However, in many woody species, after apical or internodal wounding, clumps of sclereids regenerate on the radius where fibre strands normally occur; but there is no strand formation or longitudinal elongation, whereas the regenerating phloem and xylem form longitudinal strands. Thus the component of positional control which causes longitudinal continuity is absent in the regenerating sclerenchyma ring.

Periodicity associated with diffusion–reaction systems

The diffusion–reaction system propounded by Turing (1952) was important in drawing attention to the patterning that could result from interaction between diffusible morphogens, but it seems that this system may not be greatly involved in the production of plant tissue patterns. It can give rise to more or less regular periodic patterns, such as commonly appear in the tangential component as seen in transverse section. However, most of these patterns, when considered in three dimensions, are seen to derive from strand formation which has been attributed to flux-dependent facilitation, and not to diffusion–reaction in Turing's sense.

In tissue culture, however, it is not uncommon for tracheary elements to occur in clusters or groups (Earle, 1968; Roberts, 1976) involving a degree of non-randomness, and sometimes the groups or nodules are positioned in a ring and show more or less regularity or periodicity of arrangement similar to that displayed in simulations of diffusion–reaction systems (Wetmore & Rier, 1963; Warren Wilson *et al.*, 1982). It seems likely that such patterns are indeed due to morphogen diffusion and interactions as envisaged by Turing.

A particularly striking case arises when an interrupted sclerenchyma cylinder regenerates as sclereid groups in wounded stems (Warren Wilson *et al.*, 1983). Both in transverse and in longitudinal sections these groups, though not regular, nevertheless display some consistency in the sizes of the sclereid groups and the intervals between them (Fig. 9.3a, b). Diffusion–reaction seems the most likely positional control system.

General aspects of tissue pattern control mechanisms

Wolpert (1970, 1971) has analysed positional control mechanisms into two components. First each cell is assigned positional information in terms of morphogen concentrations or some other positional value. Secondly this positional information, interacting with the cell's developmental history and genome, brings about differentiation appropriate to the position of the cell. This chapter has emphasized the mechanisms providing positional information rather than those interpreting it through appropriate differentiation, but both components are essential to tissue pattern formation.

Indeed, it seems that these two components are in some cases mutually promotive: the positional information leads to patterned differentiation, and the latter exerts a positive feedback effect that reinforces the positional information. The term 'differentiation-dependent information' has been introduced by Sachs (1978) to describe this interaction, which can promote stability and assist in regulation following damage. Two major mechanisms of positional control described above involve differentiation-dependent patterning. In the 'gradient induction' mechanism, the steepness of the gradient in auxin/sucrose concentration can be expected to increase with (a) auxin production by autolysis in differentiating xylem mother cells, (b) increase in sucrose in differentiating phloem as it joins the sucrose-rich phloem transport system of the plant, (c) increased breakdown of auxin by high IAA-oxidase activity in the differentiated phloem, and (d) increased utilization in the actively metabolic cambium and immediate derivatives. In the 'flux-dependent facilitation' mechanism of strand formation the increase in flux is itself a biophysical form of differentiation; the subsequent cell division in preferred planes, as well as the eventual differentiation of conducting tissues, can both be expected to reinforce the flux which is the positional signal. In both mechanisms, differentiation results from patterns in positional information, and in turn it promotes these patterns.

In conclusion, the possible role of different control mechanisms in the development of major vascular tissue patterns can be outlined as follows.

In the short term (a few days), strands develop. In the normal apex these are termed procambial strands and they subsequently differentiate to give primary xylem and phloem. These structures seem to correspond to the tracheary strands and sieve tube strands (sometimes called wound xylem strands and wound phloem strands) that appear within the first week after wounding of stems or roots, as well as in tissue cultures. The pattern of these strands is defined by facilitated-flow canalization of auxin transport, and they tend to extend between sources and sinks of auxin, though strongly influenced by polarity of auxin movement; sucrose sources may also affect their pathway. The strands arise a short distance beneath the external surface. Phloem differentiates on the outside, nearer the surface, and xylem on the inner side; this orientation, and the positioning of the strands a certain distance beneath the surface, are attributed to a morphogenetic gradient in auxin/sucrose concentration, which forms beneath exposed surfaces where auxin is destroyed.

In the longer term (a few weeks) a vascular cambium forms; its depth beneath the external surface is again controlled by the auxin/sucrose gradient so that it forms a layer between any previously differentiated xylem and phloem. Homeogenetic induction, mediated by this same morphogenetic gradient, ensures union of extending cambia. In the normal stem the cambium forms first within the procambial strands and subsequently extends as interfascicular cambium. In large wounds the regenerating cambium may be the first sign of vascular regeneration; it joins up with cut ends of pre-existing cambia. In some callus cultures, the vascular cambium arising from a meristematic nodule may be the first sign of vascular differentiation. In all cases, however, it is essentially a secondary meristem, and it gives rise to derivatives which form phloem (on the outer side) and xylem (inside), this orientation being attributed to the auxin/sucrose gradient.

References

Allen, J. R. F. & Baker, D. A. (1980). Free tryptophane and indole-3-acetic acid levels in the leaves and vascular pathways of *Ricinus communis* L. *Planta*, **148**, 69–74.

Aloni, R. (1980). Role of auxin and sucrose in the differentiation of sieve and tracheary elements in plant tissue cultures. *Planta*, **150**, 255–63.

Aloni, R. & Jacobs, W. P. (1977). The time course of sieve tube and vessel regeneration and their relation to phloem anastomoses in mature internodes of *Coleus*. *American Journal of Botany*, **64**, 615–21.

Andreae, W. A. & Ysselstein, M. W. van (1960). Studies on 3-indoleacetic acid metabolism. VI. 3-indoleacetic acid uptake and metabolism by pea roots and epicotyls. *Plant Physiology*, **35**, 225–32.

Ans, B. (1979). Formation d'une structure de type vasculaire primaire végétale dans un système de réaction-diffusion. *Flora*, **168**, 358–78.

Arnold, W. N. (1968). The selection of sucrose as the translocate of higher plants. *Journal of Theoretical Biology*, **21**, 13–20.

Bacon, J. S. D., MacDonald, I. R. & Knight, A. H. (1965). The development of invertase activity in slices of the root of *Beta vulgaris* L. washed under aseptic conditions. *Biochemical Journal*, **94**, 175–82.

Baker, D. A. (1978). *Transport Phenomena in Plants*. London: Chapman & Hall.

Ball, E. (1952). Morphogenesis of shoots after isolation of the shoot apex of *Lupinus albus*. *American Journal of Botany*, **39**, 167–91.

Bandurski, R. S. (1980). Homeostatic control of concentrations of indole-3-acetic acid. In *Plant Growth Substances 1979*, ed. F. Skoog, pp. 37–49. Berlin, Heidelberg & New York: Springer-Verlag.

Bandurski, R. S. & Schulze, A. (1977). Concentration of indole-3-acetic acid and its derivatives in plants. *Plant Physiology*, **60**, 211–13.

Bannan, M. W. (1955). The vascular cambium and radial growth in *Thuja occidentalis* L. *Canadian Journal of Botany*, **33**, 113–38.

Barlow, P. W. (1976). Towards an understanding of the behaviour of root meristems. *Journal of Theoretical Biology*, **57**, 433–51.

Behnke, H.-D. & Schulz, A. (1980). Fine structure, pattern of division, and course of wound phloem in *Coleus blumei*. *Planta*, **150**, 357–65.

Benayoun, J., Aloni, R. & Sachs, T. (1975). Regeneration around wounds and the control of vascular differentiation. *Annals of Botany*, **39**, 447–54.

Bertrand, C.-E. (1884). Loi des surfaces libres. *Comptes rendus hebdomadaires des Séances de l'Académie des Sciences, Paris*, **98**, 48–51.

Beslow, D. T. & Rier, J. P. (1969). Sucrose concentration and xylem regeneration in *Coleus* internodes *in vitro*. *Plant & Cell Physiology*, **10**, 69–77.

Bieleski, R. L. (1966). Accumulation of phosphate, sulfate and sucrose by excised phloem tissues. *Plant Physiology*, **41**, 447–54.

Bonnemain, J. L. (1971). Transport et distribution des traceurs après application de AIA-2-^{14}C sur les feuilles de *Vicia faba*. *Comptes rendus hebdomadaires des Séances de l'Académie des Sciences, Paris, Série D*, **273**, 1699–702.

Briggs, W. R., Steeves, T. A., Sussex, I. M. & Wetmore, R. H. (1955). A comparison of auxin destruction by tissue extracts and intact tissues of the fern *Osmunda cinnamomea* L. *Plant Physiology*, **30**, 148–55.

Brown, C. L. & Sax, K. (1962). The influence of pressure on the differentiation of secondary tissues. *American Journal of Botany*, **49**, 683–91.

Camus, G. (1949). Recherches sur le rôle des bourgeons dans les phénomènes de morphogénèse. *Revue de Cytologie et de Biologie Végétales*, **11**, 1–200.

Cheng, T. Y. (1972). Induction of indoleacetic acid synthetases in tobacco pith explants. *Plant Physiology*, **50**, 723–7.

Child, C. M. (1941). *Patterns and Problems of Development*. Chicago: University of Chicago Press.

Clowes, F. A. L. (1969). Anatomical aspects of structure and development. In *Root Growth*, ed. W. J. Whittington, pp. 3–17. London: Butterworths.

Cohen, M. H. (1971). Models for the control of development. *Symposia of the Society for Experimental Biology*, **25**, 455–76.

Comer, A. E. (1978). Pattern of cell division and wound vessel member differentiation in *Coleus* pith explants. *Plant Physiology*, **62**, 354–9.

Crick, F. (1970). Diffusion in embryogenesis. *Nature*, **225**, 420–2.

Dalessandro, G. (1973). Interaction of auxin, cytokinin and gibberellin on cell division and xylem differentiation in cultured explants of Jerusalem artichoke. *Plant & Cell Physiology*, **14**, 1167–76.

Dalessandro, G. & Roberts, L. W. (1971). Induction of xylogenesis in pith parenchyma explants of *Lactuca*. *American Journal of Botany*, **58**, 378–85.

de la Fuente, R. K. & Leopold, A. C. (1966). Kinetics of polar auxin transport. *Plant Physiology*, **41**, 1481–4.

Dircks, S. J. (1981). An anatomical study of the origin of sclereids in wounded stems – development and positional relations of sclerenchyma. *Graduate Diploma in Science Thesis, Australian National University*, 11 pp.

Earle, E. D. (1968). Induction of xylem elements in isolated *Coleus* pith. *American Journal of Botany*, **55**, 302–5.

Ede, D. A. (1978). *An Introduction to Developmental Biology*. Glasgow: Blackie.

Edelman, J. & Hall, M. A. (1965). Enzyme formation in higher-plant tissues: development of invertase and ascorbate-oxidase activities in mature storage tissue of *Helianthus tuberosus* L. *Biochemical Journal*, **95**, 403–10.

Fadia, V. P. & Mehta, A. R. (1973). Tissue culture studies on cucurbits: the effect of NAA, sucrose, and kinetin on tracheal differentiation in *Cucumis* tissues cultured *in vitro*. *Phytomorphology*, **23**, 212–15.

Fiedler, H. (1936). Entwicklungs- und reizphysiologische Untersuchungen an Kulturen isolierter Wurzelspitzen. *Zeitschrift für Botanik*, **30**, 385–436.

Forest, J. C. & McCully, M. E. (1971). Histological study on the *in vitro* induction of vascularization in tobacco pith parenchyma. *Canadian Journal of Botany*, **49**, 449–52.

Fosket, D. E. (1980). Hormonal control of morphogenesis in cultured tissues. In *Plant Growth Substances 1979*, ed. F. Skoog, pp. 362–9. Berlin, Heidelberg & New York: Springer-Verlag.

Freundlich, H. F. (1908). Entwicklung und Regeneration von Gefässbündeln in Blattgebilden. *Jahrbucher für wissenschaftliche Botanik*, **46**, 137–206.

Fry, S. C. & Wangermann, E. (1976). Polar transport of auxin through embryos. *The New Phytologist*, **77**, 313–17.

Garrod, D. R. (1973). *Cellular Development*. London: Chapman & Hall.

Gautheret, R. J. (1959). *La Culture des Tissus Végétaux*. Paris: Masson et Cie.

— (1966). Factors affecting differentiation of plant tissues grown *in vitro*. In *Cell Differentiation and Morphogenesis*, pp. 55–95. Amsterdam: North-Holland Publishing Co.

Gersani, M., Lips, S. H. & Sachs, T. (1980). The influence of shoots, roots and hormones on sucrose distribution. *Journal of Experimental Botany*, **31**, 177–84.

Giaquinta, R. T. (1980). Translocation of sucrose and oligosaccharides. In *The Biochemistry of Plants*, vol. 3, *Carbohydrates: Structure and Function*, ed. J. Preiss, pp. 271–319. New York & London: Academic Press.

Goldsmith, M. H. M. (1966). Movement of indoleacetic acid in coleoptiles of *Avena sativa* L. II. Suspension of polarity by total inhibition of the basipetal transport. *Plant Physiology*, **41**, 15–27.

Goodwin, P. B. (1976). Physiological and electrophysiological evidence for intercellular communication in plant symplasts. In *Intercellular Communication in Plants: Studies on Plasmodesmata*, ed. B. E. S. Gunning & A. W. Robards, pp. 121–9. Berlin, Heidelberg & New York: Springer-Verlag.

Grant, P. (1978). *Biology of Developing Systems*. New York: Holt, Rinehart & Winston.

Hardham, A. R. & McCully, M. E. (1982). Reprogramming of cells following wounding in pea (*Pisum sativum* L.) roots. I. Cell division and differentiation of new vascular elements. *Protoplasma*, **112**, 143–51.

276   *J. and P. M. Warren Wilson*

333
Helm, J. (1932). Über die Beeinflussung der Sprossgewebe-differenzierung durch entfernen junger Blattanlagen. *Planta*, **16**, 607–21.

Hertel, R. & Flory R. (1968). Auxin movement in corn coleoptiles. *Planta*, **16**, 123–44.

Jacobs, W. P. (1952). The role of auxin in differentiation of xylem around a wound. *American Journal of Botany*, **39**, 301–9.

— (1979). *Plant Hormones and Plant Development*. Cambridge: Cambridge University Press.

Jacobs, W. P. & McCready, C. C. (1967). Polar transport of growth-regulators in pith and vascular tissues of *Coleus* stems. *American Journal of Botany*, **54**, 1035–40.

Jacobs, W. P. & Morrow, I. B. (1957). A quantitative study of xylem development in the vegetative shoot apex of *Coleus*. *American Journal of Botany*, **44**, 823–42.

Jacobsen, B. S. & Caplin, S. M. (1967). Distribution of an indoleacetic acid-oxidase-inhibitor in the storage root of *Daucus carota*. *Plant Physiology*, **42**, 578–84.

Jaffe, L. F. (1979). Control of development by ionic currents. In *Membrane Transduction Mechanisms*, ed. R. A. Cone & J. E. Dowling, pp. 199–231. New York: Raven Press.

Janse, J. M. (1921). La polarité des cellules cambiennes. *Annales du Jardin Botanique de Buitenzorg*, **31**, 167–80.

Jeffs, R. A. & Northcote, D. H. (1966). Experimental induction of vascular tissue in an undifferentiated plant callus. *Biochemical Journal*, **101**, 146–52.

— (1967). The influence of indol-3yl acetic acid and sugar on the pattern of induced differentiation in plant tissue culture. *Journal of Cell Science*, **2**, 77–88.

Kaan Albest, A. von (1934). Anatomische und physiologische Untersuchungen über die Entstehung von Siebrohrenverbindungen. *Zeitschrift für Botanik*, **27**, 1–94.

Karzel, R. (1924). Untersuchungen über die Regeneration von Sprosspitzen. *Jahrbucher für wissenschaftliche Botanik*, **63**, 111–41.

Kessell, R. H. J. & Carr, A. H. (1972). The effect of dissolved oxygen concentration on growth and differentiation of carrot (*Daucus carota*) tissue. *Journal of Experimental Botany*, **23**, 996–1007.

Kirschner, H. & Sachs, T. (1978). Cytoplasmic reorientation: an early stage of vascular differentiation. *Israel Journal of Botany*, **27**, 131–7.

Kirschner, H., Sachs, T. & Fahn, A. (1971). Secondary xylem reorientation as a special case of vascular tissue differentiation. *Israel Journal of Botany*, **20**, 184–98.

Kny, L. (1877). Über künstliche Verdoppelung des Leitbündel-Kreises im Stamme der Dicotyledonen. *Sitzungsberichte der Gesellschaft naturforschender Freunde zu Berlin*, Sitzung vom 19 Juni, 189–92.

Lancaster, D. M. & Rowan, K. S. (1974). The behaviour of carrot root cambium in axenic culture. In *Mechanisms of Regulation of Plant Growth*, ed. R. L. Bieleski, A. R. Ferguson & M. M. Cresswell, pp. 559–64. Wellington: Royal Society of New Zealand.

LaMotte, C. E. & Jacobs, W. P. (1963). A role of auxin in phloem regeneration in *Coleus* internodes. *Developmental Biology*, **8**, 80–98.

Leopold, A. C. & Hall, O. F. (1966). Mathematical model of polar auxin transport. *Plant Physiology*, **41**, 1476–80.

Leopold, A. C. & Lam, S. L. (1962). The auxin transport gradient. *Physiologia Plantarum*, **15**, 631–8.

Loomis, W. F. (1959). Feedback control of growth and differentiation by carbon dioxide tension and related metabolic variables. In *Cell, Organism and Milieu*, ed. D. Rudnick, pp. 253–94. New York: Ronald Press.

MacWilliams, H. K. & Papageorgiou, S. (1978). A model of gradient interpretation based on morphogen binding. *Journal of Theoretical Biology*, **72**, 385–411.

Meinhardt, H. (1978). Space-dependent cell determination under the control of a morphogen gradient. *Journal of Theoretical Biology*, **74**, 307–21.

Mitchison, G. J. (1980). A model for vein formation in higher plants. *Proceedings of the Royal Society of London*, B, **207**, 79–109.

Morris, D. A. (1977). Transport of exogenous auxin in two-branched dwarf pea seedlings (*Pisum sativum* L.). *Planta*, **136**, 91–6.

— (1982). Hormonal regulation of sink invertase activity: implications for the control of assimilate partitioning. In *Plant Growth Substances 1982*, ed. P. F. Wareing, pp. 659–68. London & New York: Academic Press.

Morris, D. A. & Kadir, G. O. (1972). Pathways of auxin transport in the intact pea seedling (*Pisum sativum* L.). *Planta*, **107**, 171–82.

Mosse, B. & Labern, M. V. (1960). The structure and development of vascular nodules in apple bud-unions. *Annals of Botany*, **24**, 500–7.

Nash, L. J. & Bornman, C. H. (1972). Movement of ^{14}C-indoleacetic acid in *in vitro* cultured *Nicotiana*. *Phytomorphology*, **22**, 69–74.

Neeff, F. (1922). Über polares Wachstum von Pflanzenzellen. *Jahrbucher für wissenschaftliche Botanik*, **61**, 205–83.

Park, D. J. M. (1975). Flux induced biochemical differentiation. *Journal of Theoretical Biology*, **54**, 363–79.

Passam, H. C., Read, S. J. & Rickard, J. E. (1976). Wound repair in yam tubers: physiological processes during repair. *The New Phytologist*, **77**, 325–31.

Peel, A. J. (1974). *Transport of Nutrients in Plants*. London: Butterworths.

Phillips, R. (1980). Cytodifferentiation. *International Review of Cytology*, Supplement **11 A**, 55–70.

Pierik, R. L. M. (1980). Hormonal regulation of secondary abscission in pear pedicels *in vitro*. *Physiologia Plantarum*, **48**, 5–8.

Pilkington, M. (1929). The regeneration of the stem apex. *The New Phytologist*, **28**, 37–52.

Prat, H. (1948). Histo-physiological gradients and plant organogenesis. *Botanical Review*, **14**, 603–43.

Priestley, J. H. (1928). The meristematic tissues of the plant. *Biological Reviews of the Cambridge Philosophical Society*, **3**, 1–20.

Ray, P. M. (1958). Destruction of auxin. *Annual Review of Plant Physiology*, **9**, 81–118.

Rayle, D. L., Ouitrakul, R. & Hertel, R. (1969). Effects of auxins on the auxin transport system in coleoptiles. *Planta*, **87**, 49–53.

Robbertse, J. R. & McCully, M. E. (1979). Regeneration of vascular tissue in wounded pea roots. *Planta*, **145**, 167–73.

Roberts, L. W. (1976). *Cytodifferentiation in Plants. Xylogenesis as a Model System*. Cambridge: Cambridge University Press.

Rubery, P. H. (1980). The mechanism of transmembrane auxin transport and its relation to the chemiosmotic hypothesis of the polar transport of auxin. In *Plant Growth Substances 1979*, ed. F. Skoog, pp. 50–60. Berlin, Heidelberg & New York: Springer-Verlag.

Rubery, P. H. & Sheldrake, A. R. (1974). Carrier-mediated auxin transport. *Planta*, **118**, 101–21.

Russell, C. R. & Morris, D. A. (1982). Invertase activity, soluble carbohydrates and inflorescence development in the tomato (*Lycopersicon esculentum* Mill.). *Annals of Botany*, **49**, 89–98.

Rzimann, G. (1932). Regenerations- und Transplantationsversuche an *Daucus carota*. *Gartenbauwissenschaft*, **6**, 612–36.

Sacher, J. A., Hatch, M. D. & Glasziou, K. T. (1963). Sugar accumulation cycle in sugar cane. III. Physical and metabolic aspects of cycle in immature storage tissues. *Plant Physiology*, **38**, 348–54.

Sachs, T. (1972). The induction of fibre differentiation in peas. *Annals of Botany*, **36**, 189–97.

— (1975). The induction of transport channels by auxin. *Planta*, **127**, 201–6.

— (1978). Patterned differentiation in plants. *Differentiation*, **11**, 65–73.

— (1981). The control of the patterned differentiation of vascular tissues. *Advances in Botanical Research*, **9**, 151–262.

Sachs, T. & Cohen, D. (1982). Circular vessels and the control of vascular differentiation in plants. *Differentiation*, **21**, 22–6.

Savidge, R. A. & Wareing, P. F. (1981). Plant-growth regulators and the differentiation of vascular elements. In *Xylem Cell Development*, ed. J. R. Barnett, pp. 192–235. Tunbridge Wells: Castle House Publications.

Sevenster, P. & Karstens, W. K. H. (1955). Observations on the proliferation of stem pith parenchyma *in vitro*. II. The internal structure of stem pith cylinders of *Helianthus tuberosus* L. cultivated *in vitro*. *Acta Botanica Neerlandica*, **4**, 188–92.

Sheldrake, A. R. (1971). Auxin in the cambium and its differentiating derivatives. *Journal of Experimental Botany*, **22**, 735–40.

— (1973). Auxin transport in secondary tissues. *Journal of Experimental Botany*, **2**, 87–96.

Sheldrake, A. R. & Northcote, D. H. (1968). The production of auxin by tobacco internode tissues. *The New Phytologist*, **67**, 1–13.

Shoji, K. & Addicott, F. T. (1954). Auxin physiology in bean leaf stalks. *Plant Physiology*, **29**, 377–82.

Simon, S. (1908). Experimentelle Untersuchungen über die Entstehung von Gefässverbindungen. *Berichte der deutschen botanischen Gesellschaft*, **26**, 364–96.

Sinnott, E. W. (1960). *Plant Morphogenesis*. New York, Toronto & London: McGraw-Hill.

Sinnott, E. W. & Bloch, R. (1945). The cytoplasmic basis of intercellular patterns in vascular differentiation. *American Journal of Botany*, **32**, 151–6.

Smith, R. H. & Murashige, T. (1982). Primordial leaf and phytohormone effects on excised shoot apical meristems of *Coleus blumei* Benth. *American Journal of Botany*, **69**, 1334–9.

Snow, R. (1942). On the causes of regeneration after longitudinal splits. *The New Phytologist*, **41**, 101–7.

Sorokin, C. (1962). Inhibition of cell division by carbon dioxide. *Nature*, **194**, 496–7.

Steeves, T. A. & Sussex, I. M. (1972). *Patterns in Plant Development*. Englewood Cliffs: Prentice-Hall.

Steward, F. C., Mapes, M. O. & Mears, K. (1958). Growth and organized development of cultured cells. II. Organization in cultures grown from freely suspended cells. *American Journal of Botany*, **45**, 705–8.

Stoutemyer, V. T. & Britt, O. K. (1965). The behavior of tissue cultures from English and Algerian ivy in different growth phases. *American Journal of Botany*, **52**, 805–10.

Sussex, I. M. (1955). Morphogenesis in *Solanum tuberosum* L.: experimental investigation of leaf dorsiventrality and orientation in the juvenile shoot. *Phytomorphology*, **5**, 286–300.

Sussex, I. M., Clutter, M. E. & Goldsmith, M. H. M. (1972). Wound recovery by pith cell redifferentiation: structural changes. *American Journal of Botany*, **59**, 797–804.

Thimann, K. V. (1969). The auxins. In *The Physiology of Plant Growth and Development*, ed. M. B. Wilkins, pp. 3–45. London: McGraw-Hill.

Thompson, N. P. (1967). The time course of sieve tube and xylem cell regeneration and their anatomical orientation in *Coleus* stems. *American Journal of Botany*, **54**, 588–95.

— (1970). The transport of auxin and regeneration of xylem in okra and pea stems. *American Journal of Botany*, **57**, 390–3.

Thompson, N. P. & Jacobs, W. P. (1966). Polarity of IAA effect on sieve tube and xylem regeneration in *Coleus* and tomato stems. *Plant Physiology*, **41**, 673–82.

Thornley, J. H. M. (1981). Organogenesis. In *Mathematics and Plant Physiology*, ed. D. A. Rose & D. A. Charles-Edwards, pp. 49–65. London & New York: Academic Press.

Thornley, J. H. M., Warren Wilson, J. & Colley, E. (1980). The initiation and maintenance of polarity: mathematical aspects of a proposed control system. *Annals of Botany*, **46**, 713–17.

Thorpe, T. A. & Meier, D. D. (1973). Sucrose metabolism during tobacco callus growth. *Phytochemistry*, **12**, 443–7.

Torrey, J. G., Fosket, D. E. & Hepler, P. K. (1971). Xylem formation: a paradigm of cytodifferentiation in higher plants. *American Scientist*, **59**, 338–52.

Turing, A. M. (1952). The chemical basis of morphogenesis. *Philosophical Transactions of the Royal Society of London*, B, **237**, 37–72.

Van Fleet, D. S. (1959). Analysis of the histochemical localization of peroxidase related to the differentiation of plant tissues. *Canadian Journal of Botany*, **37**, 449–58.

Vöchting, H. (1892). *Über Transplantation am Pflanzenkörper*. Tübingen: Verlag H. Laupp'schen Buchhandlung.

Waddington, C. H. (1966). Fields and gradients. In *Major Problems of Developmental Biology*, ed. M. Locke, pp. 105–24. New York: Academic Press.

Walker, A. J., Ho, L. C. & Baker, D. A. (1978). Carbon translocation in the tomato: pathways of carbon metabolism in the fruit. *Annals of Botany*, **42**, 901–9.

Walker, A. J. & Thornley, J. H. M. (1977). The tomato fruit: import, growth, respiration and carbon metabolism at different fruit sizes and temperatures. *Annals of Botany*, **41**, 977–85.

Wardlaw, C. W. (1950). The comparative investigation of apices of vascular plants by experimental methods. *Philosophical Transactions of the Royal Society of London*, B, **234**, 583–60.

— (1952). *Phylogeny and Morphogenesis*. London: Macmillan.

— (1965). *Organization and Evolution in Plants*. London: Longmans.

Wareing, P. F. (1978). Determination of plant development. *Botanical Magazine, (Tokyo)*, Special Issue 1, 3–17.

Wareing, P. F. & Phillips, I. D. J. (1981). *Growth and Differentiation in Plants*. Oxford & New York: Pergamon.

Warren Wilson, J. (1978). The position of regenerating cambia: auxin/sucrose ratio and the gradient induction hypothesis. *Proceedings of the Royal Society of London*, B, **203**, 153–76.

— (1980). A control system for initiating and maintaining polarity. *Annals of Botany*, **46**, 701–11.

Warren Wilson, J. & P. M. (1961). The position of regenerating cambia – a new hypothesis. *The New Phytologist*, **60**, 63–73.

— (1981). The position of cambia regenerating in grafts between stems and abnormally-oriented petioles. *Annals of Botany*, **47**, 473–84.

Warren Wilson, J., Dircks, S. J. & Grange, R. I. (1983). Regeneration of sclerenchyma in wounded dicotyledon stems. *Annals of Botany*, **52**, 295–303.

Warren Wilson, J., Roberts, L. W., Gresshoff, P. M. & Dircks, S. J. (1982). Tracheary element differentiation induced in isolated cylinders of lettuce pith: a bipolar gradient technique. *Annals of Botany*, **50**, 605–14.

Warren Wilson, P. M. & J. (1963). Cambial regeneration in approach grafts between petioles and stems. *Australian Journal of Biological Sciences*, **16**, 6–18.

Wetmore, R. H. & Rier, J. P. (1963). Experimental induction of vascular tissues in callus of angiosperms. *American Journal of Botany*, **50**, 418–30.

Whiting, A. G. & Murray, M. A. (1948). Abscission and other responses induced by 2,3,5-triiodobenzoic acid in bean plants. *Botanical Gazette*, **109**, 447–73.

Wilbur, F. H. & Riopel, J. L. (1971). The role of cell interaction in the growth and differentiation of *Pelargonium hortorum* cells *in vitro*. II. Cell interaction and differentiation. *Botanical Gazette*, **132**, 193–202.

Wolpert, L. (1970). Positional information and pattern formation. In *Towards a Theoretical Biology*, vol. 3, *Drafts*, ed. C. H. Waddington, pp. 198–230. Edinburgh: Edinburgh University Press.

— (1971). Positional information and pattern formation. *Current Topics in Developmental Biology*, **6**, 183–224.

Young, B. S. (1954). The effects of leaf primordia on differentiation in the stem. *The New Phytologist*, **53**, 445–60.

Zamski, E. & Wareing, P. F. (1974). Vertical and radial movement of auxin in young sycamore plants. *The New Phytologist*, **73**, 61–9.

Zenk, M. H. & Müller, G. (1964). Über den Einfluss der Wundflächen auf die enzymatische Oxydation der Indol-3-essigsaure *in vivo*. *Planta*, **61**, 346–51.

Ziegler, H. (1975). Nature of transported substances. In *Transport in Plants. I. Phloem Transport*, ed. M. H. Zimmermann & J. A. Milburn, pp. 59–100. *Encyclopedia of Plant Physiology*, New Series, vol. 1, ed. A. Pirson & M. H. Zimmermann. Berlin, Heidelberg & New York: Springer-Verlag.

10

Positional controls in root development

P. W. BARLOW

The structure of plant organs, such as the root and shoot, is so uniform that one may suspect that some powerful organizing influence is at work whose action transcends taxonomic boundaries. This uniformity is also expressed at the cellular level. Epidermal cells, for example, which by definition constitute the outermost layer of an organ, have a rather similar appearance irrespective of whether they are on leaf or stem, root or fruit. Likewise, vascular tissue is located in the centre of roots and stems, and is always composed of at least three types of cell – phloem, xylem and parenchyma. The different tissues and their constituent cell-types thus have characteristic positions within the plant body – positions which may be defined in terms of their distance from the surface of the organ and their distance from each other.

Undoubtedly genes determine the chemical composition of any given cell and this in turn confers identity to the tissue of which the cells are a part. But it seems less likely that genes alone directly determine the pattern of tissue development. Epigenetic influences are probably involved and cause cell differentiation to adopt a particular spatial pattern. A significant epigenetic influence could be the position that a cell occupies in a growing organ. If so, then the positional specifications are likely to be of a physico-mechanical and chemical nature. I shall attempt to show that one plant organ – the root – uses such position-related properties to regulate the spatial aspect of cell differentiation.

Development of the root and its control systems

The growing root is cylindrical, joins the stem at its basal end, and tapers to a point at its apical end. The root tip consists of two rather discrete groups of cells: a cap at the extreme apex, and the root proper which the cap surmounts. These two groups are distinct on account of their origin, histology, and function. In many species there is no obvious boundary between the root and cap; such roots are said to have an 'open' construction. In other species, notably grasses, there is such a boundary; these roots are said to have a 'closed' construction.

Both the cap and the root proper may be regarded as being composed of columns of cells. The wall of each column is made primarily of pectin and cellulose and grows longitudinally with the interpolation of new wall materials. Although the cell-columns may branch, it is possible to trace their origin to a stem-cell, or founder, region. In roots with an open construction the cell-columns of cap and root proper share a common stem-cell zone. On the other hand, roots with a closed construction have separate stem-cell zones for cap and root proper. The stem-cell zone of the root proper in the closed constructional types corresponds to a region known as the quiescent centre in which the rates of cell division and metabolism are slower than in other meristematic zones. This correspondence is not so exact in open roots.

The cell-columns making up the root are organized in both the radial and longitudinal plane into distinct tissues. The outermost ring of columns is the epidermis. This encloses the cortex which in turn encloses the endodermis and the stele, the latter forming the core of the root. In each of these tissues cell maturation occurs in a longitudinal direction. However, the position of these tissues within the body of the root, and the way in which their constituent cells arise from the stem-cell zones, speaks for a control over their differentiation operating in the transverse plane for otherwise we cannot account for the presence of different tissues across the radius of the root. Our attempt to analyse cell development in the growing root will therefore be in terms of longitudinal and transverse (i.e., radial and circumferential) positional controls.

Certain aspects of root development are also subject to positional controls that operate within the body of the plant as a whole, but apart from the following few remarks no more will be said on this type of positional control. One example of this higher level of control is that lateral roots developing at different locations on the main root may differ in their gravitropic response (Dyanat-Nejad & Neville, 1972). The positional controls that influence the orientation of branch roots with respect to gravity are poorly understood but obviously have an important effect upon the structuring of the root system in the soil (Barlow, 1983). Moreover, roots with different gravitropic behaviour can differ not only in their location on the plant but also in their structure. For example, in *Ludwigia peploides*, a shrub of marshland, upward-growing adventitious roots develop from primordia that form in the dorsal side of a node of a horizontal shoot, while downward-growing roots develop from the ventral side (Ell-more, 1981). Before emergence the primordia of each type of root are indistinguishable, but after emergence the upward-growing root becomes anatomically distinct, particularly with respect to the organization of its cortex and cap, from the downward-growing root. A further example is that cell differentiation within the root is influenced by the latter's position within the environment. In roots of *Monstera deliciosa*, for instance, the frequency of trichosclereids increases, and that of rhaphides decreases, as the roots pass from an aerial to a subterranean habit (Hinchee, 1981). Similarly, aerial roots of *Cissus* sp. may grow for three metres or more without branching, but branch as soon as their tips make contact with the soil (Zimmerman & Hitchcock, 1935).

Predominantly longitudinal controls of differentiation

Meristematic activity

The cell-columns of the root elongate with the simultaneous interpolation of new protoplasm and wall material. One end of a column may be imagined as being anchored in the stem-cell zone. Growth of the column has the effect of displacing cells away from the stem-cell zone along the axis of the root. New cells are created in the column by the periodic insertion of cross-walls that divide the existing cells into two daughters. The zone where cells both divide and elongate is the meristem and is located in the distal portion of the root proper and in the proximal portion of the cap. Beyond the meristem, cells elongate only (elongation zone) and eventually cease growing (mature zone). It is as though these three zones – meristem, elongation, mature – and the changing properties of the cells within them, are defined by a local internal environment that varies along the root axis. The inherent capacity of cells to grow results in their displacement from one local environment to another, wherein their properties change accordingly. This view requires that the internal environment is steadily maintained by the root apex and is unaffected by growth. Coupled with this is the supposition that cell maturation is in part governed by the unfolding of a developmental program which includes a progressive increase in sensitivity to the internal environment and hence an increasing capacity for position-related differentiation.

A single substance, varying in activity along the root axis, could hardly specify the differentiation of each particular tissue (epidermis, cortex, etc.) unless there were some additional informational input in a radial direction. On the other hand, a positional signal could operate longitudinally along the root axis to specify properties such as growth and division that are common to all cells.

The cell division cycle and its accompanying nuclear events are generally held to be regulated by two classes of hormones, auxin (e.g., indoleacetic acid) and cytokinin. It is still not known what these hormones actually do in terms of physiological or molecular events, or whether they are the only hormones involved in the cell cycle. They may have a role that is more than simply rendering cells competent to divide and that they actually induce separate and specific division-directing functions. So far, there is no satisfactory agreement on these points (Fosket & Tepfer, 1978; Everett, Wang, Gould & Street, 1981; Wang, Everett, Gould & Street, 1981). Auxin is made in the root apex, but is also transported there from the shoot, and accumulates in the cap. Cytokinin seems to be synthesized in the root apex and transported basipetally. The quiescent centre (QC) and surrounding cells have been suggested as a site of cytokinin synthesis (Phillips & Torrey, 1971; Short & Torrey, 1972). This idea was taken up by Barlow (1976) who proposed that there are two streams, one of cytokinin the other of auxin, flowing along the root which could govern the observed pattern of cell growth, division and DNA synthesis in the apex and hence define both the location and the length of the meristem (Fig. 10.1). If cells in and around the QC are a site of cytokinin synthesis, they could be the source of a gradient, given that cells further away metabolize or otherwise attenuate the activity of this substance. Auxin, on the other hand, because of its proximally-located source, could show a gradient attenuated towards the tip. Thus, the QC would be a region where the cytokinin/auxin ratio is high and this might account for its slow rate of metabolic processes, including division itself; we may suppose that here cytokinin is present in supra-optimal amounts. Away from the QC the cytokinin/auxin ratio falls to levels that permit faster rates of cell growth and division. At a certain distance the cytokinin level falls below a threshold necessary to trigger mitosis (see Fig. 10.1). By contrast, auxin is ubiquitous and permits DNA synthesis to occur even if it is not followed by mitosis. The cytokinin/auxin ratio along the root axis therefore defines nuclear activity: the meristem is defined by a ratio that causes (or permits) the coupling of DNA synthesis, mitosis and cytokinesis, but at another ratio the coupling of these events lapses resulting in endoreduplication of DNA. (Endoreduplication may be under an additional genetic or nucleotypic control since only certain species, particularly those with a relatively low nuclear DNA content, exhibit this process.) Eventually, DNA synthesis also fails to be initiated (see Fig. 10.1) and the nuclei may even degenerate.

A gradient of cytokinin along the root axis may arise in two ways. In one, the cytokinin may diffuse away, or enter a transport stream that draws it away, from its site of synthesis. A gradient arises because of metabolism or transport. Such a scheme would require quite a high rate of cytokinin synthesis to satisfy the constant demands of division and to maintain a meristem that may be 1–2 mm long. This might appear incompatible with the apparently slow rate of metabolism in the QC, its proposed source, though Lerman (1978) has pointed to the advantage of quiescence for certain key activities. A second possibility that may not require as much synthetic activity is one in which the QC is again the site of cytokinin synthesis, but here each cell accumulates cytokinin until it is displaced by division from the QC. Further growth and divisions by these displaced cells and their progeny occur with the utilization of the original capital of cytokinin. When this is exhausted (diluted or metabolized) division can no longer occur. Thus, the longitudinal extent of the meristem is limited, in the first instance, by the distance over which cytokinin can diffuse; in the second instance the limitation is the cytokinin content of a cell at the time it is displaced from the zone in which cytokinin is synthesized.

The longitudinal extent of the meristem varies within the root (Barlow, 1976). In *Zea mays*, for example, cells in the epidermis and pericycle

Fig. 10.1 Postulated ratio of concentrations of auxin and cytokinin along the root axis. It is proposed that the cytokinin level is high in the quiescent centre (QC) and declines on either side. Above a certain value, P, all events of the mitotic cycle (cell growth, DNA synthesis, mitosis and cytokinesis) occur, but below this value mitosis and cytokinesis cannot occur. Below a lower value P', all cycling activity ceases. (After Barlow, 1976.)

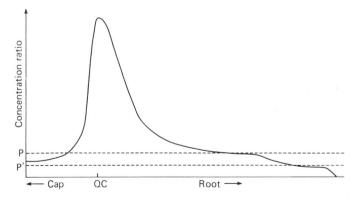

divide for a longer period (accomplishing about eight cycles of division) and over a longer distance than cells in the cortex or, more notably, those in the xylem lineages which may complete only 3–4 division cycles. Cells of the cap may complete even fewer (1–2) cycles. With a gradient model of divisional control one must postulate that this variation occurs because the longitudinal distribution of the controlling factors differs in the various tissue-related cell-columns.

Another element in the regulation of meristematic activity may reside in the balance between the rate of cell growth and the rate at which a nucleus can prepare for mitosis. Mitotic readiness in many dividing cell systems seems to depend upon the attainment of a critical cell size or cytoplasmic mass (Prescott, 1976). If the increase in cytoplasmic mass fails to keep pace with the increase in cell volume, as seems to be the case in the proximal zone of the root (e.g., Barlow, Rost & Gunning, 1982), then this system for triggering mitosis will not be activated and the probability of division will decline in parallel with the changing mass/volume relationship. Such a control may be determined by changes in the capacity to synthesize cytoplasm, in the development of the vacuome, and in the extensibility of the cell wall (Barlow, 1983; Barlow *et al.*, 1982; Lloyd & Barlow, 1982), all of which may be integrated with gradients of growth regulators and changes in the sensitivity of cells towards them.

Constructional events

Roots have a very precise manner of construction which is quite distinct in each species. This distinctiveness arises from the way in which the planes of division are regulated, particularly in and around the stem-cell zone. In the root meristems of higher plants, and in apices of those ferns which have a single apical cell (see Fig. 10.2), the planes of division in and around the stem-cell zone are periclinal and longitudinal-radial. These 'formative' divisions create initials for all the cell-columns of the root (apart from those generated later by cambial activity). Beyond this zone of formative divisions cells divide transversely to the root axis and increase the number of cells within each column (Fig. 10.2). As mentioned earlier, the number of these 'proliferative' divisions varies according to the location of the cell-column within the root.

In the water-fern, *Azolla pinnata*, formative divisions occur in a definite spatial and temporal sequence (Gunning, Hughes & Hardham, 1978) (Fig. 10.2). Indeed, so precisely regulated are these divisions that it is possible to predict where a new cell wall will form at cytokinesis with a high degree of accuracy – perhaps to less than one micrometre. Formative divisions in roots of higher plants are no less exactly placed as judged by the overall pattern they create, but probably have a more flexible positional and temporal control. No root has been analysed in the same degree of detail as that of *Azolla*, yet in the root of *Z. mays*, which is much larger though probably not very much more complex, the formative divisions that increase the number of cell-columns in the cortex exhibit some regularity. For example, the periclinal division that initiates a new cell-column at the inside of the cortex occurs in approximately 50% of the cases observed in a cell located 6–7 cells away from the most distal cell of the QC. Then, after further periclinal divisions have progressed basipetally and acropetally through the same column of inner cortical cells, the sequence begins again (Barlow, unpublished).

The manner in which the planes of division are regulated is fundamental to organogenesis (Sinnott, 1943) and is a subject covered by others in this volume (see chapters by Bopp, Korn, and Lintilhac). One possibility is that the orientation and placement of the new cell wall is a response to a pattern of mechanical stresses in the mother cell wall and its underlying cytoplasm. These stresses also interact with some element that positions the nucleus at a point in the cell where it will undergo mitosis; this element may be able to sense a shear-free plane in the cytoplasm (Miller, 1980; Lloyd & Barlow, 1982). Thus, the planes of division in various locations in the root relate to a corresponding pattern of stresses. How these stresses might be set up is one of the most intriguing problems of morphogenesis and requires more attention. Growth regulators may be involved if they exert a control over the rate of cell growth, but they would have to do this in three planes simultaneously; their flow through cells and tissues is often strongly polarized in one plane (along the root axis), though diffusion in the radial and circumferential directions may influence growth in these planes. The direction of flow of growth regulators may also directly influence the orientation of cytoplasmic elements including microtubules and hence influence the direction of cell growth and the orientation of the mitotic apparatus. Another possibility is that the chemical composition, or the structure, of the cell wall modifies the way in which it will grow. For example, the different arrangement of cellulose microfibrils in transverse and longitudinal walls may determine the growth of their respective surfaces. With regard to chemical composition, Basile (1980) has found that morphogenesis of shoots of leafy liverworts (Jungermanniales) can be modified by the inclusion of hydroxyproline in their culture medium. He proposes that this substance, when incorporated into the cell walls of the shoot can regulate the resulting form of the apex

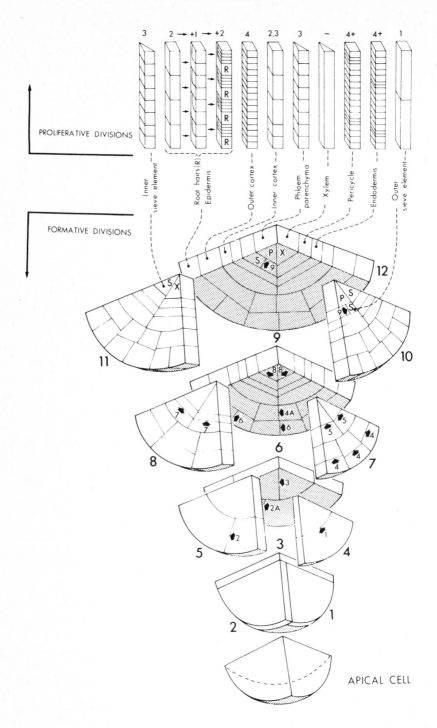

and its derivative tissues. Apical form can also be modified by other compounds, such as ammonium ions (Basile, 1980) that influence the conversion of proline to hydroxyproline. Extrapolating these effects to roots raises the possibility that different morphogenetic patterns have as their basis a spatial pattern of differences in the metabolism and structuring of cell-wall materials.

The stability of the form of the root apex might suggest that its geometry regulates its behaviour (i.e., form begets form). This stability and the predictability with which divisions occur in time and space within the apex (cf. Fig. 10.2), also suggests that it should be possible to formulate 'rules' by which divisions are specified to construct the root. So far no simple algorithm has been devised for root construction. If it was understood how cellular geometry interacts with the biochemical controls of division such an algorithm might be devised (cf. Lindenmayer, 1981). However, the notion of stable form is apparently contradicted by the observation that disturbances to the geometry of the root, such as cuts made into the apex, or removal of parts, are quickly rectified (see discussion on pp. 311–13). Moreover, root apices of certain species gradually reorganize their cellular pattern during growth (Armstrong & Heimsch, 1976). One aspect of root geometry that may contribute to the regulation of division is the interaction between the cap and the QC (Barlow & Hines, 1982). The results of certain experiments indicate that in *Z. mays* the cap physically restrains growth of the cells in the QC. Since the rate of cell growth is related to the duration of the cell cycle, this restraint has the effect of slowing the cycle in the QC. Thus, geometry and attendant physical interactions are demonstrated to play a regulatory role in development at the root apex.

Fig. 10.2. Diagrammatic representation of the production of three files of merophytes by sequential divisions of the apical cell in a root of *Azolla pinnata*. The formative divisions that give rise to the population of initials for each of the cell-types occur in meropytes 4–12; the sequence in which the divisions take place is indicated by a number (1–9) and an arrow pointing to the newly inserted wall within the merophyte. Note that epidermis (dermatogen) and outer cortex arise from a common mother cell at division 4 in merophyte 7, and inner cortex and endodermis arise from a common mother cell at division 7 in merophyte 8. The final outcome of the proliferative (transverse) divisions for each cell-type is shown at the top of the figure. The number above each column of cells refers to the number of transverse divisions per merophyte that occur in each cell-type. (Diagram modified from Gunning, 1982.)

Differentiation events in continuously produced cell-types

All roots consist of the same tissues arranged in basically the same way and it is generally assumed that a given type of tissue has a similar function in whatever species it is found. The cells within a tissue may have different modes of origin yet all develop in the same way. For example, the cortex of the root of *Azolla* consists of only two concentric layers of cells (Gunning *et al.*, 1978); both groups of cells share a similar internal structure and are thus probably identical in function, yet each has a different origin in the meristem (Fig. 10.2). Therefore, differentiation of structure and function within the tissue unit seems to be dominated more by its position within the root as a whole than by any influence handed on by the cells' ancestry.

Given the basic homology of tissues designated epidermis, cortex, etc., in what way do they differ from each other, and what is the nature of the positional control that initiates these differences? An investigation of cell differentiation in young roots of *Azolla* has gone some way to answering these questions and has also given some clue as to how the longitudinal control of differentiation may operate. The dermatogen (loosely equivalent to epidermis) and the outer cortex are examples of two groups of cells that differentiate in quite specific ways but which arise from a common ancestral cell (Fig. 10.2; Barlow *et al.*, 1982; Whatley & Gunning, 1981). The cortex is characterized by an extensive vacuole compartment (vacuome) and moderate numbers of chloroplasts with numerous thylakoid bands, while the dermatogen has a less-developed vacuome and many more, somewhat simpler, chloroplasts. The development of the vacuome is unmistakable in the two tissues and its inception can be recognized in the newest-formed cells of the cortex. A similar divergence of vacuome development occurs in inner cortex and endodermis which also arise from a common mother cell (Fig. 10.2). The vacuome in the inner cortex is similar to that of the outer cortex; the vacuome in both the dermatogen and endodermis is relatively less well developed. Likewise, the number of plastids per cell diverges rapidly upon the creation of dermatogen and outer cortical lineages; a similar divergence is also found between the inner cortex and endodermis. Thus vacuome and the plastid complement of adjacent columns of inner and outer cortex increase in a similar manner in spite of the different origin of the cells. It is possible that the cytoplasm is asymmetrically positioned in the common mother cell of dermatogen and outer cortex, and of inner cortex and endodermis, though no definite evidence for this was found (Barlow *et al.*, 1982). But if this is so, then the whole pattern of vacuome and plastid development could be determined by the partitioning of this asymmetric cytoplasm into the two newest recruits of each cell-column in each tissue. This would seem rather a hazardous system to maintain as a mechanism determining differentiation, though not an impossible one since the apical cell consistently retains a large vacuome at division and partitions less-vacuolated cytoplasm into the daughter cell (Barlow *et al.*, 1982).

The early onset of vacuome development in the cortex, the smooth pattern of its increase, and the less-vacuolate character of cells in sibling files of endodermis and dermatogen, suggest another principle of differentiation – that of homeogenetic induction. This is where a cell already embarked on a particular path of differentiation can induce a neighbouring cell to follow the same path (Lang, 1974). Cortical cell differentiation, which is characterized by extensive vacuome development, may be evidence of such a process. The onset of differentiation in cells most recently added to the cortical cell-columns may be the result of their having been penetrated by an inductive influence that causes a particular pattern of vacuome development, and perhaps of plastid growth also. Homeogenetic induction may be a general principle operating in tissue differentiation, particularly where the cells are arranged in columns. The necessary cell-to-cell communication is by means of plasmodesmata. In differentiating cells these are more frequent across the transverse walls of cells within the same column than across the longitudinal walls that link adjacent columns (for the frequency of plasmodesmata and details of their origin in roots of *Azolla*, see Gunning (1978)). An induction process also suggests that the message which passes from cell to cell is specific to the type of cell that is differentiating and carrying the message.

The inductive influence, and/or the cells' response to it, is probably under genetic control. In young differentiating regions, where proliferative divisions are still occurring, the inductive influence programs the nucleus so that the developmental pathway it directs is identical to that already established in the older cells of the same column; this is a 'feed-forward' process. In the zone of formative divisions the nuclei are not programmed so and retain a pluripotential state; they are concerned with cytoplasmic syntheses that are not directed towards the construction of any particular tissue. Thus, there must be some fundamental distinction between cells in the proliferative zone and cells in the formative stem-cell zone. Moreover, if there were not, the inductive influence might penetrate cells in the formative zone (i.e., the mother cells of the various differentiating cell-columns and their respective antecedents) and cause them to differentiate also. That the two groups of meristematic cells are distinct is shown by the finding that under conditions which inhibit cell division, cells in the zone of proliferative divisions can be caused to differentiate to completion, yet the stem-cell zone with its formative divisions does not differentiate but retains its proliferative capacity and can repopulate the meristem when conditions favourable to division return (e.g., Wilcox, 1954).

The view of longitudinal differentiation presented so far is one in which cells are irrevocably locked into a particular pathway of differentiation as a consequence of homeogenetic induction operating along the cell columns that make up the root. These various pathways probably originate during the early development of the root (irrespective of whether it arises in the embryo, or as a lateral or adventitious primordium) and are perpetuated by its continued growth. The QC, and the formative zone of ferns, continually feed cells into the homeogenetically differentiating system. Once a cell enters into a particular column of cells the feed-forward induction system causes it to differentiate to completion.

Differentiation events in discontinuously produced cell-types

Two examples will be presented to illustrate the situation where certain cells in a tissue undergo an optional step in differentiation and where the stimulus to do so is related to the cells' position. One concerns the differentiation of trichosc!ereids in the cortex of certain roots, the second concerns the development of an epidermal cell into a hair. In both cases differentiation is apparently linked to an asymmetric proliferative cell division in the proximal region of the meristem.

Trichosc! ereids are hair-like cells; they are developed in the cortex of aerial roots (as in *Monstera*, for example) and grow up and down the intercellular spaces (Fig. 10.3). They may reach lengths up to 20 times that of sibling parenchymatous cortical cells. Bloch (1946) has shown that the trichosc1ereid initial is the smaller of two daughter cells that arise at one of the later divisions in the cortex. An asymmetric division is important for trichosc1ereid development; if the division of a cortical cell is not asymmetric then no trichosc1ereid initial arises. Proximity of the initial to an intercellular air-space seems to be the trigger for the initial subsequently to extend as a hair (Fig. 10.3a).

In their illustrations, Sinnott & Bloch (1946, Figs. 1, 3) show that the newly-formed trichosc1ereid initial cells are spaced regularly along the length of the cell columns in which they lie and are separated by three intervening parenchyma cells. In the same root the products of other asymmetric divisions, the epidermal hair cells and short hypodermal cells, have only one intervening cell of the other type (i.e., non-hairs and long hypodermal cell, respectively). If all these asymmetric divisions are the final ones to occur in a particular cell-column, then this observation suggests that there are differences in the number of cell divisions that elapse between the commitment of a cell to an asymmetric division and its occurrence. In the case of the trichosclereid initials two divisions elapse, and in the case of the epidermal hair and the short hypodermal cell, one division. Commitment may consist of some sort of cytoplasmic polarization, since in many instances the small daughter has the more densely-staining cytoplasm. Thus, differences between the time of commitment and the onset of differentiation can lead to patterned development. The pattern may grow if the products of the differential division continue to divide and maintain their differentiated state. It is in this way that files of hairs, interspersed by files of hairless cells, develop on the epidermis of roots of *Ceratopteris* (Tschermak-Woess & Hasitschka, 1953), and columns of crystal idioblasts arise within the cortex of *Cissus* roots (Bloch, 1947).

In the second example, the epidermal hair, there is apparently no general, trans-species relationship between the size of the daughter cells that result from an asymmetric division and their subsequent course of differ-

Fig. 10.3. Trichosclereids (internal hairs) in an aerial root of *Monstera deliciosa*. (a) The small initial cell (I) extends (arrowed) into an intercellular space (S). (b) Later in development the nucleus (N) enters the thread of the hair. (c) Transverse section showing the hairs (there may be up to five) within the intercellular spaces. Scale bars correspond to 10 μm.

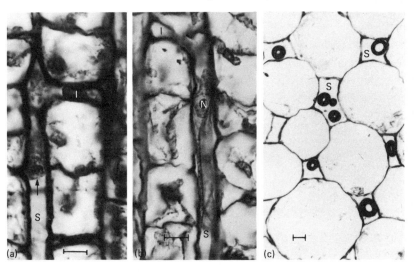

entiation: it may be either the larger or the smaller daughter that forms the hair, and this incipient hair cell may be either the proximal or distal daughter. It is probably incorrect to regard the hair as being positively induced from an essentially hairless epidermis. Rather, as experimental evidence suggests, all epidermal cells have the potentiality to form hairs. Evidence for this view comes from work with X-rays at doses which block division but not cell growth. By so treating roots of grasses (including *Z. mays*), Ivanov & Filippenko (1976) and Filippenko (1980) found that all cells of the epidermis formed hairs except those situated under the cap. In the course of their development epidermal cells seem to acquire a hair-forming zone at one end. This polarization occurs in the proximal portion of the meristem (see experiments of Filippenko (1980) who found that environmental conditions conducive to hair development act only on cells in the proximal region of meristem). Any division that occurs in the polarized cell simply causes the hair-forming potential to be confined to one of the two daughters. Division *per se* is irrelevant to hair differentiation.

The point along the outer wall at which the protuberance of the hair will emerge is opposite the cell nucleus. This was first shown by Haberlandt (1887) and later confirmed by others (e.g., Cormack, 1945). In fact, Haberlandt writes of the importance of nuclear position for differentiation of hairs and other types of cells. However, an interesting recent finding of Nakazawa & Yamazaki (1982) suggests that hair formation is independent of nuclear position. They find that the site of hair emergence is controlled by a stable (i.e., not displaced by centrifugation) zone of cortical cytoplasm or cell membrane. A detailed investigation of this system is clearly warranted. All these observations seem to suggest that a cytoskeleton (possibly involving microtubules) is a fundamental component in the regulation of certain morphogenetic processes. The relationship between the cytoskeleton and morphogenesis involves positional controls that operate at the molecular level within the cell. Ultrastructural studies that illustrate the importance of microtubule behaviour in root development have been reviewed by Gunning (1982).

Predominantly lateral controls in differentiation

Epidermis-cortex interactions

The outer cell layers of roots afford examples of lateral interactions between cells that result in a patterning of cell differentiation. Such examples are illustrative of 'heterogenetic' induction since the interaction that leads to differentiation is between cells of different tissues. One example is the differentiation of epidermal hairs on roots of white mustard, *Sinapis alba* (Bünning, 1951). Hair development is influenced by the type of contact that an epidermal cell makes with an underlying cortical cell. If an epidermal cell lies over a radial cortical cell wall it develops as a hair, but if it lies over a tangential wall it does not (Fig. 10.4a). An intercellular space is present between the radial cortical wall and the epidermis, but no such space lies between the tangential walls of epidermis and cortex. It was suggested by Bünning (1951) that this space, or the lapse of cell contact caused by the space between epidermis and cortex, is important in hair

Fig. 10.4. Hair and non-hair cells in the epidermis of roots of *Sinapis alba*. (a) Hair cell initials (arrowed) may be recognized by their densely-staining cytoplasm. They also lie over the radial walls of underlying cortical cells. Hairless cells, which are more vacuolate and hence relatively less stained, lie over the outer tangential walls of the cortical cells. (b) When the epidermis is cut away from the underlying cortex many more cells in the epidermis appear to develop as hair cell initials (arrowed). In this example the cut was made one day before the tissue was fixed. Scale bars correspond to 50 μm.

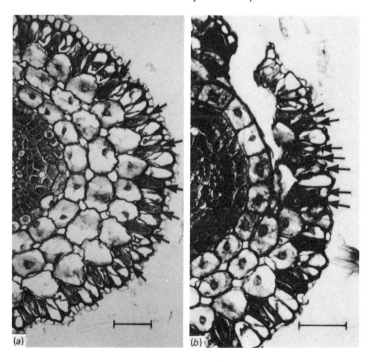

differentiation since cutting the epidermis free from the cortex causes many more of the epidermal cells to show the characteristics of hair initials (Fig. 10.4b).

Another example of heterogenetic induction, again involving hair cells, is found in *Potamogeton natans* (Tschermak-Woess & Hasitschka, 1953). Here the hair cells arise following an asymmetric division. After the hair cell initials have been established, cells of the outer cortex (hypodermis) that underlie the hairs divide, often asymmetrically, to give long and short cells; hypodermal cells underlying the non-hair epidermal cells do not divide.

These two examples provide evidence of lateral (radial) interactions between two contiguous cell layers. The result is a pattern of differentiation in one layer that reflects the pattern in the other layer. The nature of the inductive stimulus in both examples is unknown. In the case of *Sinapis* it is possible (as suggested by Bünning (1951)) that some inhibitor of hair development passes across the common tangential wall between epidermis and cortex, but no such passage occurs where cell contact is reduced by the presence of an intercellular space. This would accord with the idea that non-hair cells are restricted in their development. In *Potamogeton* the cytoplasmic asymmetry in the hair/non-hair pair of epidermal cells may influence cytoplasmic organization in the underlying hypodermal cells. A number of observations show that deposition of wall material in a cell can be influenced by the pattern of wall or cytoplasmic material in a neighbouring cell – e.g., the off-setting of transverse cell walls in meristematic tissue (Sinnott & Bloch, 1941), the coincidence of bands of secondary thickening in adjacent xylem elements (Sinnott & Bloch, 1945), and the matching-up of Casparian bands in contiguous endodermal cells.

Interactions in the stele

Complex developmental patterns in which radial, circumferential and longitudinal controls all operate are found in the stele. Here, xylem and phloem are symmetrically positioned within, and differentiate from, a matrix of parenchymatous cells. Homeogenetic induction may be partly responsible for the perpetuation of this pattern of differentiation. Evidence for this longitudinal control will be discussed first. The proposition is that a maturing xylem or phloem cell induces a similar pathway of development in younger cells of the same column. Intuitively this would seem to be a reasonable mechanism for creating the unbroken column of cells necessary for the effective functioning of a vascular system. Moreover, the vascular pattern is generally extremely constant in roots sampled over many days, or even years, of their growth (Torrey & Wallace, 1975). An alternative mechanism would be the continual elaboration of vascular elements within the undifferentiated meristematic procambial cylinder and the fitting together of their pattern with that which has already been developed. The distinction, by experimental means, between these two modes of differentiation of a stable vascular system (i.e., one that is induced by maturing tissue, the other a result of autonomous differentiation within the meristem) has been attempted by various workers. Early results, reviewed by Esau (1954), did not lead to any particularly convincing conclusions. However, Torrey (1955) in addressing this problem made use of nutrient agar medium in which to grow excised root tips of *Pisum sativum* and to observe changes in their vascular pattern. Roots of this species have a constant, symmetrical triarch vasculature. When tips which had been excised just distal to the point at which the vascular pattern was clearly established were grown in culture the majority of roots continued to show a triarch pattern. A small number of roots showed a reduction in vasculature, and became diarch or monarch. These reduced patterns arose as the result of the death of certain cells in one or two of the developing xylem arms. These observations are consistent with homeogenetic induction of the vasculature. However, as growth of the root continued, some of the monarch and diarch roots reverted to the triarch condition. This came about by initiation of new xylem strands which had no connection with pre-existing strands. Here is clear evidence that xylem elements can differentiate autonomously within the procambial cylinder. In later experiments using cultured roots from which the tip had been removed, Torrey (1957) found that a medium containing indoleacetic acid at 10^{-5} M caused the vascular system to alter to a hexarch condition; three new strands of xylem and phloem appeared *de novo* between the pre-existing strands. Reversion to the usual triarch condition occurred when the roots were transferred to auxin-free medium. An additional observation made in this work (see also Torrey 1965)) was that the degree of vascular complexity was related to the area of the procambial cylinder at the level at which the vascular system was blocked-out (Fig. 10.5). Thus, although vascular elements can arise autonomously, they only do so if the dimensions of the procambial cylinder permit this. Likewise, vascular strands may die out, presumably because the dimensions of the procambial cylinder reduce to one that cannot support extra elements. (There may be an upper limit of procambial area that can support a single vascular system. In palms, numerous independent vascular systems can co-exist within a root (Drabble, 1904); these systems may arise because the size of the procambial cylinder exceeds that capable of maintaining a single unified system.) None of the results refutes the idea that acropetal differentiation of the vascular elements is due to an inductive process, though it is fairly clear that the procambial cylinder, perhaps through some spatial limitation, plays a critical role in permitting this to occur.

The procambial cylinder is manifestly a site where differentiation of a number of cell-types is actively taking place. There may also be interactions among the processes that lead to these various pathways of differentiation. In the apices studied by Torrey this is evident from the position that new xylem strands adopt in relation to pre-existing strands (as in the conversion of a triarch to a hexarch vasculature mentioned earlier). In addition, the alterations of vascular pattern involve one symmetrical pattern converting to another. A change in symmetry was also noted by Barlow (1977a) in a study of vascular patterns in pea roots recovering from treatment with colchicine. Occasionally, asymmetric patterns arose (some by the spontaneous initiation of new xylem strands), but these eventually converted to symmetric patterns (Fig. 10.6). Such observations permit the inference that circumferential interactions determine the pattern of the vasculature. Furthermore, in *Zea*, with a polyarch vasculature, Feldman (1977) found that not all xylem elements are initiated at the same level in the root. After the establishment of the first four elements

Fig. 10.5. Range of dimensions of the procambial cylinder at the level of inception of the vascular pattern in relation to the complexity of the vascular system. The relationship is in terms of (a) procambial cylinder diameter (μm), and (b) the number of cells across a diameter. The number of roots analysed is indicated above each data point. Dimensions are from cultured root tips of *Pisum sativum* derived from the distal 0.5 mm of primary roots. (Data from Torrey, 1965.)

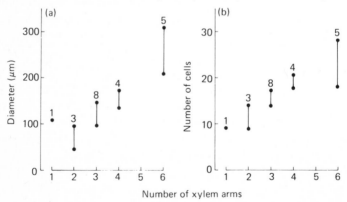

additional element arises between two pre-existing elements only when they are separated by a certain minimum circumferential distance. These observations suggest that there is some sort of limitation of sites where new xylem elements can be initiated within the procambial cylinder.

The way in which the vascular pattern is blocked out and differentiates illustrates the action of radial influences. For example, in *Pisum* the first indication of xylem development is the enlargement of cells towards the inside of the procambial cylinder and a subsequent centrifugal enlargement of cells in each of the three arms. The later stages of development, however, are centripetal with cells at the pole of the arm, the protoxylem, maturing first. The cells that enlarged first mature later as metaxylem, followed by the differentiation (as late metaxylem) of the most central cells. Such a radial sequence of differentiation along the xylem arms may have some element of homeogenetic induction. Thus, in the procambium, longitudinal influences may predominantly specify what *types* of cells will differentiate, while transverse influences may specify what *pattern* they will adopt.

If the procambial cylinder is a site of action and interaction of stimuli for differentiation, can anything more definite be said about what these factors are and how they might operate? Auxin and sucrose when supplied to cultures of callus stimulate the formation of nodules containing

Fig. 10.6. Experimentally-induced changes in vascular (xylem) patterns in roots of *Pisum sativum*. The roots were treated with colchicine and allowed to regenerate (Barlow, 1977*a*). (a) A new xylem lineage is initiated (arrow) in a previously diarch root. (b) A hexarch pattern in which the xylem arms are asymmetrically placed within the stele; later in root development (c) this pattern converted to a symmetrical pentarch pattern. Scale bar corresponds to 100 μm.

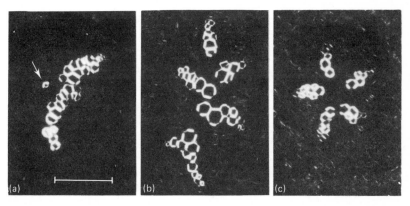

xylem, phloem, and sometimes cambium (Wetmore & Rier, 1963; Jeffs & Northcote, 1966; Aloni, 1980). Working with callus of *Syringa*, Wetmore & Rier (1963) believed that a high auxin/sugar ratio favoured xylem, while a low ratio favoured phloem. This conclusion could not be confirmed by Aloni (1980). Instead, he found that low levels of auxin resulted in the differentiation of phloem alone and he therefore proposed that auxin was the sole determinant of this cell type. In contrast, xylem differentiation required higher auxin levels. Because xylem occurred only if phloem was also present, Aloni suggested that a factor from the phloem, as well as auxin, was necessary for xylem differentiation. Since sugars are transported down mature phloem strands into the root, and developing xylem is believed to be a source of auxin (Sheldrake & Northcote, 1968), it seems possible that these two types of cell can induce their own kind by virtue of the substances they contain. These substances may be distributed within the procambial cylinder along concentration gradients (and probably amongst cells of graduated sensitivity) which interact in both the longitudinal and transverse planes. This would determine (should other conditions – e.g., procambial size – allow) the spacing of the vascular elements and also the site where additional elements will form, or where supernumerary elements will be terminated. Similar interactions may determine the site of cambium formation and of secondary xylem and phloem differentiation. Computer-aided modelling may be one approach to forming an understanding of these tissue patterns (see the chapter by H. Meinhardt). For example, the impressive diversity of stelar anatomy found in stems of Palaeozoic gymnosperms (see Fig. 4 of Sporne, 1965) may be rendered explicable by such an approach.

No one has yet satisfactorily explained why the vascular patterns of root and shoot are different, though the production of leaf primordia by the apical meristem of shoots probably has an influence on the shoot's vascular pattern. Likewise, the patterns in the transition zone, where root and shoot join, are poorly known. The root and shoot patterns merge in this zone and the pattern of differentiation here may depend on the interaction of different sets of inductive influences that originate in the root and shoot poles of the embryo.

The origin of cell patterns within the stele raises the question of how the patterns of the larger cell-groups of stele, cortex, endodermis, and epidermis arise. These tissues originate not only in roots arising in the embryo but also in root primordia that arise on pre-existing roots and stems – a rather remarkable phenomenon that speaks for a highly canalized developmental pathway. Lang (1974) suggests that the development of polarity in the early embryo is the result of heterogenetic induction emanating from maternal tissue. Once this polarity is set up other possibilities for heterogenetic interactions arise within the cluster of embryonic cells and distinct tissues begin to differentiate. Differentiation within the embryo is a topic of great interest to biologists and has been explored extensively in animal systems. Here, many heterogenetic inductive processes operate following gastrulation and cause the inception of the major tissues. In earlier developmental stages, however, simple concepts such as 'position as a morphogenetic influence' (that is, whether a cell is internal or external in the morula or blastula) have been entertained (Hillman, Sherman & Graham, 1972). There is also the segregation, by cell division, of cell surface constituents (Ziomek & Johnson, 1982) that may be important in regulating the position of cells within the morula and their subsequent response to inductive stimuli. The same principles may operate in plant embryogenesis: outside/inside positions may correspond to points along gradients of nutrients and solutes which can act as developmental switches, and differential plasmalemma properties may also be segregated at the early divisions. These divisions have a regular pattern (Souèges, 1937; Mestre, 1967) and consequently could effect such a segregating system. Once a particular tissue has been differentiated, its properties remain stable and are propagated during meristematic growth. This would correspond to homeogenetic induction, i.e., it is the reproduction of an already differentiated cytoplasm and nucleus (see also Bloch, 1947, 1948). The other term we have employed, 'heterogenetic induction', may have its material basis in cytoplasmic asymmetry and differential responses to evocators of differentiation.

Longitudinal and lateral controls of differentiation

Lateral root development

Lateral root primordia commonly originate from cells in the pericycle, although the endodermis is sometimes also involved. In ferns, laterals arise mostly from endodermal cells. The site at which primordia are initiated is related to the pattern of the vascular system. In many species the primordia arise opposite the protoxylem poles, though in polyarch roots, such as found in monocotyledonous species, they arise opposite the phloem. In diarch roots, however, the primordia arise between phloem and xylem. It is relatively easy to understand the apparent importance of vascular strands in determining the site of primordium development if the strands transport, and leak radially, trophic factors (such as sugars and vitamins) and mitogens (such as auxins and cytokinins). The relative amounts of these substances required for primordium initiation, and the amounts that can diffuse from the vascular strands into the pericycle, may determine the exact location of the primordium with respect to the vascular pattern.

Although a positional control upon primordium initiation is imposed by the vascular system, the primordia themselves interact to determine their relative spacing. These interactions operate in a longitudinal direction, and possibly transversely also. In species where primordia arise in ranks overlying a vascular strand the spatial interactions can be theoretically categorized as follows: longitudinal spacing within a rank; longitudinal spacing irrespective of rank; and circumferential spacing including the sense (handedness) in which nearest longitudinal neighbours arise. Analyses of these possible interactions have been performed by W. A. Charlton, most recently using roots of banana, *Musa acuminata*, where ranks of primordia are initiated opposite protoxylem poles (Charlton, 1982). His findings with this species serve to illustrate the principles that underlie the patterning of laterals though their physiological bases remain to be established. Within a rank the distribution of laterals departed somewhat from a statistically normal distribution. There was a deficiency of short spacings which could mean that, in addition to primordium initiation being a regular process in both time and space, a pre-existing primordium actively prevents initiation of a new primordium up to a certain distance along the same rank. A developing primordium may act as a sink for nutrients and thus prevents another primordium from developing too near it. Laterals also tended to be clustered irrespective of rank; this has been noted in some other species like *Cucurbita maxima* (Mallory, Chiang, Cutter & Gifford, 1970). Clustering may mean that there is no inhibition of primordium development across ranks; in fact the reverse would be the case if it is due to an interaction between primordia at the time they are initiated. On the other hand, clustering across the ranks may be due to the coincidence, or coming into register, of the spacing pattern along each of the individual ranks. Moreover, it is possible that changes in the physiology of the apex influence circumferential spacing since it is known that the apex exerts a 'dominance' effect whereby primordia are actively prevented from forming within a certain distance from the apex. Some of the departures from a normal distribution of laterals within a rank might also arise from periodic changes in dominance, the influence of which would operate in a longitudinal direction. Not enough is known about the controls of primordium initiation to choose among these possibilities.

The role of the primary root apex on the early growth of lateral roots has recently been re-investigated in pea by Wightman, Schneider & Thimann (1980). Their results provide evidence of a long-range control over the sequence of primordium development. These authors found that there were separate physiological controls regulating primordium formation and the subsequent emergence of the primordium through the cortex. One of the controlling factors favoured by Wightman *et al.* is cytokinin, four species of which were found in pea roots. The cytokinins had a high concentration in the apex but this declined towards the base of the root (cf. Short & Torrey, 1972). Primordium initiation was found to be less sensitive to inhibition by added cytokinin than lateral emergence. Thus, different concentration thresholds along the primary root axis were inferred to regulate the induction of these two aspects of lateral root development.

Occasionally the pattern of lateral formation is disturbed by genetic or environmental factors. Proteoid roots, where the spacing between laterals is much reduced (Purnell, 1960), may be an example of a changed endogenous physiological control over primordium initiation such as the lessening of the inhibitory influence that one primordium exerts on another in the same rank. These roots are induced by an as yet uncharacterized metabolite of microbes in the root environment (Malajczuk & Bowen, 1974), though in *Lupinus albus* proteoid roots can develop in aseptic conditions (Gardner, Barber & Parbery, 1982) thus giving evidence that the proteoid morphology is an endogenous trait whose threshold of expression is modified by the presence of microbes. Nishimura & Maeda (1982) found that root segments of *Oryza sativa* placed on a medium containing the growth regulator, 2,4-dichlorophenoxyacetic acid (2,4-D) showed alterations in the positioning and polarity of the lateral root primordia. Normally (i.e., in the absence of 2,4-D) the primordia are always initiated opposite a phloem element, but many of those formed in the presence of 2,4-D are not. Similarly, the axis of cell enlargement in the primordia is altered by the 2,4-D causing them to become callus-like structures. However, the tips of the callus-like primordia retain an ultrastructure characteristic of cells in such a position; these are future cap cells and, as we shall see, their properties may arise in response to their position at the surface of the apex. Thus, although the 2,4-D may modify (possibly by swamping) the gradient of morphogens involved in lateral formation, it does not modify certain of its limits: initiation still occurs in the pericycle and the distal limit of the primordium still terminates in a cap.

The orderly arrangement of lateral roots in the water-fern, *Ceratopteris thalictroides*, illustrates the operation of another type of factor in lateral development. In this species, as in *Azolla* (which, incidentally, does not form lateral roots) each proximal descendent (merophyte) of the apical cell undergoes a number of formative and then proliferative divisions, the products of which remain together within a merophyte (cf. Fig. 10.2). A particular cell of a merophyte in the endodermal file can become the initial for a lateral root (Chiang, 1970). This indicates that the primordium is specified by the preceding pattern of cell divisions. Superimposed on this is a relationship between further development of this primordial initial and the vascular pattern. The vasculature is diarch, yet the root is made of three 120° merophyte sectors; only two of the three merophyte sectors in each gyre of merophytes initiate a primordium. Therefore, although the cell that is the potential primordium may be determined by its heredity, its further development is dependent upon some influence from the vascular system. Moreover, the stimulation of primordium development operates in advance of any overt differentiation of vascular strands in the neighbouring stelar tissue (Chiang, 1970; Mallory *et al.*, 1970). Both the pattern of lateral primordia and the vascular system are examples of a bilaterally symmetrical system arising within a radially symmetrical meristem. They are clearly examples of positional controls operating at two different levels of root organization – the genetic and the spatial.

Diaphragms and aerenchyma

Another interesting, though little-studied, example of co-ordination of cell behaviour in the radial and longitudinal planes is evident in the development of diaphragms in the root cortex of many aquatic species (e.g., *Thalassia testudinum* – Tomlinson, 1969) (the development of diaphragm cells is also discussed by R. W. Korn in chapter 2). In *Thalassia*, diaphragms appear at regular intervals (about one millimetre apart) along the length of the mature root and are recognized as bands of small cells (2–4 cells deep) traversing the cortex which is otherwise composed of elongated cells (Fig. 10.7). They probably serve a structural function and may also transport solutes across the radius of the root. How the alternation of groups of long and short cells arises along the root axis is not known. Some species within the Eriocaulaceae, however, show short cells randomly distributed in the cortex (Tomlinson, personal communication). Thus, the aggregation of short cells as diaphragms (as in *Thalassia*) might reflect the development, during the phylogeny of aquatic plants, of some ordering principle in the transverse plane that co-ordinates the production, in the longitudinal plane, of short cells so that a cell-group of functional significance is produced.

Diaphragms occur within the air-conducting tissue, or aerenchyma, of roots of certain species. The aerenchyma itself can arise through the lysis of groups of cortical cells in response to a hormonal signal. Not all cortical cells lyse, however, and there may be some mechanism that allows

Fig. 10.7. Sections through a root of *Thalassia testudinum* to show diaphragms in the cortex. (a) Longitudinal section of a root apex showing the small cells of the diaphragm (D) traversing the mid-cortex (C). (b) Transverse section of a mature region of a root; the section is from a zone lying between diaphragms. Air spaces (S) run the length of the mid-cortex and are bounded by uniseriate radial plates of elongated cells. These plates fork or fuse. (c) Transverse section through a diaphragm in a mature region of a root. The diaphragm (D) completely occupies the cross-section of the mid-cortex. Its cells stain densely and the radial walls are thickened. The scale bars represent 250 μm. (Photographs printed from negatives kindly loaned by Professor P. B. Tomlinson, Cabot Foundation, Harvard Forest, Mass., USA.)

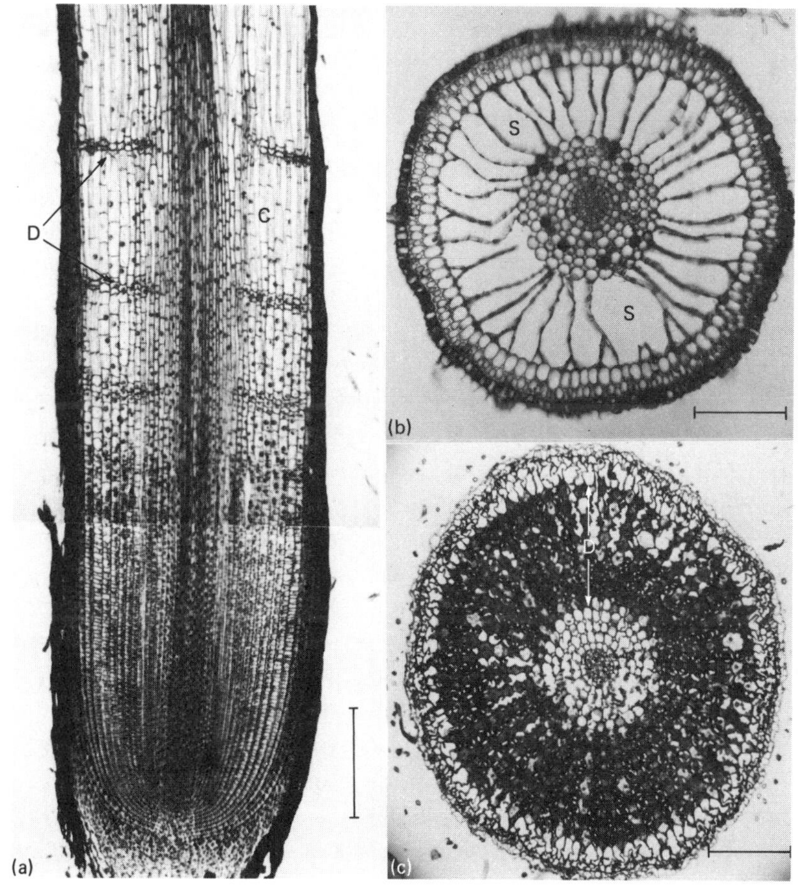

lysis to proceed in a controlled way so that a spatially-structured system of air channels results. Konings & Verschuren (1980) state that in roots of maize and willow the cells that do not undergo lysis are located opposite the xylem poles.

An experimental approach to positional controls

The examples presented in the previous section indicate that the position of a cell within a growing and differentiating zone of the root apex influences its future development. The same general idea that cell function is related to position was held by Hermann Vöchting over 100 years ago (Vöchting, 1877). In this section some experimental evidence will be presented that supports the relationship between cell position and function.

Changing a cell's position relative to the whole

The root cap is an organ with which it is possible to investigate positional effects upon cell function. In *Zea mays* the cap has well-defined zones of different cellular activities (Barlow, 1975) and is also large enough to make surgical intervention into its development practicable.

The cap continually renews itself from a meristematic stem-cell zone, yet the properties of cells at any point within it remain the same. Cells are constantly moving into different positions within the cap and acquire properties characteristic of that position: the meristem occupies the 3–4 most proximal cell-tiers; in the centre of the cap the plastids develop large starch grains (amyloplasts); at the outside of the cap the starch disappears and a mucilage forms in the dictyosome vesicles which discharge through the plasmalemma to the outside of the cell. The outer cap cells separate from each other and by their detachment from the cap expose formerly internal cells which in turn acquire the properties of outer cells. These observations are in keeping with the hypothesis that some aspects of cell differentiation are the result of a cell's position along an inside/outside transect.

One way of testing this hypothesis of positionally controlled differentiation (and thus modestly retreading the path taken by Vöchting in arriving at his conclusion) is to change a cell's position relative to the whole and then see whether it acquires properties characteristic of its new position. When a longitudinal cut is made into the cap and one half removed, cells of the cap interior now lie at the outside. A property of outer-cap that can be easily monitored is slime synthesis. Criteria by which this can be judged

are electron microscopy and a rapid uptake and metabolism of carbohydrate precursors (e.g., ^{3}H-glucose or ^{3}H-fucose) that can be recorded by
autoradiography. Shortly after the cut is made the new outside cells acquire a slime-synthesizing capacity (Fig. 10.8). The same is true of cells of
the QC exposed by removal of the root cap (Barlow, 1974; Barlow &

Fig. 10.8. Change in incorporation of D-^{3}H-glucose (a marker of slime synthesis)
by inner cap cells of *Zea mays* as a consequence of causing them to lie at the
surface of the cap. Caps were cut longitudinally and one half removed. Some roots
were labelled for two hours in ^{3}H-glucose (10 μCi/ml) immediately after the
operation (0 day) and fixed. Others were grown on in damp sphagnum for up to 3
days before being similarly labelled and fixed. (a) Mean grain densities ($\pm$ s.e.)
over cells at the cut surface ($\bullet$) are recorded as a percentage of the mean density
over cells at the natural outer surface of the cap (taken as 100% – broken line)
sampled 0–3 days after removing half the cap. Also shown is the mean grain
density ($\blacktriangle$) over inner cells of intact caps sampled over the same period. These
cells are the ones from which the inner-converted-to-outer cells are derived and
their grain density should therefore be compared with that of the inner-converted-
to-outer cells. (b) Autoradiograph of a bisected cap fed with ^{3}H-glucose on the
third day of regeneration. The blackened areas at the outside of the cap (O,
natural outside; I, inside-converted-to-outside cells) represent sites of radioactive
slime synthesis. Internal cells (asterisk) take up relatively little of the glucose. Scale
bar corresponds to 100 μm.

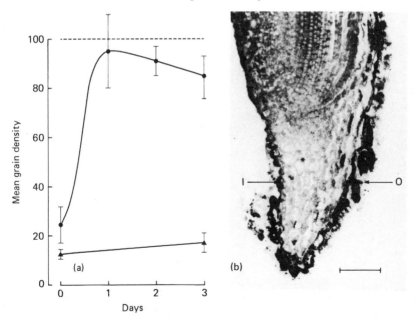

Sargent, 1978): here the normally inactive dictyosomes of the QC become filled with slime, and in this respect look like those of outer cap cells. A single longitudinal cut through the cap into the root also reveals that the resulting two half-caps regulate their internal structure since each of them differentiates as a small, but complete cap of normal appearance (Fig. 10.9). The cells at the newly-exposed surface lose their starch grains while other formerly starch grain-containing cells seem to adjust their starch content so that the new cap acquires the characteristic starch grain distribution. The pattern of differentiation in these half-caps seems to transcend the original tissue boundary since the pattern extends across the former cap/QC junction (arrowed in Fig. 10.9).

Fig. 10.9. An apex of *Zea mays* was bisected and fixed after three days of further growth. The starch grains (shown as black deposits in the photograph) are found in the cells in the centre of the tip about equidistant from each edge. The edge at the left-hand side is formed from a portion of the root meristem, while the edge at the right-hand side is formed from a portion of the original cap. Thus, within this apical tissue there has been an adjustment of the pattern of differentiation that leads to starch deposition. The pattern appears to be regulated by reference to the outer surfaces; there is no respect for the original root/cap junction (arrowed). Scale bar corresponds to 100 μm.

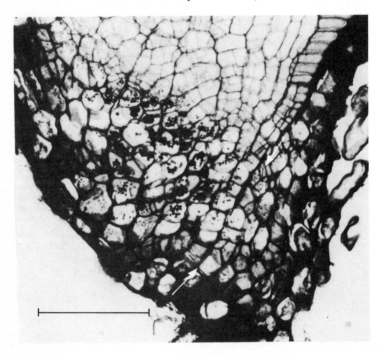

It could be argued that such cytological changes are not dependent upon position but would inevitably occur in time. For example, changes in the structure and function of the cytoplasm in a cap cell occur according to a sequence determined by the time that elapses following that cell's exit from the stem-cell zone. That this is not the case can be inferred from a number of observations. For instance, the time of onset of slime synthesis in cells at the cut surface of a bisected cap (e.g., Fig. 10.8) seems to be similar irrespective of whether they were former central, meristematic, or even QC cells; normally, each of these cells would take a different time to reach the outer cell position in the cap. Moreover, when the cap is prevented from making new cells by growing the root in a solution of colchicine for up to four days, cells are held in positions that are constant with respect to each other. Under such circumstances electron microscopy (Barlow & Sargent, unpublished) shows that meristematic cells (those distal to the stem-cell tier) do not acquire the properties of central cells, and central cells do not become like outer cells, even though they would normally have undergone such a transformation of position and properties in the time that elapses. On the other hand, nuclear DNA synthesis does continue in the presence of colchicine, and over a four-day period nuclei in the various regions of the cap acquire DNA contents uncharacteristic of those regions (Barlow, 1977*b*). Therefore, one may conclude that nuclear growth is determined according to a time-scale and not according to position, while cytoplasmic changes are position-dependent.

Similar positional controls appear to operate in the root itself. Some have been mentioned in connection with hair development (pp. 295–6). Certain properties of the epidermal layer may also be a consequence of its position at the outside of the root. Like the cap, it secretes a polysaccharide mucilage to the cell exterior, some of which forms a thick skin over the outer wall (Fig. 10.10a). If epidermis is stripped from the root, the underlying cortical cells in time acquire an appearance identical to the removed cell layer (Fig. 10.10b). When the experiment is done with *Sinapis*, hair and non-hair cells differentiate from former cortical cells and acquire a characteristic pattern (Fig. 10.10c, and compare with Fig. 10.4a). Bloch (1944) obtained comparable results in aerial roots of *Monstera* and *Philodendron*. In these roots a layer of brachysclereids is normally present beneath the epidermis. When the outer surface of the root is damaged new brachysclereids differentiate from the cortex just beneath the exposed surface. This reaction also occurs when the wound is inside the cortex, indicating that a surface is the primary determinant of sclereid differentiation, and that whether the surface is inward- or outward-facing is of little importance. These simple examples illustrate that additional developmental potentialities are normally latent in differentiated cells and may be evoked by changes in cellular position relative to the tissue as a whole. Further examples are described by J. & P. M. Warren Wilson in Chapter 9.

There are many similarities between the regenerative processes at the root surface described above and the repair of damage to the surfaces of other organs or organisms. For example, leaves (Foard, 1959), potato tubers (Borchert & McChesney, 1973), the thallus of algae (Cabioch & Giraud, 1978; Dreher, Grant & Wetherbee, 1978), and even the epidermis of mammals (Christophers, 1972) can all redifferentiate the cell-type characteristic of the surface from previously internal cells. Since regeneration of the surface is such a widespread phenomenon it may indicate that a common principle underlies the determination of cellular properties at a surface.

The converse of the experiments mentioned above is to force outer-type

Fig. 10.10. Regeneration of the epidermis of roots of *Zea mays* and *Sinapis alba*. (a) The epidermis has been peeled from a portion of the root (within bracket) of *Z. mays* and allowed to regenerate for one day. Features of normal intact epidermis (outside the bracket) are cells with a darkly-stained cytoplasm containing starch grains and with a thick, densely-stained outer cell wall. Some divisions have occurred in cells of the outer cortex from which the epidermis has been peeled. (b) Three days after epidermal peeling, a new epidermis has regenerated (within the bracket). (c) Two days after peeling epidermis from a root of *S. alba* a new epidermis has been regenerated from former cortical cells (in the zone within the bracket). Hair initials (arrowed) and non-hair cells have developed in characteristic positions (compare with other cells outside the bracket, and also Fig. 10.4a). Scale bars represent 50 μm.

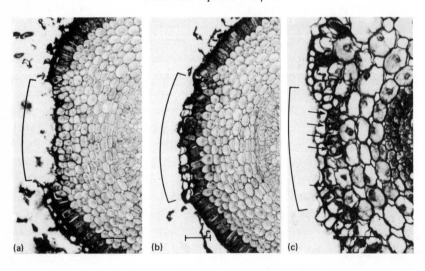

(a)   (b)   (c)

cells together and see if this converts them to inner-type cells. This often occurs in Nature with certain ontogenetic fusion phenomena (Cusick, 1966). A crucial component in the outer-to-inner cell conversion is cell-to-cell contact. For example, epidermal cells of the primordial carpel of *Catharanthus roseus* will dedifferentiate (losing both their cuticle and their dense cytoplasmic structure), divide, and redifferentiate into stigmatic transmitting tissue only if the cells of the two opposing surfaces actually come into contact. If the epidermal cells of the two surfaces are prevented from touching each other by means of a gold leaflet (which may be only 20 μm thick) inserted into the fissure between the two carpel primordia, they retain their epidermal state (Walker, 1978). Cusick (op. cit.), in his review of ontogenetic fusion, found difficulty in accepting that the developmental and morphogenetic events involved could have a purely genetic basis; as he suggests, the responses of the cells involved are more readily interpreted in terms of epigenetic processes. It may be easily imagined that in all these systems the overall differentiation of superficial and internal tissues is maintained by a positional control which, when the arrangement of the cells is disturbed, tends to impose on the remaining cells and restore the *status quo ante*. This positional control presumably uses some property of the external boundary of the tissue as a 'reference' against which the appropriate differentiation processes are regulated.

   In the above-mentioned examples it is the phenotype of the cell that is influenced by the inside/outside position. This is perhaps to be expected since all the cell-types mentioned are end-points of a developmental sequence. However, outside/inside positions may be interpreted and translated into more complex results if surgical interventions are made at an early stage of organ ontogeny. For example, a cut made at the centre of a very young capitulum of sunflower leads to ray florets and bracts (i.e., whole organs) being developed around the edge of the cut in just the same way as they would develop at the natural edge of the capitulum (Palmer & Marc, 1982). In roots the dedifferentiation (ontogenetic regression) of a segment by natural ageing, or by culture *in vitro*, can increase the developmental possibilities open to a primordium formed within it: given an appropriate stimulus, root, shoot and floral meristems can form from such primordia (reviewed by Peterson, 1975).

Regeneration

Regeneration of plant parts can be induced experimentally and serves as a system for observing the processes of development from another viewpoint. This is because the removal of a tissue or organ introduces a discontinuity into its normal developmental sequence; regeneration then recapitulates this sequence starting from a developmentally earlier stage.

The regeneration of the root cap of Z. *mays* has been described by Barlow (1974), Barlow & Sargent (1978) and Barlow & Hines (1982), and the process has been interpreted in terms of positionally-controlled cell differentiation (Barlow & Sargent, 1978; Barlow, 1981). The cap can be cleanly removed from the root apex to expose cells of the QC. Here, cell growth and division are activated, and this group of cells, which may be referred to as a 'cap anlage', gradually acquires the morphology of the lost cap tissue. The first day after decapping sees nuclear and cytoplasmic changes, such as mitosis, slime synthesis and amyloplast evolution, occurring simultaneously within the cap anlage. It is as though the normally discrete zones of meristem, central, and outer cap have all become telescoped together. With continued growth of the cap anlage the various properties of cap cells become spatially segregated. By the time the anlage reaches the size of a normal cap it may be regarded as a *de facto* cap since the various cell-types it contains occupy the correct relative positions. Furthermore, the planes of division within the regenerating apex are so regulated that the characteristic pattern of cell-columns, as well as a distinct boundary between the cap and the re-established QC, are restored (Barlow, 1974).

In animal and plant systems regeneration is usually accompanied by cell division and is termed 'epimorphosis'. However, cell division may be unnecessary for regeneration because when decapped roots of Z. *mays* are grown in solutions containing inhibitors of DNA synthesis and mitosis, such as hydroxyurea or aminouracil, a cap-like tissue is re-modelled (a process known as morphallaxis) from the remaining portion of the apex (Barlow, 1981). Soon after decapping, cells at the extreme apex seem to acquire cap-like properties (except that mitoses are absent). By the fourth day these properties have segregated spatially to result in what appears to be a cap residing in former meristematic tissue. A conclusion to be drawn from observations of both morphallactic and epimorphic regenerants is that the essence of regeneration is the reformation of a system of positional co-ordinates (in this case a 'cap-field') that specifies cell differentiation and can influence the orientation of cell growth. Possibly, one end or surface of the cap anlage corresponds to a positional reference point. When division is permitted, the regenerating cap-field grows concurrently with the growing anlage. When division is suppressed, the cap-field can still develop at the apex since the exposed surface can still act as a reference point.

Regeneration is basically a morphallactic process. But because of the intrinsic capacity of cells for growth and division epimorphosis tends to be the rule. In effect, division only amplifies the number of cells within the tissue in which morphallactic principles operate. With this realization many of the difficulties encountered in theoretical analyses of regeneration in animal systems (discussed by Maden, 1982) which have been allied to the concept of positional information appear less important.

Concluding comments

As mentioned in the previous section, the surface and the apical location probably have strong effects on the spatial pattern of differentiation of certain cells at the root apex, but the nature of the information that these positions contain is not known. The surface alone cannot be a determinant of differentiation since the superficial cells of cap and epidermis are not identical. Their shared properties may be positionally controlled; their dissimilar properties may relate to differential gene action and its inheritance from ancestral cells, as well as to properties which, though positional in origin, are evoked by qualitatively different positional signals.

Ideals of intra- and supra-cellular gradients in developing and regenerating systems and their regulation of genetic activity are attractive to biologists, possibly as a result of the influential writings of C. M. Child (1941). But whether such gradients are real or not, and if real, whether they have any role in development, has yet to be settled. Plants could contain gradients of morphogens because their cells are literally awash with solutes flowing, usually in a polarized direction, through and around them. Also, meristems and differentiating tissue could certainly serve as sources and sinks for substances in the solute stream, such as sugars and hormones, which are known to have definite effects on development. Moreover, meristems, because of their growth and metabolism, could meet the important requirement of actively maintaining a gradient. As a general scheme we propose that in the growing root apex chemical signals direct the course of cellular development and that these signals vary in a quantitative and qualitative manner, according to position. Such signals exist as gradients and it is the gradient, coupled with the cell's capacity to respond, which is informational. The substances in the gradient may be transported actively by pumps, or diffuse passively. These means of movement may differ in the different zones of the root because of different anatomical and physiological properties of the cells. These properties, in turn, may determine whether longitudinal, radial or circumferential gradients operate.

The form of the apex (as distinct from its functional components) may arise through the operation of a quite different class of positional controls. The pattern of stresses within a tissue may govern the plane of cell division and thus determine the cellular architecture of the apex. The root apex perpetuates its form during growth, and it follows that the pattern of stress must also be reproduced in parallel with the reproduction of the cells. In this way form begets form. Microtubules are important in both division and certain aspects of cell differentiation (e.g., Gunning, 1982) and the patterns which they adopt within a cell can also be handed on through successive divisions. Microtubules, cell wall structure, and architecture of groups of cells are interlocking form-generating systems. Disturbance to any one of these can result in changes in the others. However, such changes may recover the original state (as in regeneration), or lead to the new states (as during the unfolding of a morphogenetic sequence). What gives the system its stability as well as its ability to change is a challenging question that has yet to be answered.

As a tentative conclusion, I suggest that roots give evidence that form, which is primary, is specified by information resident in stress patterns; differentiation, and hence function, is secondary and results from information resident in a pattern of chemical factors that may, in the first instance, be specified by form.

References

Aloni, R. (1980). Role of auxin and sucrose in the differentiation of sieve and tracheary elements in plant tissue cultures. *Planta*, **150**, 255–63.

Armstrong, J. E. & Heimsch, C. (1976). Ontogenetic reorganization of the root meristem in the Compositae. *American Journal of Botany*, **63**, 212–19.

Barlow, P. (1974). Regeneration of the cap of primary roots of *Zea mays*. *The New Phytologist*, **73**, 937–54.

Barlow, P. W. (1975). The root cap. In *The Development and Function of Roots*, ed. J. G. Torrey & D. T. Clarkson, pp. 21–54. London, New York & San Francisco: Academic Press.

— (1976). Towards an understanding of the behaviour of root meristems. *Journal of Theoretical Biology*, **57**, 433–51.

— (1977a). New patterns of vascular development in roots of *Pisum* recovering from colchicine treatment. *Experientia*, **33**, 1153–4.

— (1977b). An experimental study of cell and nuclear growth and their relation to cell diversification within a plant tissue. *Differentiation*, **8**, 153–7.

— (1981). Division and differentiation during regeneration at the root apex. In *Structure and Function of Plant Roots*, ed. R. Brouwer, O. Gašparíková, J. Kolek & B. C. Loughman, pp. 85–7. The Hague, Boston & London: M. Nijhoff/W. Junk.

— (1983). Root gravitropism and its cellular and physiological control. In *Wurzelökologie und ihre Nutzanwendung – Ein Beitrag zur Erforschung der Gesamtpflanze*, ed. W. Böhm, L. Kutschera & E. Lichtenegger, pp. 279–94. Irdning: Verlag Bundesanstalt für alpenländische Landwirtschaft.

Barlow, P. W. & Hines, E. R. (1982). Regeneration of the root cap of *Zea mays* L. and *Pisum sativum* L.: a study with the scanning electron microscope. *Annals of Botany*, **49**, 521–9.

Barlow, P. W., Rost, T. L. & Gunning, B. E. S. (1982). Nuclear and cytoplasmic changes during the early stages of cell differentiation in roots of the water fern, *Azolla pinnata*. *Protoplasma*, **112**, 205–16.

Barlow, P. W. & Sargent, J. A. (1978). The ultrastructure of the regenerating root cap of *Zea mays* L. *Annals of Botany*, **42**, 791–9.

Basile, D. V. (1980). A possible mode of action for morphoregulatory hydroxyproline-proteins. *Bulletin of the Torrey Botanical Club*, **107**, 325–38.

Bloch, R. (1944). Developmental potency, differentiation and pattern in meristems of *Monstera deliciosa*. *American Journal of Botany*, **31**, 71–7.

— (1946). Differentiation and pattern in *Monstera deliciosa*. The idioblastic development of the trichoscleroids in the air root. *American Journal of Botany*, **33**, 544–51.

— (1947). Irreversible differentiation in certain plant cell lineages. *Science*, **106**, 320–2.

— (1948). The development of the secretory cells of *Ricinus* and the problem of cellular differentiation. *Growth*, **12**, 271–84.

Borchert, R. & McChesney, J. D. (1973). Time course and localization of DNA synthesis during wound healing of potato tuber tissue. *Developmental Biology*, **35**, 293–301.

Bünning, E. (1951). Über die Differenzierungsvorgänge in der Cruciferenwurzel. *Planta*, **39**, 126–53.

Cabioch, J. & Giraud, G. (1978). Comportement cellulaire au cours de la régénération directe chez le *Mesophyllum lichenoides* (Ellis) Lemoine (Rhodophycées, Corallinacées). *Comptes rendus hebdomadaires des Séances de l'Académie des Sciences, Paris, Série D*, **286**, 1783–5.

Charlton, W. A. (1982). Distribution of lateral root primordia in root tips of *Musa acuminata* Colla. *Annals of Botany*, **49**, 509–20.

Chiang, Y.-L. (1970). Macro- and microscopic structure of the root of *Ceratopteris pteridoides* (Hook.) Hieron. *Taiwania*, **15**, 31–49.

Child, C. M. (1941). *Patterns and Problems of Development*. Chicago: University of Chicago Press.

Christophers, E. (1972). Kinetic aspects of epidermal healing. In *Epidermal Wound Healing*, ed. H. I. Maibach & D. T. Rovee, pp. 53–69. Chicago: Year Book Medical Publishers.

Cormack, R. G. H. (1945). Cell elongation and the development of root hairs in tomato roots. *American Journal of Botany*, **32**, 490–6.

Cusick, F. (1966). On phylogenetic and ontogenetic fusions. In *Trends in Plant Morphogenesis*, ed. E. G. Cutter, pp. 170–83. London: Longmans.

Drabble, E. (1904). On the anatomy of the roots of palms. *Transactions of the Linnean Society of London (Botany)*, **6**, 427–90.

Dreher, T. W., Grant, B. R. & Wetherbee, R. (1978). The wound response in the siphonous alga *Caulerpa simpliciuscula* C. Ag.: fine structure and cytology. *Protoplasma*, **96**, 189–203.

Dyanat-Nejad, H. & Neville, P. (1972). Étude sur le mode d'action du méristème radical orthotrope dans le contrôle de la plagiotropie des racines latérales chez *Theobroma cacao* L. *Revue Générale de Botanique*, **79**, 319–40.

Ellmore, G. S. (1981). Root dimorphism in *Ludwigia peploides* (Onagraceae): development of two root types from similar primordia. *Botanical Gazette*, **142**, 525–33.

Esau, K. (1954). Primary vascular differentiation in plants. *Biological Reviews of the Cambridge Philosophical Society*, **29**, 46–86.

Everett, N. P., Wang, T. L., Gould, A. R. & Street, H. E. (1981). Studies on the control of the cell cycle in cultured plant cells II. Effects of 2,4-dichlorophenoxyacetic acid (2,4-D). *Protoplasma*, **106**, 15–22.

Feldman, L. J. (1977). The generation and elaboration of primary vascular tissue patterns in roots of *Zea*. *Botanical Gazette*, **138**, 393–401.

Filippenko, V. N. (1980). Determination of the direction of rhizodermal cell differentiation in corn. *Soviet Journal of Developmental Biology*, **11**, 189–96.

Fosket, D. E. & Tepfer, D. A. (1978). Hormonal regulation of growth in cultured plant cells. *In Vitro*, **14**, 63–75.

Gardner, W. K., Barber, D. A. & Parbery, D. G. (1982). Effect of microorganisms on the formation and activity of proteoid roots of *Lupinus albus* L. *Australian Journal of Botany*, **30**, 303–9.

Gunning, B. E. S. (1978). Age-related and origin-related control of the numbers of plasmodesmata in cell walls of developing *Azolla* roots. *Planta*, **143**, 181–90.

— (1982). The root of the water fern *Azolla*: cellular basis of development and multiple roles for cortical microtubules. In *Developmental Order: Its Origin and Regulation*, ed. S. Subtelny & P. B. Green, pp. 379–421. New York: Liss.

Gunning, B. E. S., Hughes, J. E. & Hardham, A. R. (1978). Formative and proliferative cell divisions, cell differentiation, and developmental changes in the meristem of *Azolla* roots. *Planta*, **143**, 121–44.

Haberlandt, G. (1887). Ueber die Lage des Kernes in sich entwickelnden Pflanzenzellen. *Berichte der deutschen botanischen Gesellschaft*, **5**, 205–12.

Hillman, N., Sherman, M. I. & Graham, C. (1972). The effect of spatial arrangement on cell determination during mouse development. *Journal of Embryology and Experimental Morphology*, **28**, 263–78.

Hinchee, M. A. W. (1981). Morphogenesis of aerial and subterranean roots of *Monstera deliciosa*. *Botanical Gazette*, **142**, 347–59.

Ivanov, V. B. & Filippenko, V. N. (1976). Role of asymmetric mitosis in differentiation of rhizoderm cells of festucoid grasses. *Doklady Botanical Sciences*, **230**, 389–91.

Jeffs, R. A. & Northcote, D. H. (1966). Experimental induction of vascular tissue in an undifferentiated plant callus. *Biochemical Journal*, **101**, 146–52.

Konings, H. & Verschuren, G. (1980). Formation of aerenchyma in roots of *Zea mays* in aerated solutions, and its relation to nutrient supply. *Physiologia Plantarum*, **49**, 265–70.

Lang, A. (1974). Inductive phenomena in plant development. In *Basic Mechanisms in Plant Morphogenesis. Brookhaven Symposia in Biology*, no. 25, 129–44.

Lerman, M. I. (1978). The biological essence of resting cells in cell populations. *Journal of Theoretical Biology*, **73**, 615–29.

Lindenmayer, A. (1981). Developmental algorithms: lineage versus interactive control mechanisms. In *Developmental Order: Its Origin and Regulation*, ed. S. Subtelny & P. B. Green, pp. 219–45. New York: Liss.

Lloyd, C. W. & Barlow, P. W. (1982). The co-ordination of cell division and elongation: the role of the cytoskeleton. In *The Cytoskeleton in Plant Growth and Development*, C. W. Lloyd, pp. 203–28. London: Academic Press.

Maden, M. (1982). Morphallaxis in an epimorphic system: size, growth control and pattern formation during amphibian limb regeneration. *Journal of Embryology and Experimental Morphology*, **65** (Supplement), 151–67.

Malajczuk, N. & Bowen, G. D. (1974). Proteoid roots are microbially induced. *Nature*, **251**, 316–17.

Mallory, T. E., Chiang, S.-H., Cutter, E. G. & Gifford, E. M. (1970). Sequence and pattern of lateral root formation in five selected species. *American Journal of Botany*, **57**, 800–9.

Mestre, J.-C. (1967). La signification phylogénétique de l'embryogénie. *Revue Générale de Botanique*, **64**, 273–324.

Miller, J. H. (1980). Orientation of the plane of cell division in fern gametophytes: the roles of cell shape and stress. *American Journal of Botany*, **67**, 534–42.

Nakazawa, S. & Yamazaki, Y. (1982). Cellular polarity in root epidermis of *Gibasis geniculata*. *Die Naturwissenschaften*, **69**, 396–7.

Nishimura, S. & Maeda, E. (1982). Cytological studies on differentiation and dedifferentiation in pericycle cells of excised rice roots. *Japanese Journal of Crop Science*, **51**, 553–60.

Palmer, J. H. & Marc, J. (1982). Wound-induced initiation of involucral bracts and florets in the developing sunflower inflorescence. *Plant and Cell Physiology*, **23**, 1401–9.

Peterson, R. L. (1975). The initiation and development of root buds. In *The Development and Function of Roots*, ed. J. G. Torrey & D. T. Clarkson, pp. 123–61. London, New York & San Francisco: Academic Press.

Phillips, H. L. & Torrey, J. G. (1971). The quiescent center in cultured roots of *Convolvulus arvensis* L. *American Journal of Botany*, **58**, 665–71.

Prescott, D. M. (1976). *Reproduction of Eukaryotic Cells*. New York & London: Academic Press.

Purnell, H. M. (1960). Studies on the family Proteaceae I. Anatomy and morphology of the roots of some Victorian species. *Australian Journal of Botany*, **8**, 38–50.

Sheldrake, A. R. & Northcote, D. H. (1968). The production of auxin by tobacco internode tissues. *The New Phytologist*, **67**, 1–13.

Short, K. C. & Torrey, J. G. (1972). Cytokinins in seedling roots of pea. *Plant Physiology*, **49**, 155–60.

Sinnott, E. W. (1943). Cell division as a problem of pattern in plant development. *Torreya*, **43**, 29–34.

Sinnott, E. W. & Bloch, R. (1941). The relative position of cell walls in developing plant tissues. *American Journal of Botany*, **28**, 607–17.

— (1945). The cytoplasmic basis of intercellular patterns in vascular differentiation. *American Journal of Botany*, **32**, 151–6.

— (1946). Comparative differentiation in the air roots of *Monstera deliciosa*. *American Journal of Botany*, **33**, 587–90.

Souèges, R. (1937). Les lois du développement. *Actualités Scientifiques et Industrielles*, **521**, 94 pp.

Sporne, K. R. (1965). *The Morphology of Gymnosperms. The Structure and Evolution of Primitive Seed-Plants*. London: Hutchinson.

Tomlinson, P. B. (1969). On the morphology and anatomy of turtle grass, *Thalassia testudinum* (Hydrocharitaceae). II. Anatomy and development of the root in relation to function. *Bulletin of Marine Science*, **19**, 57–71.

Torrey, J. G. (1955). On the determination of vascular patterns during tissue differentiation in excised pea roots. *American Journal of Botany*, **42**, 183–98.

— (1957). Auxin control of vascular pattern formation in regenerating pea root meristems grown *in vitro*. *American Journal of Botany*, **44**, 859–70.

— (1965). Physiological bases for the organization and development of roots. In *Encyclopaedia of Plant Physiology*, vol. 15/1, ed. W. Ruhland, pp. 1256–327. Berlin, Heidelberg & New York: Springer-Verlag.

Torrey, J. G. & Wallace, W. D. (1975). Further studies on primary vascular tissue pattern formation in roots. In *The Development and Function of Roots*, ed. J. G. Torrey & D. T. Clarkson, pp. 91–103. London, New York & San Francisco: Academic Press.

Tschermak-Woess, E. & Hasitschka, G. (1953). Über Musterbildung in der Rhizodermis und Exodermis bei einigen Angiospermen und einer Polypodiacee. *Österreichischen botanischen Zeitschrift*, **100**, 646–51.

Vöchting, H. (1877). Ueber Theilbarkeit im Pflanzenreich und die Wirkung innerer und äusserer Kräfte auf Organbildung an Pflanzentheilung. *Pflüger's Archiv für die gesamte Physiologie des Menschen und der Tiere*, **15**, 153–90.

Walker, D. B. (1978). Morphogenetic factors controlling differentiation and dedifferentiation of epidermal cells in the gynoecium of *Catharanthus roseus* I. The role of pressure and cell confinement. *Planta*, **142**, 181–6.

Wang, T. L., Everett, N. P., Gould, A. R. & Street, H. E. (1981). Studies on the control of the cell cycle in cultured plant cells III. The effects of cytokinin. *Protoplasma*, **106**, 23–35.

Wetmore, R. H. & Rier, J. P. (1963). Experimental induction of vascular tissues in callus of angiosperms. *American Journal of Botany*, **50**, 418–30.

Whatley, J. M. & Gunning, B. E. S. (1981). Chloroplast development in *Azolla* roots. *The New Phytologist*, **89**, 129–38.

Wightman, F., Schneider, E. A. & Thimann, K. V. (1980). Hormonal factors controlling the initiation and development of lateral roots II. Effects of exogenous growth factors on lateral root formation in pea roots. *Physiologia Plantarum*, **49**, 304–14.

Wilcox, H. (1954). Primary organization of active and dormant roots of noble fir, *Abies procera*. *American Journal of Botany*, **41**, 812–20.

Zimmerman, P. W. & Hitchcock, P. W. (1935). The response of roots to 'root-forming' substances. *Contributions from the Boyce Thompson Institute*, **7**, 439–45.

Ziomek, C. A. & Johnson, M. H. (1982). The roles of phenotype and position in guiding the fate of 16-cell mouse blastomeres. *Developmental Biology*, **91**, 440–7.

11

Shoot meristem development

CARL N. McDANIEL

Multicellular organisms often exhibit cellular specialization at some time during their life cycle. Thus a variety of cell types can be produced from a single cell. We identify these various cell types as specialized because they are morphologically, physiologically, or biochemically different from each other. The process by which a cell becomes different is called differentiation. When a cell has become morphologically, physiologically and/ or biochemically different in a measurable way, it is said to be differentiated. A cell has become terminally differentiated when it has achieved a specialized condition which is stable and maintained under normal conditions. Since senescence and cell death may be developmental events which are genetically regulated, one might consider differentiation, in some cases, to be complete only after a cell dies.

The regulation of cell differentiation has been a major area of research in developmental biology. However, our understanding is elementary and incomplete. The essence of regulation is the selective expression of one type of cellular specialization while leaving the other types unexpressed. How is this accomplished? Do all multicellular organisms employ similar mechanisms? Does an individual organism utilize one or several different mechanisms?

The process of differentiation has many components. During differentiation one of the early events, perhaps the first, appears to be programming of a cell or tissue for a specialized role. The programming is called determination, and a cell, or a group of cells, in which determination has taken place is said to be determined. The complete programming of a cell does not appear to take place in a single determination event. There may be several such events as the developmental fate becomes more and more specific during the differentiation process. Determination is an operationally defined concept. In general terms, a cell, or a group of cells, is determined when it exhibits the same developmental fate whether grown *in situ*, in isolation, or at a new place in or on the organism.

During differentiation cells respond to external signals which can come from the environment or other parts of the organism. The ability to develop in response to a specific signal is called competence. Thus, the competent cell must be able to recognize the signal and translate it into a response. This would imply that the cell has receptors for the signal and that the perception of the signal, or the binding of signal molecules, is linked biochemically or physically to the process regulated. As with determination, competence is defined operationally; i.e., a cell/tissue/organ is exposed to a signal and if it responds in the expected manner, it is competent.

Logically, a competent cell would be one which is also determined. Thus, it should be possible to find cells which at one time during differentiation are determined for a particular developmental fate but not capable of achieving that fate because they cannot respond to the signal which elicits that response from the cell. In other cases, determination and competence may be tightly linked thereby making it difficult or impossible to separate the processes experimentally.

In this chapter I will consider a model for shoot development in which determination and competence are important developmental events. Evidence will be presented which indicates that determination and/or competence are analysable events in shoot meristem differentiation. In the process of evaluating the importance of competence and determination, other aspects of shoot development will be considered. However, the emphasis will be on these early developmental events. For a broader treatment, as well as other aspects of shoot development, the reader should consult the recent review by Halperin (1978), various chapters in *Patterns in Plant Development* by Steeves & Sussex (1972), and other chapters of this volume.

The use of the concepts of determination and competence has been extensive in animal embryology. I believe that it will be instructive to consider that usage before considering its application to plant differentiation. As an example I will discuss briefly the differentiation of imaginal discs in the fruit fly, *Drosophila*. This model system for the differentiation of multicellular organisms clearly illustrates the concepts of determination and competence.

Drosophila imaginal disc development as a paradigm for understanding determination and competence in plants

The development of *Drosophila* imaginal discs has been clearly described by Nöthiger (1972) and the major aspects will be presented briefly here.

After a *Drosophila* egg is fertilized the zygotic nucleus divides 12 or 13 times without cytokinesis thereby creating a syncytium. At about the time of the twelfth division many of the nuclei migrate to the egg cortex and there cell membranes form around the cortical nuclei creating the cellular blastoderm (Turner & Mahowald, 1976). At this time or shortly thereafter, clusters of cells are set apart from the cells which will form larval structures. These clusters of cells are the progenitors of the imaginal discs which will give rise to the structures of the adult fly. Each disc will produce specific adult structures, e.g., eye-antennal disc, wing disc, first leg disc, etc. When the egg hatches, each disc is made up of fewer than 50 cells. During the three larval stages (instars) the number of cells increases to more than a thousand, the exact number depending upon the disc type. Although each disc type (eye-antennal, first leg, wing, etc.) has a unique morphology, the eventual adult morphology is only apparent after metamorphosis. Metamorphosis is brought about by the presence of a high concentration of ecdysone (moulting hormone) and the absence or a low concentration of juvenile hormone. Metamorphosis begins in the larval-pupal moult and is completed in the pupal-adult moult. It should be noted that ecdysone is not a trigger signal which in one short burst initiates the whole developmental program. In contrast, it must be present for a period of days. This same long-term requirement also exists for thyroxine which controls metamorphosis in amphibians (Beckingham-Smith & Tata, 1976). If ecdysone or thyroxin is withdrawn during the required period, metamorphosis ceases.

Chan & Gehring (1971) provided the early evidence that determination for disc type occurs soon after the cellular blastoderm is formed. They took cells from the posterior half and the anterior half of cellular blastoderm-stage embryos and cultured them separately in adult females to permit cellular proliferation. The proliferated cell masses were then placed in third instar larvae. After metamorphosis of the host larvae, the adult structures produced by the implanted cells were examined. The cells from the posterior half only gave rise to structures which were normally formed from discs originating in the posterior half of the embryo. Anterior half cells only gave rise to structures which were normally formed from discs originating in the anterior half of the embryo. More recently, Illmensee (1978) has transplanted blastoderm-stage cells to different locations in blastoderm embryos and the cells likewise developed according to their original positions. Thus, the future developmental fate of embryonic cells which will form imaginal discs is fixed soon after formation of the cellular blastoderm. Although the determined state is clonally propagated, determination appears to be conferred upon groups of cells and not single cells (Nöthiger, 1972).

Once the cells of a disc are determined, this condition is clonally propagated and very stable (Gehring, 1972). Disc cells can undergo cell proliferation and be maintained for years by serial transplants into the abdomens of adult females. The state of determination in an *in vivo* cultured disc piece can be assayed by placing the disc piece into the abdomen of a host larva about to undergo metamorphosis. When the host metamorphoses, the implanted disc will also metamorphose. In almost all cases the discs will produce structures characteristic of the original source of the disc. However, at a measurable frequency a disc will undergo transdetermination to form adult structures not associated with the original source disc or it may lose its ability to undergo metamorphosis altogether. For example, derivatives of an eye disc may undergo transdetermination to wing disc. This new determined state as a wing disc will then be stable and clonally propagated.

Although a group of about 10–20 cells becomes determined as a specific disc type soon after cellular blastoderm formation, these cells are not competent to respond to ecdysone by differentiating into adult structures until near the end of larval development (Nöthiger, 1972; Mindek, 1972). Discs from first or early second instar larvae which are implanted into third instars do not undergo metamorphosis when the host undergoes metamorphosis (Mindek, 1972). Rather, they undergo cellular proliferation. If they are then implanted into a second host larva, they will metamorphose with the host (Nöthiger, 1972). Thus determination and competence are distinct events in the development of imaginal discs (Fig. 11.1).

The differentiation of germ cells (progenitors of sperms and eggs) in *Drosophila* has also been extensively researched (Illmensee, 1978). The results of this work illustrate one mechanism by which cells may be determined. The cytoplasm at the posterior end of an unfertilized egg is unique and contains structures called polar granules. The nuclei which migrate into this cytoplasm are programmed by this cytoplasm to be the progenitors of germ cells. This has been convincingly demonstrated by a series of elegant cytoplasm and cell manipulations (Illmensee & Mahowald, 1974, 1976). The results of these studies are perhaps the best illustration of cytoplasmic determinants, materials localized in a specific area of a cell's cytoplasm which fix the developmental fate of future cells that come to contain these materials.

This brief review of imaginal disc development illustrates several points concerning differentiation which are central to the general theory of animal development. First, the developmental fate of a cell/tissue can be established early in its developmental history. Second, determination is a position-dependent process. Third, this determined condition can be clonally propagated. Fourth, the determined condition may be very stable but not irreversible. Fifth, although the cell/tissue may be determined, it may not be competent to respond to the developmental signal which elicits expression of the determined condition.

Fig. 11.1 Differentiation of *Drosophila* imaginal discs. Imaginal disc cells are set apart from future larval cells soon after the formation of the cellular blastoderm and undergo an initial determination event at this time. Several days later they become competent to respond to the hormone, ecdysone. During metamorphosis they will respond to ecdysone and differentiate in accordance with their determined condition, e.g., leg discs form legs, wing discs form wings, etc. The question marks indicate a hypothetical controlling input.

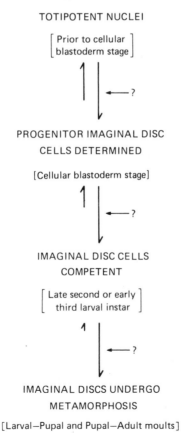

TOTIPOTENT NUCLEI

[Prior to cellular
blastoderm stage]

⟵ ?

PROGENITOR IMAGINAL DISC
CELLS DETERMINED

[Cellular blastoderm stage]

⟵ ?

IMAGINAL DISC CELLS
COMPETENT

[Late second or early
third larval instar]

⟵ ?

IMAGINAL DISCS UNDERGO
METAMORPHOSIS

[Larval—Pupal and Pupal—Adult moults]

Are these points valid for plant development? Will viewing plant development as a process similar to *Drosophila* imagnal disc development enable us to design experiments and analyse data so as to advance our understanding of plant development? Obviously these are open questions, but ones we should consider. In the remainder of the chapter I will present my opinions concerning determination and differentiation in plants as well as some of the data which led me to adopt these opinions.

Meristem development

Model

Initiation of apical meristems

During plant embryogeny the embryo proper becomes polar, establishing the root-shoot axis, and thereby forms two very different types of apical meristems. We have virtually no information on the developmental events associated with this organization, but we do know that once organized, these meristems are very stable. In angiosperms these patterns of organization have not been shown to be interconvertible: i.e., a shoot apical meristem does not become a root apical meristem or vice versa (Halperin, 1978). However, cell derivatives of each type can become organized into the other type: i.e., shoots can be rooted and roots can produce shoots.

Studies of apical meristem initiation outside normal embryogenesis have indicated that hormones may be involved in this process. The classic work from Skoog's laboratory (Skoog & Miller, 1957) on organogenesis in tobacco callus shows that different ratios of exogenous auxins and cytokinins lead to different types of meristematic organization. A high auxin to cytokinin ratio leads to root meristem formation while a low auxin to cytokinin ratio leads to shoot formation. Thus, callus cells interpret the exogenous hormone concentrations (or modify them in some unknown way which is then interpreted) and form aggregates which have self-perpetuating, three-dimensional organization.

Related to the work of Skoog's laboratory are the more recent studies on plant regeneration where just one medium is not sufficient to bring about organogenesis. A good example of this work which has been put in a developmental context is on regeneration of alfalfa (*Medicago sativa*) by Walker, Yu, Sato & Jaworski (1978) and Walker, Wendeln & Jaworski (1979). They have shown that pre-treatment of callus with hormones is required for subsequent organorgenesis in hormone-free medium. A four-day pretreatment with high kinetin and low levels of the synthetic auxin

2,4-dichlorophenoxyacetic acid (2,4-D) leads to root formation, while four days with low kinetin and high 2,4-D leads to shoot formation if the callus is subsequently placed on hormone-free medium. Regeneration will not occur on the hormone-containing medium and small pieces of callus (less than 105 μm in diameter containing 1 to 12 cells) are not capable of responding to pre-treatment (they are incompetent) by undergoing organogenesis. However, if the small aggregates are permitted to grow to a larger size, then they become competent and can be induced to regenerate. This work illustrates two developmental concepts. First, the history of a cell/tissue influences its developmental fate and second, competence is an acquired condition.

The work of Wochok & Sussex (1974, 1975) on a lower plant also indicates that hormones are involved in the establishment of apical meristem type. In *Selaginella willldenovii*, a member of the Lycopsida, there are two angle meristems at the base of each leaf which can develop either as a root or as a shoot meristem. Under normal circumstances the ventral meristem forms a root and the dorsal meristem forms a shoot. However, if the dorsal meristem is exposed to an appropriate concentration of auxin at an early point in its development, it will form a root. Failure to be exposed to auxin will lead to development as a shoot. However, the *Selaginella* apical meristems may not be as stable as apical meristems of angiosperms since a root meristem grown in culture medium lacking auxin can reorganize as a shoot meristem (Wochok & Sussex, 1976). This work illustrates two basic developmental concepts. First, there is a window during which a developmental signal can be received and second, this developmental signal fixes the future developmental fate of the tissue in a stable but not necessarily irreversible way.

Developmental fate of apical meristems

Once organized the two types of apical meristems have very different developmental fates. The root apical meristem will give rise to a root and lateral roots but will itself differentiate into no other structures. On the other hand, the shoot apical meristem may undergo further differentiation to become a thorn, a tendril or a flower. That is, the shoot apical meristem may express more than one pattern of organization. The type of growth expressed can be unlimited (continued production of leaves) or limited (production of a tendril, thorn, or flower).

The limited growth patterns of thorn or tendril production have been studied by only a few authors. In *Parthenocissus inserta* tendril differentiation begins soon after the cells form this axillary meristem since they do not normally, and could not experimentally, be made to form shoots (Millington, 1966). When the plant is floral, the presumptive tendril meristem differentiates into an inflorescence and frequently one observes meristems that differentiate as part tendril and part inflorescence. It would appear that the floral pattern supersedes the tendril pattern which has superseded the vegetative shoot pattern. Grapevine tendrils, *in vivo* and *in vitro*, will form an inflorescence if treated with exogenous cytokinin (Srinivasan & Mullins, 1978, 1979). The differentiation of thorns for two species has been described (Blaser, 1956; Bieniek & Millington, 1967). Unfortunately, little is known about the developmental events associated with the transition from leafy shoot to thorn or tendril, or from tendril to inflorescence.

In contrast to thorn and tendril development, flower development has been studied extensively. There have been numerous excellent reviews written on flowering (e.g., Lang, 1952, 1965; Zeevaart, 1976; Bernier, Kinet & Sachs, 1981a, b) as well as reviews on specific aspects of flowering (e.g., Chouard, 1960; Lang, 1961; Gifford, 1964; Popham, 1964; Nougarède, 1967; Bernier, 1971; Chailakhyan, 1975; Miginiac, 1978; Lang, 1980). My intent is not to review the total literature but rather to consider primarily aspects of the literature which indicate that determination and/ or competence are analysable states during floral differentiation.

Conversion from vegetative to floral growth

The preponderant approaches for the analysis of floral development have been physiological and morphological ones where photoperiodic plants have been the primary subjects (Bernier *et al.*, 1981a, b). The literature on photoperiodic induction is immense and illustrates the complexity of the situation. There are a host of photoperiodic types: long-day plants, short-day plants, long-short-day plants, day-neutral plants, long-day plants which first require a chilling period, etc. This bewildering array of response types presumably represents a complex group of developmental control mechanisms. Conceivably, however, the complexity of response types may be superimposed on a simpler set of basic developmental controls.

Perhaps all of the complexity observed in the regulation of flowering has obscured some basic generalizations which would enable us to investigate the problem more fruitfully. The brilliance of Mendel's analysis of pea genetics was not so much in that it explained all that was known about inheritance. Rather, in a relatively simple way, it explained a large number of observations which previously appeared as a complex array of virtually unexplainable observations. We now know that Mendel was wrong on several counts and that exceptions to his general rules are many and extremely important. However, without Mendelian genetics it would not have been possible to develop any degree of sophistication in the study of inheritance.

I believe that, although extremely complex, differentiation must have some basic principles and generalizations which are rather simple. Simple algorithms or sets of rules can be used to describe what appear to be complex biological patterns (Lindenmayer, 1978). We must seek to discern the underlying generalizations which will then enable us to deal with complexity and exceptions.

Taking this point of view, I will discuss flowering as a developmental process which will be called floral differentiation. Floral differentiation is a process which includes all events/processes occurring in the meristem or the cells of the meristem which pertain to the production of the flower (inflorescence) as well as its maturation and senescence. Thus, 'floral differentiation' is an all-inclusive process. However, I shall concern myself with the events which occur up to floral induction.

The flowering processes we study today are the products of millions of years of evolution. The hypothetical primitive angiosperm was a small evergreen tree or shrub which grew in a moist tropical area and its flowers were borne singly at the ends of leafy branches (Cronquist, 1968). If we knew the archetype for floral differentiation, we might be able to discern the basic developmental process upon which evolutionary changes were made. The archetype is unknown but we do know that in primitive and highly evolved angiosperms the flower is a terminal structure. Thus, a leaf-producing meristem that ultimately flowers terminates its growth by producing a series of modified leaves which, in essence, consume the meristem. One might speculate that the developmental processes involved in the conversion from leaf production to flower production are very old and are perhaps common to all angiosperms.

For the newly-evolved angiosperm which grew in the moist tropics, flowering at any time during the year would be a successful reproductive strategy. Delaying the production of flowers until the plant was large enough to provide adequate food for successful reproductive activities would also appear to be of adaptive advantage for the primitive angiosperm. Modifications to the archetype for floral differentiation would be selected for and preserved if they bestowed upon the plant a survival advantage which enhanced the probability of reproductive success. For example, as the primitive angiosperms evolved to invade the Temperate Zone, reduction of the juvenile period and control by environmental conditions (temperature, photoperiod) would have been adaptive modifications of the flowering process. Over millions of years the archetype for floral differentiation would be modified not only by inserting or adding-on new steps, but also by inserting control points in the original sequence. When considering flowering in herbaceous annuals, a highly evolved group of angiosperms, it is certainly not surprising that we have compiled a complex data base which has not been easily interpreted.

Perhaps some insight may be gained by considering a basic sequence of development events which might be involved in the early stages of floral differentiation (Fig. 11.2) (McDaniel, 1980*a*). During embryogeny the shoot apical meristem is organized and programmed to form leaves, lateral branches and stem tissues. This is a stable state which is developmentally regulated. The regulation of vegetative growth is not well understood and certainly merits more consideration than it has been given in the past. Once achieved, the vegetative state of shoot differentiation may persist either for a long period or only briefly. After some time interval a change in the meristem cells will occur. This change may be autonomously controlled by the genome of the plant or may result from interaction between the plant and the environment. The change results in a programming of the meristem and its cells for the production of a flower. The meristem is now determined for floral differentiation but a flower will not be produced until the meristem becomes competent to respond to the developmental signal(s) which elicit flower development. Even though the meristem may be able to translate the flowering signal into flower development, flowering may still not occur. An inhibitor(s) may, for one of many reasons, interfere with reception of the signal. Nutritional conditions may not be conducive for flower development or the signal may not reach the meristem cells. A flower will be formed by a meristem when the meristem is competent to respond to the signal and when the signal which elicits flower development is present in physiologically active amounts. It should be noted that the signal (floral stimulus) may be endogenously controlled or it may be produced only when the external environmental conditions are permissive, e.g., an inductive photoperiod.

In this model I assume that there is a developmental signal, probably hormonal in nature, which elicits the development of a flower from a competent meristem. The nature of this developmental signal or floral stimulus is unknown (Bernier *et al.*, 1981*a, b*). It may be several substances, the ratio of two substances, or other more complex arrangements. For simplicity I will write as if there were a developmental signal knowing full well that this is an oversimplification of current observations.

If the above basic model approximates reality, then much of the complexity observed in the regulation of flowering in angiosperms should be explicable by inserting various control points along the sequence of developmental events. This, of course, assumes that floral differentiation is an obligatory sequence of events. This is not a trivial assumption, and its validity can be challenged. For example, we know it is possible to produce embryos in tissue culture without first forming a flower (Steward, Mapes, Kent & Holsten, 1964), but we should also recognize that the developmental relationship between flower formation and embryogenesis is far from clear. Thus, assuming floral differentiation is an obligatory sequential process is reasonable but not necessarily true. However, to assume that floral differentiation is a facultative sequential process would imply that the history of the meristem and its cells is irrelevant as far as flower formation is concerned. These possibilities need to be considered in a broader context and this will be done when I consider possible ways for controlling the developmental events in shoot differentiation.

If the *Drosophila* model is of general predictive value, then determination and competence would be two types of developmental events which should be part of the floral differentiation process. In imaginal disc differentiation, determination for the types of adult structures to be formed (leg, wing, etc.) and the competence to respond to the developmental signal, ecdysone, could be separated experimentally. However, in most differentiation processes, determination and competence have not been separated experimentally. One might legitimately ask whether the *Drosophila* example is an anomaly and whether these two terms refer to the same developmental event. One could hypothesize that determination is the programming of a cell or tissue for a specific program of differentiation and competence is the linking of this program to a developmental signal. However, it has not been possible in any system to separate the various components which establish and permit the expression of a specific developmental program. For this reason the above hypothetical explanations of competence and determination will be employed with the knowledge that these explanations are purely speculative. Although this limitation should be recognized, the use of this conceptual framework may prove to be valuable in interpreting as well as designing experiments.

Experimental evidence for the model

What is the experimental evidence which indicates that floral differentiation is an obligatory sequence of developmental events where determina-

TOTIPOTENT CELLS

(Early embryogeny)

CELLS DETERMINED/COMPETENT FOR SHOOT MERISTEM ORGANIZATION

(Early embryogeny)

VEGETATIVE MERISTEM

(Early to late heart stage embryo)

MERISTEM CELLS DETERMINED FOR LIMITED GROWTH PATTERN

(During embryogeny or later; species specific)

MERISTEM CELLS COMPETENT TO RESPOND TO THE DEVELOPMENTAL SIGNAL

WHICH ELICITS FLOWER, TENDRIL OR THORN DEVELOPMENT

(During embryogeny or later; species specific)

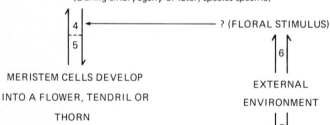

? (FLORAL STIMULUS)

MERISTEM CELLS DEVELOP

INTO A FLOWER, TENDRIL OR

THORN

(After embryogeny)

EXTERNAL

ENVIRONMENT

Fig. 11.2. Model for the development of the shoot apical meristem of angiosperms. The shoot apical meristem is formed during embryogeny from totipotent cells. The vegetative meristem may at some time express a limited growth pattern and thereby terminate its existence as a meristem. During this differentiation there may be numerous developmental events which are regulated. Some possible developmental events are indicated. Bidirectional arrows indicate that the processes are reversible. The numerals indicate possible developmental control points as discussed in the text. The question marks indicate a hypothetical controlling input.

tion and competence are among these events? Most of the work on flowering neither supports nor contradicts the proposed model. This situation has resulted primarily because most flowering studies have been concerned with photoperiodic induction (Bernier *et al.*, 1981*a*), and equally important because researchers have generally not designed experiments to investigate determination or competence. Despite this situation there is a limited amount of evidence which can be interpreted as consistent with this view of floral differentiation.

In the model proposed in Fig. 11.2 there are seven possible control points for the regulation of flowering. One, an organized meristem is not present. Two, the meristem is undetermined. Three, the meristem is incompetent. Four, the floral stimulus is not detected by the meristem. Five, the floral stimulus does not reach the meristem. Six, the plant cannot detect regulating environmental conditions. Seven, environmental condition(s) is not permissive. Let us consider these seven proposed control points and discuss examples of floral differentiation where control appears to be at one of these places. It should be noted that a species may utilize more than one of these points for controlling the time when flowers will be formed. Although in the examples presented I will consider there to be just one control point, this is obviously not true in some cases. To ascertain which control points are being utilized, a species/variety must be studied throughout its life cycle.

There are many examples of control at point seven. Plants like *Pharbitis nil* (Kujirai & Imamura, 1958), *Chenopodium rubrum* (King, 1972) and others (Lang, 1965; Zeevaart, 1976) will flower almost immediately after germination if given an inductive photoperiod. In these plants all of the other developmental events required for flowering have occurred during embryogeny or early growth of the seedling.

Control point six appears to be the place of regulation in some plants that require vernalization ('The acquisition or acceleration of the ability to flower by a chilling treatment' (Chouard, 1960)) and in some plants that exhibit a juvenile period during which floral induction is not possible. *Byrophyllum daigremontianum* is a long-short-day plant which has a long juvenile phase. Zeevaart (1962) has grafted juvenile plants onto flowering plants and the juvenile plants flowered as rapidly as mature plants. Thus, the meristem is capable of receiving and responding to the floral stimulus but the juvenile leaves appear to be unresponsive to the inductive photoperiod.

Some biennial varieties of henbane (*Hyoscyamus niger*), sugar beet (*Beta* sp.), carrot (*Daucus* sp.), and cabbage (*Brassica* sp.) require vernalization as well as long-day photoperiodic induction following the cold period (Chouard, 1960; Bernier *et al.*, 1981*a*). It is the shoot tip which is vernalized, and the vernalized state can be stable and clonally propagated. However, the shoot meristem of these varieties can be made to flower without being chilled if it is grafted to a stock of the same species which is flowering or in some cases to other species which are flowering. Again the meristem is competent to respond to the flowering signal but the leaves are apparently insensitive to photoperiodic induction. Vernalization may change the meristem so that it produces leaves which are sensitive to the photoperiod. Thus, an unvernalized meristem produces incompetent leaves. This interpretation is strengthened by observations with *Cichorium intybus* (Chouard, 1960) (cold-requiring, long-day plant) where it has been shown that it is the new leaves produced by the vernalized meristem which sense the photoperiod.

The work on floral inhibitors (Lang, Chailakhyan & Frolova, 1977; Lang, 1980) and nutrient translocation (King & Zeevaart, 1973; Sachs & Hackett, 1977) provides evidence that flowering in some plants may be controlled by a disruption of the translocation of the flowering signal or by interfering with its reception by a competent meristem. The leaves play an important role in these processes (inhibitors, nutrients), and thus the interface with floral differentiation is complex (Sachs & Hackett, 1977; Lang, 1980). However, some results from grafting and/or leaf removal experiments can be interpreted to indicate that the leaves play a direct role in inhibiting flowering. Defoliation of the annual long-day plant *Hyoscyamus niger* eliminated photoperiodic dependence such that the plant flowered in long days, short days, and continuous darkness (Lang, 1980). In *Perilla crispa* using leaf removal and grafting, King & Zeevaart (1973) provided evidence consistent with the idea that assimilate transport patterns can keep the floral stimulus from reaching competent meristems. Flowering of day-neutral *Nicotiana tabacum* cv. Trapezond can be prevented by grafting either *Nicotiana sylvestris* (long-day) or *Hyoscyamus niger* (long-day, annual strain) onto the tobacco plant and maintaining the plants under short-days (Lang *et al.*, 1977; Lang, 1980). These observations indicate that some plant species employ control points four and five for regulating the time of flowering.

As discussed earlier it has not been possible to distinguish clearly, through experimental procedures, an initial determination event from the competence event except in the differentiation of *Drosophila* imaginal discs. Thus, I cannot present with complete confidence examples where determination and competence are clearly distinguishable. Recognizing this limitation, I will discuss one situation where a meristem appears to be determined but incompetent and several cases where the meristems are at least incompetent. Finally, a possible example of an undetermined meristem which becomes determined and then competent will be considered.

In sunflower, *Helianthus annuus* (a day-neutral plant), the terminal shoot meristem normally produces about 14–16 leaf pairs followed by a shift to the development of reproductive structures (Habermann, 1964; Habermann & Sekulow, 1972). Grafting of shoot tips from plants of various physiological ages (plants which had produced different numbers of leaf pairs) onto seedling stock does not increase the total number of nodes produced by the grafted meristem. The total number of leaf pairs (produced before grafting + produced after grafting) is always about 14–16. However, grafting seedling terminal meristems onto the tops of stocks of increasing physiological age causes a reduction in the number of leaf pairs produced by the scion. For example, a seedling grafted onto a stock which has produced six leaf pairs will produce about ten leaf pairs while a seedling grafted onto a stock which has produced nine leaf pairs will produce about six leaf pairs. Rooting shoot tips of plants of various physiological ages neither increases nor decreases the total number of leaf pairs produced by the meristem. The shoot tips of plants which had produced three leaf pairs at the time of rooting produced an additional 12.5 leaf pairs while ones that had produced nine leaf pairs before rooting produced 6.7 after rooting. These data indicate that the amount of pre-reproductive growth can be reduced but not increased. Thus, the sunflower terminal meristem is programmed for a limited growth pattern in the young seedling, i.e., determined for floral development. The expression of this determined state is usually preceded by a period of growth, but by placing the meristem in a 'physiologically-older' environment, this growth period can be reduced and in the extreme case almost eliminated (Habermann & Sekulow, 1972). It would appear that competence is acquired by the meristem during this growth period or after being placed at the top of another more mature plant.

In black currant, *Ribes nigrum*, axillary buds flower when the plant is exposed to inductive short days only if they are separated by about 20 nodes from the roots (Schwabe & Al-Doori, 1973). The axillary buds of a shoot with eight nodes will not flower in short days, but if such a young shoot is grafted to the top of a shoot with more than 20 nodes, its axillary buds will flower. With leaf removal experiments Schwabe & Al-Doori (1973) demonstrated that the eight basal leaves were not capable of promoting flowering in the eight basal buds but these basal leaves could induce flowering in the uppermost buds of a 26-node plant. These results indicate that the basal buds may be incompetent to respond to the flowering signal which is produced by the basal leaves.

In the day-neutral, perennial plant, *Geum urbanum*, axillary buds will flower after a chilling period, but under normal environmental conditions very young and older axillary buds as well as the terminal meristem remain vegetative (Chouard, 1960). Vernalized flower buds form numerous axillary branches when cut or exposed to high temperatures (30–35°C) and all of these, as well as successive generations of buds, flower. If vernalized axillary buds with their subtending leaf are rooted, all of the secondary axillary buds flower. In some rare cases adventitious buds form on the petiole of the rooted leaf and these buds are always vegetative. These observations demonstrate that vernalization acts on the meristem to make it competent to respond to the floral stimulus. This competence is stable and clonally propagated, but not transmitted to unvernalized meristems.

In day-neutral *Nicotiana tabacum* the terminal shoot meristem differentiates into a single flower after giving rise to the cells which produce the rest of the above-ground plant. The conversion from vegetative to floral growth pattern is precisely regulated and endogenously controlled. That is, the number of nodes varies as a function of light intensity, day length and temperature but is very uniform under one set of environmental conditions (Seltman, 1974; Thomas, Anderson, Raper & Downs, 1975; McDaniel & Hsu, 1976a, b). For example, two populations of greenhouse-grown plants of *N. tabacum* cv. Wisconsin 38, raised at different times, produced 35.4 ± 1.0 (S.D.) nodes ($n = 30$, range = 33–38) and 33.7 ± 1.5 (S.D.) nodes ($n = 26$, range = 31–36). Under normal conditions the terminal meristem will always form a flower but flower formation is not an obligatory developmental event. By continually rooting a shoot, the shoot meristem can be made to remain vegetative indefinitely (McDaniel, 1980b). Why does the meristem normally differentiate into a flower? Two lines of experimental investigation indicate that the cells of the meristem undergo stable developmental changes which lead to flower formation.

The first line of evidence comes from a series of experiments on the development of the axillary buds of *N. tabacum* cv. Wisconsin 38 (McDaniel & Hsu, 1976a, b; McDaniel, 1978). *N. tabacum* axillary buds below the inflorescence are arrested after the production of seven to nine leaf primordia, and at the time of anthesis of the first flower appear to be morphologically vegetative as indicated by dissection and serial sections. Decapitation, rooting, and grafting experiments were used to measure a bud's developmental potential (number of nodes produced by a meristem before production of a flower). These experiments clearly demonstrate that there are two populations of axillary buds in terms of their developmental potential. Buds in population A develop in accordance with their new environment thereby showing no 'memory' of their original position. For example, buds from the fourth node below the inflorescence produced about 21 nodes *in situ* but about 38 nodes when rooted and about 34 nodes when grafted to the base of the main axis. Buds in population B develop according to their original position. For example, buds from the second node below the inflorescence produced about 19 nodes *in situ*, about 18 nodes when rooted and about 22 nodes when grafted to the base of the main axis.

On a single plant the populations are not intermixed along the main axis. Rather, B buds are always directly below the inflorescence and separated from A buds by a single internode (Fig. 11.3). For example, the following numbers of nodes were produced by the five buds directly below the inflorescence from the plant in Fig. 11.3 when these buds were rooted and grown to maturity; 1st bud – 18 nodes, 2nd bud – 17 nodes, 3rd bud – 37 nodes, 4th bud – 38 nodes, 5th bud – 38 nodes. B buds exhibit the same growth potential when grown *in situ*, when grown independently, and when grafted to and grown at a new place on the plant. Thus, they fit the classical definition of developmentally determined tissue. Since their growth is restricted to producing a limited number of leaves and then a flower, it is reasonable to say they have undergone an early determination event in floral differentiation. That is, like the cells in the *Drosophila* embryo which have become determined several hours after fertilization, these determined meristem cells must proliferate and undergo additional changes (perhaps the acquisition of competence) before the determined state is expressed.

The second group of studies demonstrates that determination for floral development occurs not only in an organized meristem but also at the tissue or cell level (Aghion-Prat, 1965; Wardell & Skoog, 1969; Tran Thanh Van, 1973; Chailakhyan, Aksenova, Konstantinova & Bavrina, 1975; Hillson & La Motte, 1977). These researchers (and others) have shown that cells from cultured internode segments or tissue explants of day-neutral *N. tabacum* will form *de novo* floral meristems if the explants are from or near the inflorescence area. Similarly-treated explants taken from more basal areas of the plant form, in almost all cases, only vegetative buds. These results dramatically illustrate that cells in the inflorescence area of some day-neutral tobaccos are determined and competent for flower formation.

These several examples have provided evidence which indicates that floral differentiation may be an obligatory sequence of developmental events. Floral differentiation may be interrupted by a block at some point in this sequence of events. By considering the classical response types (photoperiodic plants, day-neutral plants, plants requiring vernalization, plants with a juvenile phase) it has been possible to show where a particular species may be blocked in floral differentiation. This way of categoriz-

Fig. 11.3. Positions of determined and undetermined *N. tabacum* axillary buds. The plant is developmentally at anthesis (opening) of the first flower. 1st, 2nd, etc., indicate positions of axillary buds below the inflorescence. The boundary between determined and undetermined buds varies from one plant to another but has never been observed below bud four. (Reproduced from McDaniel (1978) with permission.)

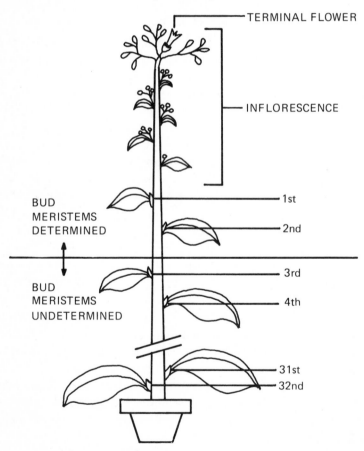

ing the flowering response in a plant may clarify some otherwise confusing observations. For example, in some plants that require vernalization, an unvernalized plant will flower if it is grafted to a flowering donor. In other species 'vernalization' is not graft-transmissible. These observations make sense if in the former case vernalization acts on the meristem so that it then produces photoperiodically sensitive leaves, and if in the latter case vernalization acts on the meristem so that it is made competent to respond to the floral stimulus. I believe the model can be useful not only for interpreting published observations but also in designing experiments.

Although evidence has been presented which indicates that floral differentiation can be controlled by regulating an event in the differentiation processes, the underlying mechanisms of regulation have not been considered. The various processes may be regulated by similar mechanisms or they might be regulated by quite diverse mechanisms. At this time speculation can be made based upon our understandings of biochemistry, biophysics, and molecular biology as well as upon regulatory mechanisms which have been shown to be operative in other developmental systems. However, hard evidence directly related to the regulation of differentiation in multicellular organisms has been difficult to obtain, and it is not abundant.

Regulation of determination/competence

As noted by the early developmental biologists, a cell differentiates in accordance with its position. This is a trite and yet profound statement. Flowers develop as modified shoot apical meristems and not as modified root apical meristems. Trichomes develop as leaf epidermal derivatives and not as root epidermal derivatives. Root hairs form as root epidermal derivatives and not shoot epidermal derivatives. In flower development stamens form after petals, and carpels form after stamens. These and many other observations indicate that position is of ultimate developmental significance.

Does the position of a cell/tissue lead to stable, clonally propagated changes which influence a cell's developmental future, or does position provide an environment that elicits a response which is only expressed in that environment? These opposing points of view can be stated in another way. Does the history of a cell/tissue influence in a predictable way its developmental future, or is a cell/tissue totally plastic and its future developmental expression only a function of its current environment? For animal cells it has been experimentally demonstrated in numerous cases that a cell's position leads to stable, clonally propagated changes which influence the cell's developmental fate (Graham, 1976; Graham & Wareing, 1976). The same unequivocal statement cannot be made for plant cells. This could mean that developmental processes are fundamentally different in the two types of organisms. Perhaps, however, the major difference between animal and plant cells lies not in the developmental processes that occur but in their stability and/or timing.

In order to establish that a specific position or environment has influenced a cell's developmental future, the cell must express a predictable pattern of development in different environments. But in order to accomplish this, the programming for the future pattern of development must be stable enough to withstand the manipulations involved in the experiment. Plant cells and tissues from many species exhibit a high degree of instability or plasticity. Some regenerate or produce a callus when manipulated *in vitro* (Murashige, 1974), while other cells will differentiate into unexpected cell types. For example, mesophyll cells of *Zinnia elegans* when cultured will differentiate into cells with secondary wall thickenings similar to tracheary elements and this differentiation can occur without DNA replication or cell division (Fukuda & Komamine, 1980).

Grafting of cells or small pieces of tissue has been extremely valuable in studying animal development (Nöthiger, 1972; Graham, 1976; Graham & Wareing, 1976; Illmensee, 1978). However, grafting of plant cells and tissues has not been exploited as fully. Some of the interesting cell and tissue types are not easily accessible: e.g., early embryonic tissues, provascular cells, prosporogenous cells, etc. Additionally, it has been extremely difficult to graft small numbers of cells successfully because of wounding responses and/or the subsequent lack of growth (Ball, 1950). Thus, many of the manipulations employed by animal-developmental biologists have not been successful with plants either because the manipulations have not been possible or because the cells respond rapidly or not at all in their new environment. It would appear that if a plant cell's history is important, the 'memory' of this history is easily erased or modified in many instances. Thus, it will be difficult, if not impossible, in some cases to know how to interpret the results of experiments in which plant cells are manipulated.

Unlike the development of animal cells, determination and competence have not been demonstrated to be involved in the differentiation of any specific type of plant cell. Sachs (1978) has suggested that the differentiation of some plant cell types (vascular cells and guard cells) is differentiation dependent; i.e., 'the processes of differentiation are responsible for the orderly distribution of the signals which in turn control continued differentiation'. Thus a cell does not become determined and competent for a particular cell type as prerequisites for responding to the developmental signal which elicits the development of the particular cell type. Rather cells are plastic and only after entering a pathway of differentiation as a result of their position do they become programmed for that pathway. Continued expression along a particular pathway is contingent upon continued differentiation in a unique environment which itself is intimately related to the particular type of cellular differentiation. This is an appealing hypothesis which permits plant cells to maintain a high degree of plasticity through almost the whole differentiation process. The inability of plant cells to move may have made this type of differentiation effective and of adaptive advantage over a more rigid program.

It is common knowledge that many plant species have the ability to form new plants from tissue pieces (Murashige, 1974). Single cells from some species can even be manipulated in culture to form entire plants (Vasil & Hildebrandt, 1965) while other species will form somatic embryos in culture (Steward *et al.*, 1964). Regeneration is not a universal trait, some groups and species being quite refractory; e.g., legumes, cereals and many woody species. However, some successes with these refractory groups have been achieved recently by culturing embryonic tissues. These results indicate that during development plant cells may undergo stable changes which lead to the loss of totipotency. Unfortunately, little is known about these changes as they have been virtually impossible to study.

Despite the fact that determination and competence have not been clearly separated from morphological differentiation in plant cellular specialization, there are some examples which suggest that the history of organized groups of plant cells does, in a predictable way, influence their developmental future. Vegetative shoot meristems cultured without leaf primordia will continue to grow as vegetative meristems (Smith & Murashige, 1970) and meristem pieces organize complete meristems (Sussex, 1952; Steeves & Sussex, 1972). Meristems which undergo phase change like English ivy (*Hedera helix*) form adult meristems which are stable and revert to juvenile-like forms, in most cases, only after considerable manipulation (Hackett, 1976; Wareing & Frydman, 1976). In plants which require vernalization it is the shoot tip which perceives the cold treatment (Chouard, 1960; Bernier *et al.*, 1981*a*). The vernalized state once established is often stable and clonally propagated; i.e., derivatives of the meristem are vernalized. Determination for floral differentiation in day-neutral tobacco is also a stable state which is clonally propagated

(Konstantinova, Aksenova, Bavrina & Chailakhyan, 1969; Tran Than Van, 1973; McDaniel, 1978). Cultured floral meristems exhibit autonomous development and complete flowers can be produced *in vitro* (Blake, 1966; Hicks & Sussex, 1970).

Fern leaf primordia appear to pass through a sequence of developmental stages in which their developmental potential is fixed (Haight & Kuehnert, 1971). Isolated, young leaf primordia can develop as shoots, but if they are isolated at progressively older stages, this developmental potential is reduced to leaf plus a bud and then finally to only a leaf. Leaf primordia of angiosperms have their developmental fate fixed quite early in organogenesis since excised primordia are only capable of forming leaves and not complete shoot meristems (Steeves, Gabriel & Steeves, 1957; Mauseth, 1977).

The foregoing gives one the firm impression that the differentiation of plant cells is most assuredly position dependent. In some cases a cell's positional history is 'remembered' but often it is easily erased. In general it appears that organized groups of cells have this 'stable memory' quality while individual cells do not. But how is position known? Positional models for plant development are discussed in Chapter 1 of this volume. The possible application to plant development of Wolpert's (1969, 1971) model of positional information has been considered by Holder (1979). In a nutshell, we do not know how positional information is communicated, but these authors and others have proposed a variety of ideas which should be considered and tested experimentally.

One way in which a cell's position can be important is for a cell to divide so that the daughter cells contain different cytoplasm. If the cytoplasmic differences result in unique types of differentiation, position has become relevant because of the existence of a polar distribution of cytoplasmic entities. As discussed earlier, this is the one known mechanism for regulating determination in animals – determination of *Drosophila* germ cells being the best example (Illmensee, 1978). This may also be the mechanism by which a *Fucus* egg divides to form two different cell types (Quatrano, 1978; Chapter 6 of this volume). Perhaps this is the mechanism by which the fertilized egg of an angiosperm establishes two cell types, suspensor cell and embryo-proper cell, at the time of the first division.

We know that the embryo-proper cell will go on to produce a polar embryo with a root and shoot axis. How is the root-shoot axis established and what are the developmental parameters involved in organizing the shoot apical meristem? At this point our only information comes from studies of *de novo* meristem organization in culture. Some of these studies have been discussed earlier (Skoog & Miller, 1957; Walker *et al.*, 1978, 1979). One is left with the strong impression that the classical hormones, auxins and cytokinins, are prime candidates as regulators for meristem organization. The study by Wochok & Sussex (1976) of cultured *Selaginella willdenovii* roots showed that a root meristem in this primitive plant can form a shoot meristem if auxin is not in the culture medium. Perhaps in an early stage of meristem evolution auxin was involved in initiating as well as stabilizing the meristem. Subsequently, other mechanisms may have evolved for meristem stabilization while the early role of hormones in the organization process remains.

Studies on the role of hormones in the flowering process also indicate their importance in regulating shoot meristem development (Lang, 1961, 1965; Zeevaart, 1976; Bernier *et al.*, 1981*b*). Auxins, cytokinins, ethylene and gibberellins have been reported to be involved. Each hormone has been shown in some species to stimulate and in other species to inhibit the flowering process. Thus, it is far from clear how these hormones are functioning to regulate floral differentiation or whether there are commonalities in the roles played by each hormone in various species.

Other than the classical plant hormones, there is little information on specific types of molecules which may be involved in shoot development, but nutrients have been implicated. Sachs & Hackett (1977) have proposed that the nutritional status of a meristem could be the means by which its development is regulated. The fact that the quantity of light received by a plant can dramatically influence flowering indicates that nutritional status may play a key regulatory role. It is known that many plants grow vegetatively under low light intensities but do not flower. This has been demonstrated for *N. tabacum* cv. Wisconsin 38 (Wardell, 1976; Sachs, Hackett, Ramina & Maloof, 1978). Certainly the nutritional status of a plant dramatically influences its development but the nature of the influence is obscure.

Earlier in this section the importance of stability was considered but nothing has been said about how stability of a determined or competent state might be achieved. The work of Meins and colleagues on cytokinin habituation in tobacco has provided a model system (Meins & Binns, 1979). They have shown that the requirement for exogenous cytokinin in the culture medium can be lost (cytokinin habituation) and that this loss can be controlled by manipulating the culture environment. Habituation is clonally propagated but reversible. They propose that cytokinins 'either induce their own production or block their own degradation and that it is this positive feedback relationship that maintains the habituated state' (Meins & Binns, 1979). It is possible that cells and tissues may employ similar positive-feedback relationships to stabilize states of differentiation like determination and competence.

More recent work indicates that cytokinin habituation may represent physiological changes which are associated with tissue differentiation. Different tissues exhibit tissue-specific phenotypes in relation to habituation (Meins & Lutz, 1979). That is, pith tissues are not immediately habituated when put in culture, while stem-cortex tissues are. They have also observed that pith tissues from different parts of the plant exhibit different capacities for habituation (Meins, Lutz & Binns, 1980*a*; Meins, Lutz & Foster, 1980*b*). Thus, pith tissue from the base of the plant rarely habituates, while pith tissue higher up the stem habituates readily. In addition, they have found that small explants (less than about 20–30 mg) do not habituate.

These are very interesting observations which may relate to topics discussed earlier. Ease of cytokinin habituation positively correlates with determination/competence for floral differentiation in terms of position along the main axis. Is this correlation meaningful? Walker *et al.* (1979) found that competence for alfalfa organogenesis required cell clumps of at least 105 μm, and Meins *et al.* (1980*b*) also reported a size limitation for habituation. Do these size limitations relate to the notion that determination/competence is conferred upon populations of cells and not single cells? This is certainly an intriguing thought.

Conclusion

Multicellular organisms produce a variety of tissues and organs which contain a host of specialized cell types. Early in the differentiation process these cells and tissues acquire the capacity to respond uniquely to developmental signals. Developmental biologists have employed the concepts of determination and competence in an attempt to deal with the early events in the differentiation process. These concepts have become crucial components of the general theory for animal development but they have not become central in our thinking about plant development (Wareing, 1978; Meins & Binns, 1979).

I have taken the position that determination and competence are developmental concepts which will be useful as we seek to understand the early developmental events in shoot meristem organization and differentiation.

I have taken this point of view because we must focus on and think about the early events in plant development if we are to broaden our understanding of this process. In this regard determination and competence may provide us with a theoretical basis from which significant advances can be made.

When all is said and done, shoot meristem development is a topic with more unknowns than knowns. Most of the work is still ahead of us. It is my hope that this book, this chapter and the simple model proposed for shoot development will not only be of value to those studying plant development, but will also stimulate a few, creative, hard-working persons to focus their attention on some of the many challenging problems alluded to in this volume.

I would like to thank Professors Wesley Hackett and Roy Sachs (University of California – Davis) for their support and interest during the writing of this chapter. I am indebted to Professor Harry Roy and Ms Susan Singer (both of Rensselaer Polytechnic Institute), and to Professor Ian Sussex (Yale University) for critically reviewing the manuscript.

References

Aghion-Prat, D. (1965). Néoformation de fleurs *in vitro* chez *Nicotiana tabacum* L. *Physiologie Végétale,* **3,** 229–303.
Ball, E. (1950). Isolation, removal and attempted transplants of the central portion of the shoot apex of *Lupinus albus* L. *American Journal of Botany,* **37,** 117–36.
Beckingham-Smith, K. & Tata, J. R. (1976). The hormonal control of amphibian metamorphosis. In *The Developmental Biology of Plants and Animals,* ed. C. F. Graham & P. F. Wareing, pp. 232–45. Philadelphia: Saunders.
Bernier, G. (1971). Structural and metabolic changes in the shoot apex in transition to flowering. *Canadian Journal of Botany,* **49,** 803–19.
Bernier, G., Kinet, J.-M. & Sachs, R. M. (1981*a*). *The Physiology of Flowering,* Vol. 1. *The Initiation of Flowers.* Boca Raton: CRC Press.
– (1981*b*). *The Physiology of Flowering,* Vol. 2. *Transition to Reproductive Growth.* Boca Raton: CRC Press.
Bieniek, M. E. & Millington, W. F. (1967). Differentiation of lateral shoots as thorns in *Ulex europaeus. American Journal of Botany,* **54,** 61–70.
Blake, J. (1966). Flower apices cultured *in vitro. Nature,* **211,** 990–1.
Blaser, H. W. (1956). Morphology of the determinate thorn-shoots of *Gleditsia. American Journal of Botany,* **43,** 22–8.
Chailakhyan, M. K. (1975). Forty years of research on the hormonal basis of plant development – some personal reflections. *The Botanical Review,* **41,** 1–29.

Chailakhyan, M. K., Aksenova, N. P., Konstantinova, T. N. & Bavrina, T. V. (1975). The callus model of plant flowering. *Proceedings of the Royal Society of London*, B, **190**, 333–45.

Chan, L. & Gehring, W. J. (1971). Determination of blastoderm cells in *Drosophila melanogaster*. *Proceedings of the National Academy of Sciences of the USA*, **68**, 2217–21.

Chouard, P. (1960). Vernalization and its relations to dormancy. *Annual Review of Plant Physiology*, **11**, 191–238.

Cronquist, A. (1968). *The Evolution and Classification of Flowering Plants*. Boston: Houghten Mifflin.

Fukuda, H. & Komamine, A. (1980). Direct evidence for cytodifferentiation to tracheary elements without intervening mitosis in a culture of single cells isolated from the mesophyll of *Zinnia elegans*. *Plant Physiology*, **65**, 61–4.

Gehring, W. J. (1972). The stability of the determined state in cultures of imaginal disks in *Drosophila*. In *The Biology of Imaginal Disks. Results and Problems in Cell Differentiation*, vol. 5, ed. H. Ursprung & R. Nöthiger, pp. 35–58. Berlin: Springer-Verlag.

Gifford, Jr., E. M. (1964). Developmental studies of vegetative and floral meristems. In *Meristems and Differentiation. Brookhaven Symposia in Biology*, no. 16, 126–37.

Graham, C. F. (1976). The formation of different cell types in animal embryos. In *The Developmental Biology of Plants and Animals*, ed. C. F. Graham & P. F. Wareing, pp. 14–28. Philadelphia: Saunders.

Graham, C. F. & Wareing, P. F. (1976). Determination and stability of the differentiated state. In *The Developmental Biology of Plants and Animals*, ed. C. F. Graham & P. F. Wareing, pp. 45–54. Philadelphia: Saunders.

Habermann, H. M. (1964). Grafting as an experimental approach to the problem of physiological aging in *Helianthus annuus* L. In *Proceedings of the XVIth International Horticultural Congress*, vol. IV, pp. 243–51. Gembloux: J. Ducolot S.A.

Habermann, H. M. & Sekulow, D. B. (1972). Development and aging in *Helianthus annuus* L. Effects of the biological *milieu* of the apical meristem on patterns of development. *Growth*, **36**, 339–49.

Hackett, W. P. (1976). Control of phase change in woody plants. *Acta Horticulturae*, **56**, 143–54.

Haight, T. H. & Kuehnert, C. C. (1971). Developmental potentialities of leaf primordia of *Osmunda cinnamomea*. VI. The expression of P_1. *Canadian Journal of Botany*, **49**, 1941–5.

Halperin, W. (1978). Organogenesis at the shoot apex. *Annual Review of Plant Physiology*, **29**, 239–62.

Hicks, G. S. & Sussex, I. M. (1970). Development *in vitro* of excised flower primordia of *Nicotiana tabacum*. *Canadian Journal of Botany*, **48**, 133–9.

Hillson, T. D. & La Motte, C. E. (1977). *In vitro* formation and development of floral buds on tobacco stem explants. *Plant Physiology*, **60**, 881–4.

Holder, N. (1979). Positional information and pattern formation in plant morphogenesis and a mechanism for the involvement of plant hormones. *Journal of Theoretical Biology*, **77**, 195–212.

Illmensee, K. (1978). *Drosophila* chimeras and the problem of determination. In *Genetic Mosaics and Cell Differentiation. Results and Problems in Cell Differentiation*, vol. 9, ed. W. T. Gehring, pp. 51–69. Berlin: Springer-Verlag.

Illmensee, K. & Mahowald, A. P. (1974). Transplantation of posterior polar plasm in *Drosophila*. Induction of germ cells at the anterior pole of the egg. *Proceedings of the National Academy of Sciences of the USA*, **71**, 1016–20.

– (1976). The autonomous function of germ plasm in a somatic region of the *Drosophila* egg. *Experimental Cell Research*, **97**, 127–40.

King, R. W. (1972). Timing in *Chenopodium rubrum* of export of the floral stimulus from the cotyledons and its action at the shoot apex. *Canadian Journal of Botany*, **50**, 697–702.

King, R. W. & Zeevaart, J. A. D. (1973). Floral stimulus movement in *Perilla* and flower inhibition caused by noninduced leaves. *Plant Physiology*, **51**, 727–38.

Konstantinova, T. N., Aksenova, N. P., Bavrina, T. V. & Chailakhyan, M. K. (1969). On the ability of tobacco stem calluses to form vegetative and generative buds in culture *in vitro*. *Doklady Botanical Sciences*, **187**, 82–5.

Kujirai, C. & Imamura, S. (1958). Über die photoperiodische Empfindlichkeit der Kotyledonen von *Pharbitis nil* Chois. *Botanical Magazine (Tokyo)*, **71**, 408–16.

Lang, A. (1952). Physiology of flowering. *Annual Review of Plant Physiology*, **3**, 265–306.

– (1961). Auxins in flowering. In *Encyclopaedia of Plant Physiology*, vol. XIV, ed. W. Ruhland, pp. 909–50. Berlin: Springer-Verlag.

– (1965). Physiology of flower initiation. In *Encyclopaedia of Plant Physiology*, vol. XV/1, ed. W. Ruhland, pp. 1380–536. Berlin: Springer-Verlag.

– (1980). Inhibition of flowering in long-day plants. In *Plant Growth Substances 1979*, ed. F. Skoog, pp. 310–22. Berlin: Springer-Verlag.

Lang, A., Chailakhyan, M. K. & Frolova, I. A. (1977). Promotion and inhibition of flower formation in a day neutral plant in grafts with a short-day plant and a long-day plant. *Proceedings of the National Academy of Sciences of the USA*, **74**, 2412–16.

Lindenmayer, A. (1978). Algorithms for plant morphogenesis. *Acta Biotheoretica*, **27** (Supplement), 37–81.

McDaniel, C. N. (1978). Determination for growth pattern in axillary buds of *Nicotiana tabacum* L. *Developmental Biology*, **66**, 250–5.

– (1980*a*). A model for floral differentiation. *Plant Physiology*, **65** (Supplement), 101.

– (1980*b*). Influence of leaves and roots on meristem development in *Nicotiana tabacum* L. cv. Wisconsin 38. *Planta*, **148**, 462–7.

McDaniel, C. N. & Hsu, F. C. (1976*a*). Position-dependent development of tobacco meristems. *Nature*, **259**, 564–5.

– (1976*b*). Positional information in relation to aging. *Acta Horticulturae*, **56**, 291–8.

Mauseth, J. D. (1977). Cytokinin- and gibberellic acid-induced effects on the determination and morphogenesis of leaf primordia in *Opuntia polyacantha* (Cactaceae). *American Journal of Botany*, **64**, 337–46.

Meins, Jr., F. & Binns, A. N. (1979). Cell determination in plant development. *BioScience*, **29**, 221–5.

Meins, Jr., F. & Lutz, J. (1979). Tissue-specific variation in the cytokinin habituation of cultured tobacco cells. *Differentiation*, **15**, 1–6.

Meins, Jr., F., Lutz, J. & Binns, A. N. (1980*a*). Variation in the competence of tobacco pith cells for cytokinin-habituation in culture. *Differentiation*, **16**, 71–5.

Meins, Jr., F., Lutz, J. & Foster, R. (1980*b*). Factors influencing the incidence of habituation for cytokinin of tobacco pith tissue in culture. *Planta*, **150**, 264–8.

Miginiac, E. (1978). Some aspects of regulation of flowering: role of correlative factors in photoperiodic plants. *Botanical Magazine (Tokyo)*, Special Issue No. 1, 159–73.

Millington, W. F. (1966). The tendril of *Parthenocissus inserta:* determination and development. *American Journal of Botany*, **53**, 74–81.

Mindek, G. (1972). Metamorphosis of imaginal discs of *Drosophila melanogaster*. *Wilhelm Roux' Archives für Entwicklungsmechanik der Organismen*, **169**, 353–6.

Murashige, T. (1974). Plant propagation through tissue cultures. *Annual Review of Plant Physiology*, **25**, 135–66.

Nöthiger, R. (1972). The larval development of imaginal disks. In *The Biology of Imaginal Disks. Results and Problems in Cell Differentiation*, vol. **5**, ed. H. Ursprung & R. Nöthiger, pp. 1–34. Berlin: Springer-Verlag.

Nougarède, A. (1967). Experimental cytology of the shoot apical cells during vegetative growth and flowering. *International Review of Cytology*, **21**, 203–351.

Popham, R. A. (1964). Developmental studies of flowering. In *Meristems and Differentiation. Brookhaven Symposia in Biology*, no. 16, 126–37.

Quatrano, R. (1978). Development of cell polarity. *Annual Review of Plant Physiology*, **29**, 487–510.

Sachs, R. M. & Hackett, W. P. (1977). Chemical control of flowering. *Acta Horticulturae*, **68**, 29–49.

Sachs, R. M., Hackett, W. P., Ramina, A. & Maloof, C. (1978). Photosynthetic assimilation and nutrient diversion as controlling factors in flower initiation in *Bougainvillea* (San Diego Red) and *Nicotiana tabacum* cv. Wis. 38. In *Photosynthesis and Plant Development*, ed. R. Marcelle, H. Clijsters & M. Van Poucke, pp. 95–101. The Hague: Junk.

Sachs, T. (1978). Patterned differentiation in plants. *Differentiation*, **11**, 65–72.

Schwabe, W. W. & Al-Doori, A. H. (1973). Analysis of a juvenile-like condition affecting flowering in the black currant (*Ribes nigrum*). *Journal of Experimental Botany*, **24**, 969–81.

Seltman, H. (1974). Effect of light periods and temperatures on plant form of *Nicotiana tabacum* L. cv. Hicks. *Botanical Gazette*, **135**, 196–200.

Skoog, F. & Miller, C. O. (1957). Chemical regulation of growth and organ formation in plant tissue cultured *in vitro*. *Symposia of the Society for Experimental Biology*, **11**, 1118–31.

Smith, R. H. & Murashige, T. (1970). *In vitro* development of the isolated shoot apical meristem of angiosperms. *American Journal of Botany*, **57**, 562–8.

Srinivasan, C. & Mullins, M. G. (1978). Control of flowering in the grapevine (*Vitis vinifera* L.). Formation of inflorescences *in vitro* by isolated tendrils. *Plant Physiology*, **61**, 127–30.

– (1979). Flowering in *Vitis:* conversion of tendrils into inflorescences and bunches of grapes. *Planta*, **145**, 187–92.

Steeves, T. A., Gabriel, H. P. & Steeves, M. W. (1957). Growth in sterile culture of excised leaves of flowering plants. *Science*, **126**, 350–1.

Steeves, T. A. & Sussex, I. M. (1972). *Patterns in Plant Development*. Englewood Cliffs: Prentice-Hall.

Steward, F. C., Mapes, M. O., Kent, A. E. & Holsten, R. D. (1964). Growth and development of cultured plant cells. *Science*, **143**, 20–7.

Sussex, I. M. (1952). Regeneration of the potato shoot apex. *Nature*, **170**, 755–7.

Thomas, J. F., Anderson, C. E., Raper, Jr., C. D. & Downs, R. J. (1975). Time of floral initiation in tobacco as a function of temperature and photoperiod. *Canadian Journal of Botany,* **53,** 1400–10.

Tran Thanh Van, M. (1973). Direct flower neoformation from superficial tissue of small explants of *Nicotiana tabacum* L. *Planta,* **115,** 87–92.

Turner, F. R. & Mahowald, P. A. (1976). Scanning electron microscopy of *Drosophila* embryogenesis I. The structure of the egg envelopes and the formation of the cellular blastoderm. *Developmental Biology,* **50,** 95–108.

Vasil, V. & Hildebrandt, A. C. (1965). Differentiation of tobacco plants from single isolated cells in microcultures. *Science,* **150,** 889–92.

Walker, K. A., Wendeln, M. L. & Jaworski, E. G. (1979). Organogenesis in callus tissue of *Medicago sativa.* The temporal separation of induction processes from differentiation processes. *Plant Science Letters,* **16,** 23–30.

Walker, K. A., Yu, C. P., Sato, S. J. & Jaworski, E. G. (1978). The hormonal control of organ formation in callus of *Medicago sativa* L. cultured *in vitro. American Journal of Botany,* **65,** 654–9.

Wardell, W. A. (1976). Floral activity in solutions of deoxyribonucleic acid extracted from tobacco stems. *Plant Physiology,* **57,** 855–61.

Wardell, W. L. & Skoog, F. (1969). Flower formation in excised tobacco stem segments. I. Methodology and effects of plant hormones. *Plant Physiology,* **44,** 1402–6.

Wareing, P. F. (1978). Determination in plant development. *Botanical Magazine (Tokyo),* Special Issue No. 1, 3–17.

Wareing, P. F. & Frydman, V. M. (1976). General aspects of phase change, with special reference to *Hedera helix* L. *Acta Horticulturae,* **56,** 57–69.

Wochok, Z. S. & Sussex, I. M. (1974). Morphogenesis in *Selaginella.* II. Auxin transport in the root (rhizophore). *Plant Physiology,* **53,** 738–41.

– (1975). Morphogenesis in *Selaginella.* III. Meristem determination and cell differentiation. *Developmental Biology,* **47,** 376–83.

– (1976). Redetermination of cultured root tips to leafy shoots in *Selaginella willdenovii. Plant Science Letters,* **6,** 185–92.

Wolpert, L. (1969). Positional information and the spatial pattern of cellular differentiation. *Journal of Theoretical Biology,* **25,** 1–47.

– (1971). Positional information and pattern formation. *Current Topics in Developmental Biology,* **6,** 183–224.

Zeevaart, J. A. D. (1962). The juvenile phase in *Bryophyllum daigremontianum. Planta,* **58,** 543–8.

– (1976). Physiology of flower formation. *Annual Review of Plant Physiology,* **27,** 321–48.

12

Positional information and plant morphology

D. J. CARR

The term positional information was coined by Wolpert (1968, 1969, 1975) (although the idea had occurred almost a century ago to Vöchting and Driesch [see Driesch, 1908]) to imply the existence of unknown mechanisms which cause cells in an embryo, an organ or a small animal to respond to their position along a developmental axis or within an aggregate of cells. Such responses could be elicited in normal development or in regeneration (in its widest sense) following damage. But positional information is also held to account for the localization of capacities for organ formation (e.g., tentacles and buds in hydroids, legs in insects). It is to this more morphological meaning of the term that this chapter is devoted.

Morphological positional information in plants is demonstrated in two ways: firstly, by observations which show that pieces of the plant, morphologically similar but from different locations in the plant body, differ in performance during regeneration; secondly, in that shoot or root meristems, although not apparently differing among themselves in cellular organization, can give rise to shoots or roots that differ in their morphology or physiological behaviour depending on the position the meristem occupied in the plant body. Evidence will be presented for morphological positional information, demonstrated in three ways, taking examples from three systems, experimental and natural: regeneration, rejuvenation, and flowering and reproductive meristems. These observations suggest the existence along the plant of morphogenetic gradients to which cells or organs are subject and which determine their development when conditions permit. We will begin with a few words about gradients and the general role that they play in the response of an organ to position.

Gradients

The gradients which cells assess for positional information are associated with axes of polarity, axial or radial. Cells in callus tissue lack both polarity and positional values, although they retain antigenic tissue specificities (Raff, Hutchinson, Knox & Clarke, 1979). In shoots, new buds form from superficial tissues, often from epidermal cells, frequently from a single epidermal cell. In roots, on the other hand, buds originate endogenously, usually from pericycle tissues. These different origins of buds reflect different positional values along the radii of roots and shoots, associated with their own difference in tissue organization. In two essays Prat (1948, 1951) discusses pattern formation in histogenesis (e.g., the 'dermograms' of grass leaf epidermis) and morphology in terms of physico-chemical gradients. Bünning (1953, 1965) also deals with the origin of pattern in terms of stimulatory or inhibitory fields or gradients around structures which either produce morphogens or deplete the surroundings of them. These theories explain the origin of pattern as a direct result of the gradients or fields, making them directly responsible for the appearance or non-appearance of particular cell or organ types. Despite the supposed linearity of such gradients, there is a comparative rarity of anomalies in differentiation (e.g., intermediate cell types or aberrant organ types, teratological malformations such as occasionally occur in floral organs) resulting from them, and no explanation is provided of this fact. The difference between gradient theories and positional information theory is that, while gradients are primarily held by the latter theory to elicit a cellular or morphological response, it is the position the cell or meristem occupies in the gradient that determines which of a number of courses of differentiation it will take. The gradient itself is not directly responsible for differentiation but provides differences which are interpreted by the cells according to their positions, above or below a threshold value (Lewis, Slack & Wolpert, 1977) along the gradient. Once determined, the morphological patterning remains because the cells or meristems remain committed to a particular course of development, at least for some relatively long period, or until strongly influenced by other determining factors. Let us now turn to the evidence for positional information in plant morphology.

Position and regeneration

Regeneration from non-meristematic plant parts

Differences in regenerative performance of initially non-meristematic explants that are equivalent, except in initial position in the plant body, must depend on some initial differences in cellular potentiality quite diff-

erent from overt cellular differentiation itself. The resumption of meristematic activity denotes a reversal of at least a part of the differentiation
process. If, preceding the cell divisions which lead to regeneration, the cells
of explants from different initial positions in the plant undergo de-differentiation to the same extent, then the remaining differences between them,
which lead to different performances, are differences due to positional
information. It is evident that in preparing explants for regeneration, any
differences due to their positions in standing gradients of metabolizable
chemicals in the intact plant must lapse, leaving only such relatively stable
states of determination as will result in different regenerative performances.
It is expected that the collapse of such standing gradients would be more
rapid than the onset of de-differentiation. Correlative phenomena, in which
the growth of parts of the plant is governed by other parts with which they
are in connection, and from which governance they escape when the
connection is severed, are not here considered as due to positional information, although the status of the governing and governed parts may be
positionally determined, as for example a terminal and a lateral shoot
apical meristem. The dominance relationships which result in correlative
growth phenomena demonstrate the existence of fields of influence, interpreted by cells in terms of their position within the fields (e.g., a leaf
primordium in the field of an apical meristem which has just initiated it).

An excellent example of morphological positional information is provided by the (largely unpublished) work of Bauer (1963) on regeneration
from pieces of the very young sporophytes of the moss *Physcomitrium
pyriforme*, and those of the hybrid *P. pyriforme × Funaria hygrometrica*.
Pieces of older sporophyte yield only protonema but pieces of younger
ones, still growing by means of an apical cell, regenerate differentially
according to their proximity to the foot or to the apical end of the sporophyte (Fig. 12.1). Apical fragments regenerate callus at both ends and this
callus tissue gives rise immediately to new apogamous sporophytes. Intermediate and basal (foot-end) fragments consist of developmentally older
cells and regeneration from them proceeds initially *via* protonema formation. Between the apical and the intermediate regions there is a differentiating zone (B2 in Fig. 12.1) the behaviour of which in regeneration is
intermediate, i.e., some new seta tips are formed and some protonema, the
balance depending on the composition of the culture medium. Even here,
the protonema produced differs from normal protonema in eventually
giving rise, not to normal gametophyte buds, but directly to new sporogonia. In other words, information for sporophyte formation is transmitted
into the cells of the protonema, just as it is into the cells of the callus tissue

which apical segments produce, irrespective of the de-differentiation processes which precede their formation. Bauer (loc.cit.) comments 'we can conclude that those regions of the developing sporogonium which are still mainly embryonic possess a specific state of differentiation in their tissues which is preserved through and beyond regeneration... New properties which the cells (of the differentiating zone of the sporophyte) have acquired in the process of differentiation also persist throughout the regeneration process'.

Fig. 12.1. Course of regeneration in fragments of sporogonia of *Physcomitrium pyriforme* × *Funaria hygrometrica*. A, apical cell and immediately adjacent cells; B_1, meristematic zone which eventually gives rise to the capsule; B_2, meristem contributing to the seta, at a later stage forming the apophysis; C, extension zone; D, fully differentiated part of the seta; E, foot. 1, Protonema with gametophytic buds; 2, caulonema producing apogamous sporophytes; 3, sporophytes produced directly and on caulonema; 4, sporophytes produced directly on callus tissue. (Modified from Bauer, 1963.)

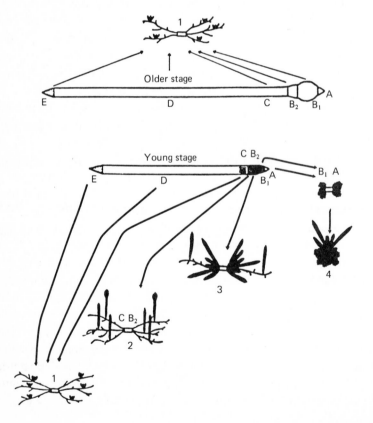

The instability of the positional information of the differentiating zone is attested by the fact that repeated transfer of the regenerating cultures to new media quickly eliminates the capacity to form sporophytes on the protonema. But this capacity can also be stabilized by transfer of the initial regenerating protonema to relatively dry (3–4%) agar medium, on which sporophytes continue to be formed, as Bauer (1963) says, through 'numerous subcultures'. The capacity to form sporophytes is also stabilized by addition of small amounts of glucose to the medium, reminiscent of the morphogenetic effects of sugars described by Allsopp (1953, 1954) and Hirsch (1975). Bauer maintains that the sporogonium-forming factor cannot be 'a hormone-like substance which passes from the (initial developing) sporogonium to the protonema... The factor must have the property of self-reproduction. It is able to propagate itself in every protonema cell, even though it is in the "wrong" physiological place'.

Similar gradients in regenerative performance have been found by a number of authors in the leaves and stems of bryophytes. In mosses regeneration proceeds by means of caulonema, from which vegetative buds arise. The caulonema are formed from epidermal cells. Basal stem segments produce more protonema axes than apical ones (Gay, 1971). Inner cells of the stem are unable to regenerate and epidermal cells, especially of the upper part of the stem, produce shoots directly (Lersten, 1961). Leaves of Polytrichales produce protonema only from the large cells over the veins which support the chlorophyllose lamellae; lamellae are absent from the basal part of the leaf, the cells of which, as well as those of the wing of the leaf, are incapable of producing protonema (Gay, loc. cit.).

In the higher plants, differences in regenerative performance have also been demonstrated to depend on the position in the whole plant from which the regenerating plant piece is taken. The most striking example, that of an apparent axial gradient in the quality of regeneration (i.e., whether vegetative or reproductive) from pieces of internode or leaf, taken from flowering plants of *Nicotiana tabacum* var. Wisconsin 38, is outlined in the chapter by Nozeran (see also the chapter by McDaniel). The observations have been repeated, using the tobacco variety Trapezond, by Chailakyan (1982) and his co-workers. Similar, but less striking, axial gradients in regeneration performance have been demonstrated by Chlyah, Sqalli-Khalil & Chlyah (1980) for internodes of flax (but interpreted by the authors as an ageing phenomenon), and by Margara (1965) for internodes of flowering stems of *Cichorium intybus*. In the latter case flowers were formed only by some of the uppermost segments, but when cultivated on a basal medium supplemented with 5% sucrose, all segments, apical and basal, produced flowers. According to Szweykowska (1978), who lists several Russian references relevant to this point, flower initiation cannot be obtained from tissues excised from young vegetative plants or from lower segments of the stem.

Gradients in the quality of regeneration have been demonstrated by Němec (1911) and Oehlkers (1956) in the leaf of *Streptocarpus wendlandii* (Gesneriaceae). Mature plants consist of the hypocotyl and a single, enormously enlarged cotyledon. An inflorescence is formed, following vernalization for 4–8 weeks, from a residual meristem at the top of the hypocotyl. Němec (1911) showed that cuttings from the basal parts of leaves of plants in flower, or able to flower, regenerated only inflorescences while, at the base of veins of cuttings made further towards the leaf tip, vegetative shoots formed. On cuttings made from regions in between, buds of intermediate character were produced. Oehlkers (1956) (whose article makes no reference to the earlier experiments of Němec and others who have experimented with this and related plants) made cuttings from the leaf each consisting of a lateral vein together with its attendant distal piece of lamina and a portion of the longitudinally halved midrib. Cuttings from non-vernalized plants generated either vegetative buds or none at all. Basal cuttings made from leaves of plants cold-treated for 4–8 weeks produced inflorescences directly (Fig. 12.2). Cuttings from further up the leaf generated buds which produced some leaves and then inflorescences. More distal cuttings produced either vegetative shoots or none at all.

Gradients of regenerative performance have been claimed, in a brief report by Stewart (1922), to be shown by the root of *Acanthus*. A root about 30 cm long was cut into segments about 7.5–10 cm long which were laid horizontally. Buds formed on the upper surface, but only one or two developed per segment. 'The shoot which develops from the youngest piece which bears the root apex is juvenile in character whereas the growth formed on the oldest portion shows much more adult characters.' These 'characters' were expressed in the leaf spectrum of the developed shoot, the 'juvenile' character being shown in the production of leaves like those of seedlings and then a set of transitional leaves, whereas a shoot of more 'adult' character proceeded directly to the formation of 'adult' leaves. It would be of interest to repeat these experiments, and also to remove the shoot apex from the youngest segment to see whether it exercises a hormonal influence on the character of the regeneration.

There is a considerable body of published information on the loss of ability of stem cuttings to form roots with height of insertion of the cuttings on the main stem of trees. Capacity to root is thus a 'juvenile' character (Doorenboos,1965; see also the chapter by Nozeran), retained by the tissues of the earliest internodes. There is evidence of the stability of 'positional information' which determines differences in rooting ability between juvenile and adult tissue. Cuttings of *Eucalyptus grandis* will root readily so long as they are taken from seedlings up to about the tenth node. Cuttings taken above this node do not root. Capacity for rooting appears to be retained at the lower nodes for many years. Pieces of stem tissue (including the cambium) from lower nodes of *E. grandis* patch-grafted to an adult stem retained their ability to form roots when pieces of the stem were treated as cuttings (Paton & Willing, 1973). The difference in rooting ability is therefore inherent in juvenile tissue (even when it is not primary) and is not lost during a sojourn in 'adult' surroundings.

Fig. 12.2. Diagram of a leaf of *Streptocarpus wendlandii* × *grandis*. Scale along the midrib in centimetres. Cuttings each including 12 cm of midrib were made from a leaf of a plant vernalized for 8 weeks. Data on types of regeneration from cuttings at the locations of their origin: ⊗, inflorescences without leaf production; ⊙, inflorescences preceded by leaves; ○, leaves only; ●, neither leaves nor inflorescences; ×, cuttings in soil; *, cuttings in sand plus Belvitan (β-indolyl acetic acid). (After Oehlkers, 1956.)

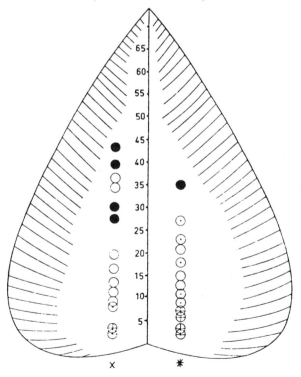

Regenerative performance of meristems

In higher plants, in addition to an axillary bud, a leaf may subtend one or more accessory shoot meristems. The fate of these meristems – or their potentiality in regeneration – appears to be determined at their inception and is often different from that of the axillary bud in the same axil. In eucalypts, the axillary bud is *sylleptic* (as re-defined by Tomlinson & Gill, 1973), i.e., it usually grows out during the current year's flush of growth, or it abscises without much extension. The accessory buds are *proleptic* and undergo a period of 'relative dormancy', which may last for many years, before growing out. This difference in bud behaviour is attributable to the different positional relationships of the two kinds of meristems in the leaf axil.

In terms of vegetative growth, the most striking example of differential performance determined by the position of shoot meristems is that shown by *Araucaria excelsa* (Massart, 1923) which is responsible for the rigorously geometrical, 'pagoda-like' architecture of trees of that species. Figure 12.3 shows the three kinds of terminal bud meristems, A, B, and C. A is the terminal bud meristem which gives rise to the main stem and its leaves. Buds terminating main branches (first-order branches) are called B. Buds terminating second-order branches are called C. In the leaf axils there are other meristems, a, b, and c, which remain suppressed. Experiment shows that buds of type a, borne on the main stem, give rise only to erect (orthotropic) stems, those of type b (on the first-order branches) only to branches which, in their plagiotropic behaviour and dorsiventrality, resemble first-order branches, while type c buds, on second-order branches (branchlets), can give rise only to branchlets which they resemble in their diageotropic behaviour, with their tips curved downwards.

As had already been shown by Vöchting (1904), the tip of the main stem, treated as a cutting, grows on like a main stem; branches treated as cuttings remain branch-like in their growth, while cuttings made from branchlets grow like branchlets. Repeated removal of the tip of the main stem causes buds of type a, irrespective of their position on the upper or lower part of the main stem, to grow out, replacing the lost 'leader', but none of these grow into branches. Nor do type b buds grow out to form branchlets following removal of all the branchlets from a branch. Buds of type B are formed in the axils of leaves on the main stem, at the end of its annual increment of growth, i.e., in place of buds of type a. Similarly, buds of type C are formed on branches at regular intervals in place of buds of type b. The spacing between the different types of buds is determined by the vigour of growth. The subsequent potentialities of these different kinds of buds are evidently determined at their inception, since regeneration performance was independent of the time at which decapitation of the main stem or branches, or branchlet removal, was carried out. It is therefore the position of these meristems in relation to growth activity just completed prior to their initiation which determines their fate, a concept explored in the chapter by McDaniel (q.v.). Very similar phenomena involving apices of the main shoot branches of the flowering plant, *Phyllanthus urinaria*, are dealt with in the chapter by Nozeran. Massart (1923) considered the precise and unchangeable fate of each kind of apical meristem to be due to a stable determination not unlike that of parts of a mosaic animal egg.

Fig. 12.3. The types of vegetative buds on *Araucaria excelsa*. For explanation see text. (After Massart, 1923.)

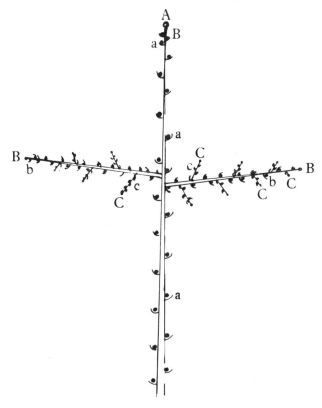

Even in plants of a growth pattern much less precisely determined than that of *Araucaria excelsa*, there is evidence of positionally-determined differences in the performance of otherwise morphologically identical shoot meristems. Among other examples, Dostál (1967) cites his early experiments on *Scrophularia nodosa* and *Circaea lutetiana* (Dostál, 1911) in which the plants were cut up into segments, each with a node bearing a pair of leaves and their axillary buds. Basal segments produced runners or tubers, while apical segments produced flowers, if the plants from which they were taken were flowering. Intermediate segments produced transitional shoots which could go on to produce flowers. It would be useful and interesting to devise and carry out experiments (e.g., reciprocal grafts) to test whether the difference in performance is really due to different informational contents of the bud meristems or whether it is due to different states of induction of the subtending leaves. Dostál (1911) found that darkening or removing a subtending leaf allows its axillary bud (excised together with a piece of stem) to develop as an erect leafy shoot, irrespective of its position on the (still leafy) part of the shoot. On the other hand, buds excised from stolons always develop as stolons. Other examples of this kind are dealt with in the chapter by McDaniel.

Carr, Carr & Jahnke (1982) have recently described an apparently irrevocable performance restricted to basally-located shoot meristems in eucalypts. In the axils of the cotyledons of the majority of species of eucalypts, and in some, in the axils of the next few pairs of seedling leaves, woody swellings called lignotubers appear. They are restricted to these nodes and those of shoots derived from them, and are never formed at later nodes on the seedling or adult stem. Each tuber is formed beneath multiple accessory buds (Carr, Jahnke & Carr, in press) which continue to proliferate, so that the surface of the developed lignotuber bears a large number of buds. If a young, basal leaf axil of a seedling is excised and cultured on a suitable medium, the accessory buds continue to proliferate indefinitely and some grow out to form shoots which may be rooted as cuttings. After some months, lignotubers are formed at the basal nodes of these cuttings, so long as they derive from a lignotuberous species, but not otherwise. Similarly, from a fully-formed lignotuber, buds may grow out which also form lignotubers at their basal nodes. In some species such shoots may be plagiotropic and develop as runners or stolons, with paired lignotubers at the nodes. The lignotubers of some species of eucalypts (especially some members of the series Corymbosae) give rise to underground rhizomes from which leafy shoots arise which form lignotubers at their basal nodes. Evidently, the information inherent in the initial accessory meristems of potentially lignotuberous nodes is inherited by meristems derived from them. It is not lost during a period of growth as a thin rhizome several centimetres long, during which the apical meristem produces opposite pairs of scale leaves. Months or years later, when the rhizome thickens and becomes woody, a pair of lignotuberous swellings appears at each node (C. J. Lacey, personal communication). The extent of transmission of the information for lignotuber formation depends on whether the secondary shoots, produced from accessory buds, are ortho-tropic, in which case only a few nodes may form lignotubers, or plagio-tropic, in which case the information appears to be transmitted clonally to an indefinite number of nodes. Exactly what influences determine whether a secondary shoot from the lignotuber will be orthotropic or plagiotropic is not yet known.

Position and rejuvenation

There is evidence that the future performance of an apical meristem may be influenced by a sojourn in a region of the shoot different from that in which it (or its immediate progenitors) arose. This evidence comes mainly from attempts to 'rejuvenate' adult parts of the plant to enable them to regain the capacity for rooting, when treated as cuttings. Cuttings made directly from adult parts of plants maintain adult characteristics when rooted and grown on as separate plants. But if an adult piece of a plant is grafted to a juvenile part of the same plan and allowed to develop for a time, before being removed and treated as a cutting, it may assume juve-nile characteristics (Doorenbos, 1965). Shoots taken from adult parts of eucalypts are not capable of forming roots but show symptoms of reju-venation (i.e., of becoming capable of forming roots) when grafted basally and allowed to grow for a while. This is demonstrated by allowing the grafted shoot to grow for some months then cutting it back, thus forcing newly-formed accessory buds to grow out (Martin & Quillet, 1974; Cauven & Marian, 1978). As cuttings, these new shoots will form roots, whereas the original (grafted) cuttings or epicormic shoots derived from them would not. Using a somewhat different technique, in which the shoots to be grafted were obtained from accessory buds developed by adult nodes of *Eucalyptus grandis* cultured on a nutrient medium, Paton, Willing & Pryor (1981) have demonstrated a similar rejuvenation, mea-sured in terms of rooting ability. The phenomenon is reminiscent of trans-determination of *Drosophila* imaginal discs (Hadorn, 1966), by which the potential developmental performance of some cells of a disc is changed by repeated transfer to the morphogenetic field of a non-homologous disc (see also the chapter by McDaniel).

The potential performance of apical meristems may be influenced by a prolonged period of suppression, or a long sojourn within stem tissues (see also the work of Ducreux on algae, reported on p. 384 in the chapter by Nozeran). As already pointed out, accessory buds of eucalypts may remain suppressed or 'dormant' for many years. The term dormancy is here relative, for the bud meristems located in the cambial region not only add to their bud strands as the girth of stem increases, but the meristems themselves proliferate (Cremer, 1972). When the accessory buds grow out as epicormic shoots, the form and other properties of the leaves they produce is influenced by the duration of their sojourn within the bark. If the accessory buds grow out on the subtending, adult stem in the year of their production or a year later, the leaves formed, although small, differ in no essential way from the surrounding adult leaves. However, on epicormic shoots that grow out following a sojourn of several years within the bark, at least the first few pairs of leaves are juvenile in character.

This is well seen in *Eucalyptus albida* Maiden & Blakely, the juvenile leaves of which are opposite, oval and very glaucous, while the adult leaves are sub-decussate, lanceolate, green and glossy (Fig. 12.4). To some degree the 'juvenization' of the accessory bud is quantitative; the longer the sojourn within the bark, the longer the sequence of juvenile leaves produced by the epicormic shoot. Epicormic shoots from the oldest accessory buds, near the base of the trunk, therefore retain their juvenility as long as would a seedling stem, whereas epicormic shoots growing out high on the stem, or on high branches, and therefore derived from relatively young accessory buds, pass quickly through the juvenile stage and then produce adult foliage.

The bud apical meristem is not the only part affected by its sojourn within the bark. Späth (1912) pointed out that sylleptic shoots lack the basal pair of – usually opposite or sub-opposite – scale-like leaves, or prophylls, which are produced on proleptic shoots. Each accessory bud of a eucalypt produces a pair of leaf primordia which are retained throughout the life of the bud. The shape and other properties of the two leaves developed from these primordia on an epicormic shoot reflect the duration of the sojourn of the bud within the bark, or alternatively the position of the bud in relation to the overall extension and lateral growth of the tree. If an adult accessory bud grows out within a year or two of its formation, its basal pair of leaves resemble small versions of the adult

Fig. 12.4. Young epicormic shoots of *Eucalyptus albida*. (1) Three pairs of shoots from successively younger parts of an adult region: (A) six years old, (B) three years old, (C) one year old; all from the same plant. All the shoots developed simultaneously, following removal of the top of the plant. The 'adult' leaves are sub-opposite, lanceolate, green and glossy (asterisks in B and C). 'Juvenile' leaves are opposite, oval and glaucous (A). Basal pairs of leaves (arrowed) in (A) are oval and quite glaucous, in (B) are oval and somewhat glaucous, and in (C) are lanceolate and green. (2) Part of a two–three-year-old stem with adult leaves and epicormic shoots. Basal leaves (arrowheads) on shoot (E) are oval and sub-glaucous, the later leaves are green and lanceolate. In contrast, the shoots labelled (D), although only a year older, have several pairs of glaucous ovate leaves. Scale, millimetres.

leaves. If the same kind of bud has remained within the bark for 4–5 years or more, on growing out the basal leaves take on a resemblance to juvenile leaves. It should be noted that some of the differences (e.g., wax secretion) between juvenile and adult leaves of eucalypts are expressed at the cellular level: the stomata of juvenile eucalypt leaves are always smaller – sometimes very much smaller – than those of the adult leaves (Carr & Carr, in preparation). Their sojourn within the bark therefore changes not only the leaf shape but also the potential performance of the cells of the basal pair of leaf primordia, as well as that of leaves subsequently produced by the reactivated apical meristem of the epicormic shoot.

Reproductive and floral meristems

Angiosperms

In general, the vast literature on the physiology of flowering has little to say on certain morphological problems, such as why the capacity to form flowers is normally restricted to a relatively few shoot apical meristems, and why there is so great a variety of species-specific arrangements of these meristems on the plant. Examples of precocious flowering (paedogenesis) are not rare either in nature (von Denffer, 1950) or induced by daylength treatment of seedlings of some plants. Observations of paedogenesis leave the impression that all shoot apices are potentially floral, but they offer no explanation of why flowering normally follows a vegetative phase, and why the capacity of a shoot meristem to become reproductive (whether as a flower or as an inflorescence) is a function of its position (on this point, see also the chapter by Nozeran). The same comments may be made concerning the disposition of apices which, on the same plant, can give rise only to male flowers or only to female flowers. Where the sexes are thus separated on the same plant, they may be in separate inflorescence or within the same inflorescence. In a few cases however, this disposition can be altered by daylength or hormonal treatment.

The sequence of parts of flowers affords another instance of positionally-determined morphology (see also the chapter by Meinhardt). Under the influence of an apical field, primordia which appear initially similar or even identical develop differently, the earliest ones as bracts or hypsophylls, the later ones in sequence as perianth parts, corolla, stamens and the gynoecium (Heslop-Harrison, 1964). This pattern can be considered as derived from a compressed strobilus, in which male organs appear on the axis below female organs. In the cryptogams the order is often reversed: in *Selaginella*, the microsporangia are at the top, the megasporangia at the bottom of the strobilus.

In Compositae there is a great variety of arrangements of sterile pseudo-male or pseudo-female, functionally male or female, and hermaphrodite flowers in the inflorescence. In a few species, the female flowers are situated on the axis of the inflorescence below the (functionally) male or bisexual flowers. In *Calendula*, a series of bracts is followed by a series of outer bisexual (ray) florets with ligulate (petal-like) corollas, then by the disc florets which are sterile or pseudo-female. Even the outer and inner fruits developed are of different shapes. The physiology of germination of these dimorphic (in some species trimorphic) fruits may also differ (as it does, for example, in *Xanthium*). Longitudinal bisection, or injury to the centre, of a sunflower capitulum (flower head) at a sufficiently early stage of development may result in the production of twin capitulae in which involucral bracts and ray florets are formed on the cut edges (Palmer & Marc, 1982). The determination of the flower primordia as ray florets evidently depends on their peripheral position, since the edges of slits made either experimentally or caused by temporary boron deficiency in the middle of the developing capitulum become occupied by ray florets, an effect not duplicated by conventional hormonal treatments (Palmer & Marc, 1982).

Apical meristems may pass through a sensitive stage in which they are competent, if conditions are favourable, to become reproductive, but after which they remain vegetative. Flowers in soybean (*Glycine max*) are formed from meristems subtended by bracts on short lateral shoots, each borne in a leaf axil. Those who have experimented with Biloxi soybeans will be aware that, in dissecting the axillary shoots following short-day treatment of the plants, the floral stimulus will have affected only certain axillary meristems which were receptive at the time when the floral stimulus reached them (see Hamner, 1969). We have, of course, no idea what determines how a meristem gains or loses receptivity, but perhaps the concept of 'competence' can be extended to explain the restriction of potentiality to flower to only certain meristems whose distribution is a characteristic of the plant species. In a right-handed clone of *Lemna perpusilla* the first daughter frond always appears in the right-hand meristematic pocket, the flowers always in the left-hand pocket (Witztum, 1979). Chouard & Tran Thanh Van (1962) have shown that, in the herbaceous perennial *Geum urbanum*, flowers are formed by lateral meristems low on the plant. Flowers are formed at increasingly high nodes with increase in the duration of vernalization. Unnatural vernalization lasting a year induces even the terminal meristem to form a flower.

In one group of plants, the palms (Corner 1966), the location of potentially reproductive meristems shows a remarkable diversity, evidently genetically determined, but physiologically unresearched. Only experiment could determine how far positional information is involved in the precise and species-specific location of the inflorescences in these plants.

Gymnosperms and cryptogams

In conifers the position of male and female cones is also subject to relatively strict rules. In *Pinus sylvestris*, for instance, at the onset of sexual maturity (5–10 years) the tree initially produces some female cones on strong leading shoots. Male cones are then produced basitonically on lower branches, female cones being produced acrotonically. Later on, male cones form higher on the annual increments of the branches, and on branches of higher order, eventually on the leading shoots. 'Since the male condition arises first on the older branches a stage is frequently reached, in younger trees, in which the lower, older branches are predominantly or entirely male and the upper branches female or vegetative' (Wareing, 1959). The positional rules in conifers are maintained even when cones are induced experimentally by treatment with gibberellins (Pharis, 1978).

Similar fairly rigid rules govern the placement of reproductive organs in lower plants. The position, whether terminal or lateral, of an apical cell that gives rise to the archegonium, sharply divides the liverworts into two main series, differing also in their more foliose or more thallose form. In mosses, with the exception of *Sphagnum*, the antheridia and archegonia are formed at the tip of the stem but not by the apical cell itself, which, in some species (e.g., *Polytrichum* spp.) can continue the growth of the stem through the antheridial cup. In the Australian moss *Cyathophorum*, the archegonia are formed from meristematic discs which are arranged on the stem in two ranks, associated with the two ranks of dorsal leaves.

Positioning and perennial reproductive meristems

In some plants, certain positions may be set aside for continued production, year after year, of reproductive structures. The spur shoots or dwarf shoots of, e.g., rosaceous fruit trees represent perennial sites for the formation of flowers. The shoot apex of the spur shoot remains vegetative and can, with the removal of correlative inhibition, convert into the apex of a spur-bearing long shoot. But flowering is restricted to lateral meristems borne in the axils of leaves on the spur shoots. Similar perennial or long-lived inflorescences are found in *Hoya carnosa* and in other members of Asclepiadaceae (e.g., *Dischidia merrillii*, according to Goebel, 1931). In *Hoya* the inflorescences are axillary and apparently leafless. In fact, the flowers are subtended by bracts, which in *Hoya* take the place of the foliage leaves of the rosaceous spur shoot. It was A. P. de Candolle who first noted the perennial inflorescences of *Hoya carnosa*; Goebel (loc. cit.) was of the opinion that flower production by a *Hoya* inflorescence probably ceased when a fruit was formed, a circumstance which, in the absence of suitable pollinators, did not occur in cultivation. According to Bruce Gray (personal communication) fruits are infrequently set by Australian species of *Hoya* in the wild and those illustrated in Jones & Gray (1977) are laterally placed on the inflorescence, i.e., were succeeded by further flower production. In some species of *Hoya* the inflorescence branches are thus perennial sites, positionally determined, of flower formation.

In the genus *Banksia*, the position of the inflorescence is exactly specified: it terminates a main shoot or a short lateral, which may be an epicormic shoot from a stem 3–4 years old. However, although the position is specified, the inflorescence may not begin to develop for several years. In many species, following cessation of growth, the main shoot terminates a dormant bud. From axillary buds just below it, several shoots may grow out to form a whorl of branches, each of which will eventually terminate in the same way as its parent shoot. The apical bud eventually grows out to form an inflorescence (Fig. 12.5 A). Usually only one inflorescence is formed at the terminal site; sometimes more than one is formed. Exceptionally, and especially in cultivation, inflorescences may proliferate indefinitely from the terminal site, so that eventually inflorescences are formed over the whole dome and extend down its flanks (Fig. 12.5 B,D). This phenomenon (as observed in *B. oblongifolia* Cav.) demonstrates vividly that the terminal site is positionally determined for inflorescence formation. Any shoot meristem remaining at that site, after outgrowth of the whorl of branches, may develop as an inflorescence. In aberrant plants of *B. oblongifolia* the whole 'inflorescence position' may be taken over and covered with cones or their remains. When that happens, residual meristems on the stem below (i.e., epicormic buds remaining from the former leaf axils) may give rise to inflorescences, the 'information' spilling over, as it were, from the now fully-occupied 'inflorescence

position'. The plant thus becomes cauliflorous (Fig. 12.5 C,E). It is interesting that each cauliflorous cone of *Banksia oblongifolia* is subtended by a normal complement of hypsophylls (bracts) (Fig. 12.5 E), unlike proleptic branches to which epicormic shoots in *Banksia* may give rise. In a number of species (e.g., *B. baueri* R.Br., *B. quercifolia* R.Br. and *B. oreophila* A. S. George) the cones are borne on short lateral branches and are rarely terminal to a main branch (George, 1981). These short laterals may be epicormic. In *B. candolleana* Meissner epicormic buds on old stems give rise to a succession of short laterals, each terminating in a cone.

Cauliflory

The phenomenon of cauliflory has received scant attention from physiologists, although Klebs (1911) and, before him, Haberlandt (1893) mentioned it and developed nutritional hypotheses to explain it. We must therefore consult books on ecology and plant geography for descriptions and analyses of this particular phenomenon.

Simple cauliflory (as in *Theobroma*, *Diospyros* spp., *Drypetes*, the Australian *Syzygium cormiflora*, etc.) is the production of flowers anywhere on the stem, on the trunk and branches, as well as on the twigs. The flowers may also be restricted to the trunk, as in the South American cannonball tree, *Coroupita guianensis*, and the Javanese *Aristolochia barbata*, *Cynometra cauliflora*, etc. In *Saurauia cauliflora* and *Durio testudinarum* (illustrated in Menninger, 1967) the flowers are restricted to the base of the trunk. 'The cauliflory of *Stelechocarpus burahol*, a small tree in the family Annonaceae, is most peculiar: the female flowers emerge in clusters from thick warty growths on the trunk while the smaller male flowers appear in the axils of leaves, which soon abscise, on the twigs' (Schimper

---

Fig. 12.5. Inflorescences of *Banksia oblongifolia*. (A) A main branch (running up from the left) three years old, has terminated in a bud and given rise to three other branches. From the terminal bud a young cone is developing. (B) Four cones, one old (pointing to the left), two flowering and one young, have developed at a single terminal 'inflorescence position'. (C) Cauliflorous cones developed on the main stem below a terminal 'inflorescence position'. The arrow indicates the hypsophylls of a new, as yet undeveloped, cone. (D) A set of cones at a terminal 'inflorescence position'. Small arrows indicate primordia of future cones. (E) Cauliflorous cones on an old leafless stem bearing a terminal 'inflorescence position' entirely covered with cones, old and new. Large arrow, the hypsophylls subtending a cauliflorous cone. Small arrows, a cluster of potential cones.

& Faber, 1935). A similar case of specific and separate location of shoot meristem function is that of *Ficus ribes*, a small Javanese rainforest tree, which shows 'a sharp separation of foliage-bearing and fruiting branches (receptaculae). The latter emerge from the base of the trunk, grow to 2.5 m in length, and are pressed closely to the soil or even enter it and then form short, normal roots' (Schimper & Faber, loc. cit.). In another rain-forest species, *Annona rhizantha*, the fertile branches remain below ground and only their tips, bearing the flowers, emerge from the soil.

All these instances – and examples could be multiplied – demonstrate the great diversity of morphological arrangements for specifying the position of potentially reproductive meristems in different plants. It seems very probable that these meristems carry positional information which determines their privileged status in reproduction, but as yet we have nothing but descriptions and the most elementary of hypotheses, mostly teleological, to explain them and their abundance in the tropical rain-forest.

Topophysis and cyclophysis

The results of experiments such as those of Vöchting on *Araucaria excelsa* led Molisch (1922) to write: 'in many plants shoots are not of identical nature but differ according to the positions they occupy and these differences persist when the shoots are treated as cuttings or graft partners'. He coined the term 'topophysis' for this phenomenon. Olesen (1978), attempting to clarify the definition, re-defined it (omitting Molisch's specific reference to the position of the shoot) as 'the phenomenon that scions, buddings and cuttings, for some time after grafting, budding or rooting, maintain the branch-like habit they had as shoots on the plant'. Both authors agree, however, in ascribing this behaviour to a property of the shoot meristems, especially of woody plants. But many of the phenomena mentioned in this chapter concern non-meristematic tissues in herbaceous plants. The neutral term 'positional information' subsumes the term topophysis and refers to a wider set of morphological phenomena of the same kind.

Molisch (loc. cit.) cites the experiences of horticulturists in propagating such plants as roses, and extends the concept of topophysis to them also. 'If long, non-flowering shoots are used as sources of buds for propagation, pauciflorous plants result. If buds are obtained from short, flower-bearing shoots, in general, the resulting plants have short, floriferous branches.' More recently, it has been shown that grafts from the lower branches of *Picea abies* flower poorly, in contrast with grafts from the top branches

(Dormling, 1976). Molisch (loc. cit.) says that 'cuttings from flowering shoots of *Sparmannia africana* yield stocky, almost dwarf but floriferous plants, while cuttings from strong, leafy shoots grow strongly and flower rarely. Sucker shoots from the roots of trees like plums ... in contrast to seedlings, yield plants with a striking tendency to form root suckers. Used as graft scions, the water-shoots of apples and pears develop into trees which retain the nature of water-shoots, strongly growing and with little tendency to flower and set fruits'. According to Büsgen & Münch (1929), the twigs and needles at the top of an old spruce or silver fir look quite different from those of young trees of about man's height. 'If a leading shoot of a spruce is grafted on to young spruce, a tree is obtained which resembles exactly in the structure of its branches and twigs and in the form of its needles the top of an old tree cut off and stuck in the ground. An artificially rejuvenated tree of this sort also bears cones more plentifully than a young one of the same age grown from seed.'

Seeliger (1924) introduced the term 'cyclophysis' to denote characters determined by age or course of life, supposing that persistent morphological changes result from internal readjustments taking place with age. There seems no way in which one can dissociate such changes from positional effects since inevitably the older branches are positioned differently from the younger ones. Some authors have sought to re-define cyclophysis as 'the process of maturation of the apical meristem' (Fortanier & Jonkers, 1976). Wareing (1959) has distinguished between phase change (i.e., juvenile to adult) and ageing, the latter being allegedly reversible, but not the former. Olesen (1978) maintains that cyclophysis is independent of topophysis and that the first is due to ageing or maturation of meristems, the second to pure position on the plant. At least some degree of rejuvenation of 'older', 'more mature' meristems has been shown to be possible (by transferring them to 'younger', 'less mature' regions of the plant). Moreover, topophytic effects are not always as irreversible as they appear to be in *Araucaria excelsa*. In *A. cunninghamii*, for instance, a branch may replace the lost leader (Massart, 1923).

Topophytic and cyclophytic phenomena are subject to reversal to an extent which varies from plant to plant. Even if the difference between them is real, both can be considered as expressions of forms of determination differing only in degree of stability. This determination extends not only to meristems but also to non-meristematic tissue. In such tissues the positionally-determined potentialities elicited in regeneration may appear to resemble, or perhaps to be an extension of, the state of determination of the meristem they derived from.

Conclusion

Holder (1979) sees the 'primary meristematic cells as those comprising the fields in which positional information is specified', other tissues derived from them having interpreted their positional values by descent. Holder's view is an attempt to reconcile positional information in plants with the tenets applied to animals, in which 'the case for universality is strengthened by the fact that all fields are about the same size, usually less than 1 mm or 50 cells in the maximum dimension' (Wolpert, 1975). Unfortunately, this cannot account for the determination of pattern in such organs as leaves which initially have their own meristems and morphogenetic fields (the problem of positionally-determined patterns in organs was recognized by Wolpert (1969) as a difficult and 'central' one), nor can it account for the distribution of differential morphogenetic capacities (e.g., the expression of juvenile and adult characters, or the localized change from vegetative to reproductive behaviour) within the whole plant. Nevertheless, some of the main elements of morphological patterning appear to be set within quite small regions, such as the tip of a shoot, a leaf primordium, a flower primordium, etc.

Most animals have a 'closed' development, unlike plants, and the ends of a gradient which the cells interpret in terms of their position within it must be either continuously separated as the plant grows, or else there are 'units' of gradient within the plant (such as might coincide with episodic increments of growth of a perennial plant). Within such a newly-developing unit of growth (? a 'phytomer') a new morphological gradient is set up; there will be temporal changes in the nature of these gradients as the plant expands and during the development of the region of the 'unit' gradient itself. Such a system is evident in *Araucaria excelsa*: the unit gradient switches towards the end of an annual increment of growth of the main shoot (perhaps due to a feedback from the increment itself, as envisaged in the chapter by Meinhardt) so that a newly-forming lateral shoot meristem interprets its position as that of a B-type bud, whereas during the previous growth it would have interpreted the same position as that of an a-type bud (see Fig. 12.3). The same sort of switching occurs periodically on the branches, when the field of the B-type bud gives rise to lateral meristems of C-type instead of the commoner b-type. These periodic switchings are thus directly related to the rhythm of growth which is in turn governed by both internal and external factors. Only colonial animals, such as corals, have a comparable 'open', episodic development.

The concept of positional information reduces the dependence of morphogenesis on size (Wolpert, 1968, 1969, 1975).Thus, morphological positional information allows for the development, even by dwarfed plants (e.g., bonsai), of a full complement of morphological behaviour. Mere size or vigour, for instance, does not affect the positional determination of the different kinds of shoot apical meristems in *Araucaria excelsa*. Wareing (1959) has concluded, on the basis of experiments with *Betula verrucosa*, that 'the attainment of size is the primary factor determining the transition to the adult condition and that it is not important whether this size is attained as a result of continuous growth or by the normal periodic succession of seasonal growth increments'. This conclusion is at variance with both paedogenesis and the attainment of sexual maturity by 'bonsai' plants, whether in nature or in cultivation.

Gradients, whether physico-chemical or hormonal, may provide signals for the establishment of the determined state. Auxin has been cited by proponents (e.g., Webster, 1971) of positional information in animal development as a model morphogen which moves polarly and elicits different kinds of responses. But the movement of auxin is measured in metres not microns and in some responses (e.g., correlative inhibition) its effects do not diminish with distance. If auxin is to be implicated in positional information, critical studies, including measurements of auxin levels, must be made within formative fields such as those close to shoot apices. So far, the only data we have come from crude application of auxin (sometimes in large quantities in lanolin) or estimations of auxin contents, not at a cellular or primordium level, but of whole organs or even whole plants (e.g., *Lemna*).

Other gradients may be implicated. Wareing (1959) as suggested that male cones of *Pinus sylvestris* arise in regions where vigour is least (i.e., where 'ageing', by his definition, is greatest). He suggests that nutritional competition between such regions and dominant shoots is to the advantage of the latter. This is an interesting correlation, and nutritional gradients may play a part in the determination of certain meristems of future male cones; but it is not a universal explanation of the distribution of male cones in conifers, nor of inflorescences in other plants in general.

Many of the experiments and observations mentioned in this chapter are old, but there seems no reason to doubt the validity of the main conclusions reached by their authors. What is needed is an expansion of experimental work, especially the performance of critical experiments to confirm these conclusions and to test the concept put forward – that of morphological positional information.

References

Allsopp, A. (1953). Investigations on *Marsilea* 3. The effects of various sugars on development and morphology. *Annals of Botany*, **17**, 447–63.

— (1954). Investigations on *Marsilea* 4. Anatomical effects of changes in sugar concentration. *Annals of Botany*, **18**, 449–61.

Bauer, L. (1963). On the physiology of sporangium differentiation in mosses. *Journal of the Linnean Society of London (Botany)*, **58**, 343–51.

Bünning, E. (1953). *Entwicklungs- und Bewegungsphysiologie der Pflanzen*, 3rd edn. Berlin, Göttingen & Heidelberg: Springer-Verlag.

— (1965). Die Entstehung von Mustern in der Entwicklung von Pflanzen. In *Encyclopaedia of Plant Physiology* vol. 15/1, ed. W. Ruhland, pp. 383–408. Berlin, Heidelberg & New York: Springer-Verlag.

Büsgen, M. & Münch, E. (1929). *The Structure and Life of Forest Trees*, 3rd edn. (English translation by T. Thomson). London: Chapman & Hall.

Carr, D. J., Carr, S. G. M. & Jahnke, R. (1982). The eucalypt lignotuber: a position-dependent organ. *Annals of Botany*, **50**, 481–9.

Cauven, B. & Marian, J. N. (1978). La multiplication végétative des *Eucalyptus* en France. *Annales de Recherches Sylvicoles*, **1**, 141–75.

Chailakyan, M. Kh. (1982). Hormonal substances in flowering. In *Plant Growth Substances 1982*, ed. P. F. Wareing, pp. 644–55. London: Academic Press.

Chlyah, H., Sqalli-Khalil, M. & Chlyah, A. (1980). Dimorphism in bud regeneration in flax (*Linum usitatissimum*). *Canadian Journal of Botany*, **58**, 637–41.

Chouard, P. & Tran Thanh Van, M. (1962). Nouvelles recherches sur l'analyse des mécanismes de la vernalisation d'une plante vivace, le *Géum urbanum* L. (Rosacées). *Bulletin de la Societé Botanique de France*, **109**, 145–7.

Corner, E. J. H. (1966). *The Natural History of Palms*. London: Weidenfeld & Nicolson.

Cremer, K. W. (1972). Morphology and development of the primary and accessory buds of *Eucalyptus regnans*. *Australian Journal of Botany*, **20**, 175–95.

Denffer, D. von (1950). Blühhormone oder Blühhemmung? Neue Gesichtspunkte zur Physiologie der Blütenbildung. *Die Naturwissenschaften*, **37**, 296–301, 317–21.

Doorenbos, J. (1965). Juvenile and adult phases in woody plants. In *Encyclopaedia of Plant Physiology*, vol. 15/1, ed. W. Ruhland, pp. 1222–35. Berlin, Heidelberg and New York: Springer-Verlag.

Dormling, I. (1976). Topophysiseffekten hos klonförökad gran (*Picea abies* (L.) Karst.). Stockholm: Avdelingen för Skoglig ekofysiologie Skogshögskolan.

Dostál, R. (1911). Zur experimentellen Morphogenesis bei *Circaea* und einigen anderen Pflanzen. *Flora*, **103**, 1–53.

— (1967). *On Integration in Plants*. Cambridge, Mass.: Harvard University Press.

Driesch, H. (1908). *The Science and Philosophy of the Organism*. London: Black.

Fortanier, E. J. & Jonkers, H. (1976). Juvenility and maturity of plants as influenced by their ontogenetical and physiological ageing. *Acta Horticulturae*, **56**, 7–43.

Gay, L. (1971). Correlative systems controlling regeneration on gametophytes of *Polytrichum juniperinum* Willd. *Zeitschrift für Pflanzenphysiologie*, **66**, 1–11.

George, A. S. (1981). The genus *Banksia* L.f. (Proteaceae). *Nuytsia*, **3**, 239–473.

Goebel, K. (1931). *Blütenbildung und Sprossgestaltung*. Jena: Fischer.

Haberlandt, G. (1893). *Eine botanische Tropenreise*. Leipzig: Engelmann.

Hadorn, E. (1966). Dynamics of determination. In *Major Problems of Developmental*

Biology, ed. M. Locke, pp. 85–104. New York: Academic Press.

Hamner, K. (1969). *Glycine max* (L.) Merrill. In *The Induction of Flowering*, ed. L. T. Evans, pp. 62–89. Melbourne, Australia: MacMillan.

Heslop-Harrison, J. (1964). Sex expression in flowering plants. In *Meristems and Differentiation. Brookhaven Symposia in Biology*, no. 16, 109–25.

Hirsch, A. M. (1975). The effect of sucrose on the differentiation of excised fern leaf tissue into either gametophytes or sporophytes. *Plant Physiology*, **56**, 390–3.

Holder, N. (1979). Positional information and pattern formation in plant morphogenesis and a mechanism for the involvement of plant hormones. *Journal of Theoretical Biology*, **77**, 195–212.

Jones, D. L. & Gray, B. (1977). *Australian Climbing Plants*. Sydney: Reed.

Klebs, G. (1911). Uber die Rhythmik in der Entwicklung der Pflanzen. *Sitzungsberichte Heidelbergische Akademie der Wissenschaften, mathematische-naturwissenschaftliche Klasse*, **23**.

Lersten, N. R. (1961). A comparative study of regeneration from isolated gametophyte tissues in *Mnium. Bryologist*, **64**, 37–47.

Lewis, J., Slack, J. M. W. & Wolpert, L. (1977). Thresholds in development. *Journal of Theoretical Biology*, **65**, 579–90.

Margara, J. (1965). Comparaison *in vitro* du développement de bourgeons de la tige florifère de *Cichorium intybus* L. et de l'évolution de bourgeons néoformés. *Comptes rendus hebdomadaires des Séances de l'Académie des Sciences de Paris*, **260**, 278–81.

Martin, B. & Quillet, G. (1974). Bouturage des arbres forestiers au Congo. *Revue de Bois et de Forêts Tropicales*, **155**, 15–33; **156**, 39–61; **157**, 21–39.

Massart, J. (1923). La coopération et le conflit des reflexes qui déterminent la forme du corps chez *Araucaria excelsa* R.Br. *Memoires de l'Académie Royale des Sciences de Belgique*, **5**, 3–33.

Menninger, E. A. (1967). *Fantastic Trees*. New York: Viking Press.

Molisch, H. (1922). *Pflanzenphysiologie als Theorie der Gärtnerei*, 5th edn. Jena: Fischer.

Němec, B. (1911). Weitere Untersuchungen über die Regeneration. IV. *Československa Akademie věd, Cisaře Frantiska Josefa. Bulletin Internationale, Classe Science, Mathématique et Médicine*, **16**, 238–56. (Cited in Priestley & Swingle, 1929).

Oehlkers, F. (1956). Veränderungen in der Blühbereitschaft vernalisierter Cotyledonen von *Streptocarpus*, kenntlich gemacht durch Blattstecklinge. *Zeitschrift für Naturforschung*, **11b**, 471–80.

Olesen, P. O. (1978). On cyclophysis and topophysis. *Silvae Genetica*, **27**, 173–8.

Palmer, J. H. & Marc, J. (1982). Wound-induced initiation of involucral bracts and florets in the developing sunflower inflorescence. *Plant & Cell Physiology*, **23**, 1401–9.

Paton, D. M. & Willing, R. R. (1973). Inhibitor transport and ontogenetic age in *Eucalyptus grandis*. In *Plant Growth Regulators*, pp. 126–32. Tokyo: Hirokawa Publishing Co.

Paton, D. M., Willing, R. R. & Pryor, L. D. (1981). Root-shoot gradients in *Eucalyptus* ontogeny. *Annals of Botany*, **47**, 835–8.

Pharis, R. P. (1978). Interaction of native and exogenous plant hormones in the flowering of woody plants. In *Regulation of Developmental Processes in Plants*, ed. H. R. Schütte & D. Gross, pp. 343–60. Saale: Halle.

Prat, H. (1948 and 1951). Histo-physiological gradients and plant organogenesis. Parts I & II. *The Botanical Review*, **14**, 603–43; **17**, 693–746.

Priestley, J. H. & Swingle, C. F. (1929). Vegetative propagation from the standpoint of plant anatomy. *United States Department of Agriculture Technical Bulletin*, no. 151, 1–98.

Raff, J. W., Hutchinson, J. F., Knox, R. B. & Clarke, A. E. (1979). Cell recognition: antigenic determinants of plant organs and their cultured cells. *Differentiation*, **12**, 179–86.

Schimper, A. F. W. & Faber, F. C. von (1935). *Pflanzengeographie auf physiologischer Grundlage*, 3rd edn, vol. 1. Jena: Fischer.

Seeliger, R. (1924). Topophysis und Zyklophysis pflanzlicher Organe und ihre Bedeutung für die Pflanzenkultur. *Angewandte Botanik*, **6**, 191–200.

Späth, H. L. (1912). *Der Johannistrieb*. Berlin: Paul Parey.

Stewart, L. B. (1922). Note on juvenile characters in root and stem cuttings of *Acanthus montanus*. *Transactions of the Botanical Society of Edinburgh*, **28**, 117–18.

Szweykowska, A. (1978). Regulation of organogenesis in cell and tissue culture by phytohormones. In *Regulation of Developmental Processes in Plants*, ed. H. R. Schütte & D. Gross, pp. 219–35. Saale: Halle.

Tomlinson, P. B. & Gill, A. M. (1973). Growth habits of tropical trees: some guiding principles. In *Tropical Forest Ecosystems in Africa and South America: a Comparative Review*, ed. B. J. Meggers, E. S. Ayensu & W. D. Duckworth, pp. 129–43. Washington D. C.: Smithsonian Institution Press.

Vöchting, H. (1878). *Über Organbildung im Pflanzenreich*, vol. 1, pp. 240–1. Bonn: Cohen.

— (1904). Über die Regeneration der *Araucaria excelsa*. *Jahrbucher für wissenschaftliche Botanik*, **40**, 144–55.

Wareing, P. F. (1959). Problems of juvenility and flowering in trees. *Journal of the Linnean Society of London (Botany)*, **56**, 282–9.

Webster, G. (1971). Morphogenesis and pattern formation in hydroids. *Biological Reviews of the Cambridge Philosophical Society*, **46**, 1–46.

Witztum, A. (1979). Morphogenesis of asymmetry and symmetry in *Lemna perpusilla* Torr. *Annals of Botany*, **43**, 423–30.

Wolpert, L. (1968). The French flag problem: a contribution to the discussion on pattern development and regulation. In *Towards a Theoretical Biology*, vol. 1. *Prolegomena*, ed. C. H. Waddington, pp. 125–33. Edinburgh: Edinburgh University Press.

— (1969). Positional information and the spatial pattern of cellular differentiation. *Journal of Theoretical Biology*, **25**, 1–47.

— (1975). The development of pattern: mechanisms based on positional information. *Advances in Chemical Physics*, **29**, 253–67.

13

Integration of organismal development

R. NOZERAN

Our aim is to seek out elements that will allow us to analyse the chain of events that constitutes the morphogenetic cycle of a plant. The phanerogams provide the basic experimental material of this study, but we shall, when necessary, draw parallels between them and findings from experiments with other plants (vascular cryptogams and thallophytes).

The occurrence of many of the steps in morphogenesis depends, in some way, upon ecological factors (e.g. photoperiod and ambient temperature). But our main objective is to identify large-scale events which play a role in morphogenesis, and more particularly to focus attention on the various internal correlations whose interplay determines that process. These correlations are positional controls *par excellence* in plant development since they can operate over considerable distances in the plant body and over long periods during the life-cycle.

First we shall describe the complex nature of whole-plant morphogenesis, or as we shall sometimes call it, the 'morphogenetic progression'. Then, we shall cite examples that help us to understand some of the individual internal elements in this sequence.

Our approach will be based essentially on morphological observations. Should it require justification, we can quote Darwin (1859, p. 415) who wrote in his master work *The Origin of Species* that morphology 'is the most interesting department of natural history, and may be said to be its very soul'. We shall also try to avoid the pitfalls of a static approach.

We have made mention of the plant's morphogenetic progression. This may be defined as the integration of the ontogenetic and physiological changes in each plant part within the sequence of its growth from a seed to its ultimate senescence and death. We believe that the integrated nature of separate developmental processes in space and time is an irrefutable fact and that the discovery and characterization of these various processes must constitute a base for our analysis. Thus, qualitative and quantitative data relative to the plant's life-cycle must be sought out, estimated, and understood as fully as possible. Such data may be obtained through direct observations. For instance, leaves may have a quite different appearance depending on which part of the plant has formed them; their morphological characteristics, both qualitative and quantitative, may be easily documented. But in addition, differences in activities in a part of an individual may also be detected indirectly through their different effects on other plant parts. As we shall see, detecting these differences requires the use of experimental artifices that often involve the removal of a specific portion of the plant. Because of this we are reminded that the part itself, by its contribution to the modification of the activities in other parts, creates, *ipso facto*, new conditions in its own environment. These will in turn react upon the plant and influence its further development, eventually leading it to a new mode of functioning.

What we have just outlined briefly is an enormously complex system; but it forms the fundamental basis for the existence of the chains of events which – physiologically and morphologically – take part in the construction of the plant's life-cycle.

GENERAL TRAITS OF THE MORPHOGENETIC PROGRAM

A plant's morphogenetic progression may be characterized by three groups of phenomena: the masking and the unmasking of morphogenetic capabilities, and the return to an initial morphogenetic state. In vascular plants, these three groups of phenomena involve the aerial as well as the underground systems (i.e. shoots and roots). We will examine them in both systems.

The aerial system

Loss of expression of morphogenetic capabilities

The production of juvenile-type leaves during the phase following seed germination characterizes the juvenile state, and is lost upon reaching the adult state. This is more strongly expressed in some species than in others.

Other morphogenetic capabilities expressed in the young seedling may cease to manifest themselves in the parts formed by the older individual. This may be shown in experiments which test, for instance, the rooting capacity of fragments excised from plants at various stages in their life. Thus, cuttings from plants such as *Hevea brasiliensis* (Gregory, 1951),

Cupressus dupreziana (Franclet, 1969; Nozeran, 1978) and *Citrus*, show an increasingly weaker rooting capacity in the course of their development, to the point of complete loss. The same is true of various other properties such as the ability to give rise to tissue cultures or to bud neoformations from organs, or fragments of organs, e.g. in *Hedera helix* (Privat, 1958).

Expression of latent vegetative morphogenetic potentialities

At the same time that some morphogenetic potentialities may be masked, others not previously expressed may reveal themselves. One of the most obvious illustrations is offered by plants with dimorphic vegetative branches such as *Coffea arabica* (Carvalho, Krug & Mendes, 1950) and *Phyllanthus amarus* (Bancilhon, 1969). Individuals of these species consist of an erect orthotropic axis that arises directly from the embryo and which, *after a certain period of development*, gives rise to plagiotropic axes of quite different structure. This type of growth is perpetuated through cuttings: the meristem from which the plagiotropic branch grows functions indefinitely in that mode, and in this sense is quite distinct from that of the erect axis (see also Chapter 12, p. 356–9).

New physiological processes may develop in the course of morphogenesis following germination. Such is the case, for instance, in the progressive acquisition of halophyte traits in *Suaeda* (Billard *et al.*, 1976; Boucaud & Ungar, 1978). Similarly, the manifestation of dormancy in buds is progressive and related to the age of the plant, as in *Picea excelsa* (Dormling, 1973) and *Pseudotsuga menziesii* (Goublaye de Nantois, 1980). It is likely that this may be related to the gradual acquisition of frost-resistance after a more sensitive young phase (as in *Metasequoia glyptostroboides*).

Return to an initial type of situation

The loss and acquisition of morphogenetic potentialities cause the development of vegetative structures which are qualitatively different from those characterizing the young seedling. This situation is the starting point of yet another new stage, flowering. Here, processes occur which lead to seed formation and favour a return to the initial state in the life-cycle, i.e., rejuvenation. In this respect, two remarks should be made. The first concerns the contradiction between the essentially ephemeral nature of the flower, one of the most rapidly senescing parts of the plant, and the fact that it is here that processes best suited to a return to a juvenile stage are located. The second is the realization that this phenomenon is not solely dependent upon fertilization. Indeed, the embryos of apomictic seeds, for instance those of various *Gramineae*, arise from the nucellus and are indistinguishable from embryos arising from the fertilized egg.

The underground system

Although investigations of the underground root system have been fewer, the available data indicate that phenomena comparable to those which we have just described for the aerial system occur here also. In certain plants, the expression of hidden potentialities in the initial stages may be easily observed. The most obvious example is provided by individuals from species with dimorphic roots. Roots arising from the seed are orthogravitropic but bear lateral roots that show indefinite plagiotropic growth. Such a situation exists in *Theobroma cacao* (Charrier, 1969; Dyanat-Nejad, 1971), in *Cupressus dupreziana* (Franclet, 1969; Nozeran, 1978), and in *Hevea brasiliensis*.

Similarly, the well-known example of suckers (stems exhibiting juvenile characters) developing from roots, illustrates that roots also may give rise to a structure characteristic of an initial ontogenic situation.

Modulations in the proposed general scheme

The scheme we propose raises various issues. One concerns the ambiguity of the adjective 'young'. This word is commonly used in describing plant growth, especially in perennial plants (annual shoots, for example, are said to be 'young'). However, if we take into account the morphogenetic potentialities of these newly-formed parts, we see that they may reveal growth processes different from those in the seedling; these result from 'losses' or 'unmasking' of growth processes, and consequently may be evidence of 'ageing'.

At first sight it may appear that the program which leads from youth (in the case of a phanerogam, from seed germination) to the senescence and death of the individual, corresponds to a continuous, linear process. However, a closer look leads us to consider it in a different, non-linear perspective.

We might illustrate linear development towards senescence in a tree, for instance, through the gradual build-up of an imbalance between branch development and mechanisms for ensuring the adequate partitioning of photosynthates and nutrients throughout the organism (Borchert, 1976; Chouard, 1977). Likewise, in annual plants it may be the imbalance (at least in some cases) between the needs of the individual and certain ecological conditions which accounts for its short life. But the apparent linearity of development cannot conceal the processes which modulate it. Elements of discontinuity do exist and are localized in specific parts of the plant. They may intervene in the natural propagation of an individual that otherwise would senesce and die. Here again we may mention suckers – rejuvenated shoots borne on roots – that allow a return to an initial mode of functioning characteristic of seedlings. The same occurs with bulbils which, in certain plants, are located at the level of, or near, the flowers. We may also mention the possibilities offered by another type of organ, the leaf, which usually enjoys a more limited destiny. We know that cells from this organ, doomed to senescence and death, are capable of giving rise to new young individuals (e.g. *Bryophyllum* and *Streptocarpus* spp.). Discontinuity of development occurs in these cases through new individuals and does not rely on sexual reproduction. There is, nevertheless, a true reversion to the juvenile state.

If we confine ourselves to the development of the individual proper, other discontinuities may be seen. We have just mentioned that the unfolding of morphogenesis involves the formation of leaves and flowers, the lives of which are limited. In a perennial plant, however, other localized phenomena of this type may be found. They relate, for example, to those processes which determine the onset of inactivity and eventual death of the meristems of branches (as in *Theobroma cacao*), and also the tendency of meristems to re-orientate their morphogenesis so that they develop in a new way, e.g. as tendrils (in *Vitis*), or thorns (in *Gleditsia triacanthos* and in various *Citrus* spp.), or as short branches with no developmental future (such as the brachyblasts of *Pinus* – Fig. 13.2a).

In some cases, it is the entire terminal portion of the growing branch which dies; this happens in various woody plants (e.g. *Ulmus, Vitis, Gleditsia triacanthos*) during autumn in temperate countries. After a period of diapause secondary branches replace the senescent axes and in so doing reveal that in their proximal region there are morphological (the shape and size of their leaves) and physiological (an aptitude toward rhizogenesis) characteristics which resemble those of young structures. Thus, through the development of this particular type of branch a partial rejuvenation provides the means of slowing down the inexorable course towards senescence and ultimate death in the perennial plant.

A border case is of special interest. It concerns branches called 'stump shoots', which arise from the base of the trunk of trees and shrubs, i.e. from that part which developed at the outset of morphogenesis. These shoots, issuing from the 'oldest' part of the plant, exhibit very obvious juvenile characteristics. In *Quercus*, for instance, cuttings from these shoots easily take root (Kazandjian, 1977), whereas branches, although newly-formed in the crown of the tree and which are therefore chronologically 'younger', do not.

This juvenile quality in 'stump shoots' may, in nature, confer exceptional longevity on individuals of certain species (e.g. *Sequoia sempervirens*) since a failing principal axis may be supplanted, after hundreds of years of development, by one or several of the branches arising at its base. This characteristic is applied to rejuvenate old plantations (for example, *Olea europaea* in Tunisia) (Kechaou, 1971).

Another type of partial return to initial processes has also been observed in woody plants living mainly in intertropical regions, and is a consequence of the 'flushing' (rhythmic growth) of their axes. The apical meristem periodically stops functioning, but when it resumes activity it returns to a former mode of functioning. Thus, periodical phenomena intervening within a primary meristem may result in a partial rejuvenation which, in other species, is acquired only by the generation of adventitious meristems.

All these observations show that the morphogenetic progression does not unfold in a smooth, linear fashion. It occurs with discontinuities, and with a partial return to initial processes.

DETERMINISM IN THE MORPHOGENETIC PROGRESSION

Now that the broad characteristics of the morphogenetic progression have been described, we will consider some issues regarding its determinism. One notion, that of external environmental factors, will guide our considerations. We will examine them in order to underline certain generalities and to call attention to some aspects that are little understood. We will also stress their impact on the evolution of the internal environment.

Any cell, group of cells, or organ, creates around itself a specific environment which influences the activities of other cells, group of cells, or organs. The new qualities gained in turn modify the environment of other cells, including those which initially triggered the changes, and so the process goes on. Thus, we face a complex system of interactions about which we know little. We will nevertheless present here some of the most recent findings.

The external environment and morphogenesis

As already mentioned, it is well established that external factors are very important for morphological movement. For example, flowering, one of the decisive cross-roads in the life-cycle of higher plants, may be achieved only under a very specific set of environmental conditions. We wish here to draw attention to *in vitro* culture conditions as another external morphogenetic factor which may yield a wealth of interesting results.

Culture conditions create a physiologically restrictive environment. When buds from the adult part of various woody or herbaceous species are placed in culture they acquire, following successive explantations from the stems on which they arose, a morphogenesis that increasingly resembles that of the young seedling (Fig. 13.1). Such juvenile structures have been obtained from adult individuals of various species of *Vitis* (Fig. 13.1a), *Solanum* (Fig. 13.1b), and *Ipomoea* (Fig. 13.1c). This 'young' morphology is attended by correspondingly 'young' physiological properties. One such attribute is the ability of these miniature individuals to give rise to neoformed roots. Various *Citrus* spp. offer a typical illustration. It is difficult to root cuttings of these species from vegetative portions of the adult individual, though rhizogenesis occurs with ease on fragments of stems grown *in vitro* (Bouzid, 1975). This rejuvenation phenomenon is caused by a restricted food supply, and the young morphogenetic structure persists for as long as it is maintained. For over twenty years a *Vitis* clone has been kept in this state through successive explants.

It is also possible to disturb the natural order of morphogenetic events through culture methods. In *Dahlia*, for instance, individuals which are morphologically juvenile can be made to flower (Watelet-Gonod & Favre, 1981); similar results have been reported in certain *Vitis* genotypes (Favre & Grenan, 1979). Moreover, in certain circumstances, 100% tuberization of juvenile *Solanum tuberosum* may also be obtained (Nancy-David, 1977). Thus, *without* passing through each and every step in their normal development, it is possible to obtain individuals which *simultaneously* exhibit young and adult characteristics.

An individual usually reacts to its environment in accordance with the stage it has reached in its life-cycle. External factors can trigger certain contradictory elements of the internal mechanisms of the plant's morphogenetic progression; they determine which process will occur, and sometimes spectacularly affect its unfolding (as in the examples above). But there is more. Individuals belonging to certain species, probably because of their exposure to ecological conditions *unusual* for them, are capable of

R. Nozeran

Fig. 13.1. Various plants presenting a juvenile morphogenesis and miniaturized when cultivated *in vitro*: (a) vine; (b) potato; (c) sweet potato.

exhibiting a stable and *unusual* morphogenesis, even when removed from the conditions which gave rise to it. We have called such plants 'variants'; without any modification to their stock of genetic information, they function differently from the normal (Nozeran, 1968; Nozeran & Bancilhon, 1972; Nozeran & Rossignol-Bancilhon, 1977). This example may explain the existence in a few gymnosperms of individuals called *Retinospora* which, whether they arise from cuttings or from seeds, exhibit juvenile foliage and sterility for a long, or even an 'indefinite', period (Schaffalitzky de Muckadell, 1954).

The experience of a different external environment can thus induce new interactions between the plant and its surroundings and may completely change its destiny. The environment may, for instance, exert its effect via a leaf, but only when certain processes are at work in that leaf. We refer to the ability of an organ, or cell, to perceive an 'impulse' from its environment, as 'competence'. This is a state which also forms an integral element in the morphogenetic progression.

The internal environment and morphogenesis

Certain aspects of the morphogenetic progression affect various parts of the plant and have intrinsic consequences for their future (i.e. they feed-forward). Others – and maybe sometimes also the same ones – exert an influence on other parts whose new mode of functioning may then initiate retro-actions on the inducing part (i.e. they feed-back). We shall next discuss these two groups of phenomena.

Intrinsic phenomena

Surprisingly, an interruption in the functioning of certain plant parts seems to have important implications for their future morphogenesis. It can also be shown that large modifications are accompanied by a reduction in the number of meristematic cells involved in the morphogenetic process.

Interruptions

When we examine the development of higher plants we notice that some meristems when compared with the initial type issuing from the seedling, manifest new and hitherto unrevealed morphogenetic potentialities. We have already cited (p. 379) as examples, plagiotropic branches, tendrils, thorns, and brachyblasts of *Pinus* (Fig. 13.2a). These often constitute

morphogenetic dead-ends. By contrast, other meristems, whose development asserts the individual's perennial nature, may give rise to a bud that
shows a more or less definite diapause, or dormancy. The branches to
which they give rise initially show juvenile characteristics in their proximal part.

It has been possible to examine this state of diapause experimentally. In
the alga *Chara*, Ducreux (1975) was successful in interrupting the
functioning of the 'branch' apical cells and he was able to induce a new
start in these cells. The resulting branch varied more or less from the
normal, and even regained the form characteristic of a newly-germinated
plantlet. Another interesting experiment was conducted by Espagnac
(1973) on the fern, *Nephrolepis biserrata*. Shoots of this species exhibit a
marked dimorphism: one form consists of short axes bearing chlorophyl-
lous leaves, the other consists of leafless stolons of varying length.
Meristems of the latter, as long as they remain attached to the mother

Fig. 13.2. *Pinus pinea*. (a) Shoot fragment bearing a short branch (brachyblast)
(br); its terminal meristem normally stops functioning after having produced two
needle leaves (nl). (b) The same, but following experimental treatment and
resumption of the terminal meristematic activity of the brachyblast; the latter then
builds a shoot (sh) bearing juvenile leaves (jl). (c) Young seedling presenting
juvenile leaves (jl) above the cotyledons (cot).

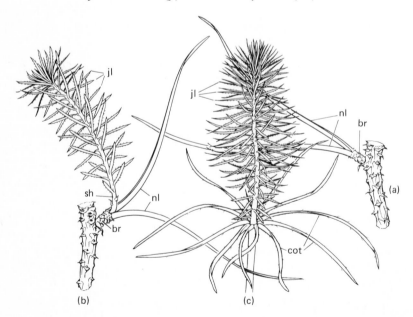

plant, produce the stolon form indefinitely. But if the distal end of such an axis is excised and then planted, growth is interrupted for at least twenty days. Following this period, growth resumes, but the resulting form is different: the meristem now produces a short, foliar axis quite similar to that arising from the germinating zygote. Unquestionably, the experimental protocol induces a diapause in the meristem which is followed by a noticeable rejuvenation of its processes.

There are comparable examples in the phanerogams. Experiments conducted in the phytotron in Gif-sur-Yvette, France, with *Citrus* show that if plants are grown at 5°C, following a period of normal development at a temperature of around 30°C, their development totally ceases. Several months later, some meristems resume activity and give rise to branches with certain characteristics of a seedling. Similarly, conditions may be created in *Pinus* by treatments such as drastic pruning of the aerial parts of the tree, or *in vitro* cultivation (Lahondère & Neville, 1959; Chaperon, 1979; David, David, Faye & Isemukali, 1979) which allow the terminal meristem of a brachyblast to resume growth. This leads to the formation of a branch (Fig. 13.2b) which may exhibit morphological characteristics of the seedling (Fig. 13.2c); it also shares other juvenile physiological traits, such as easy multiplication from cuttings.

These examples demonstrate that the functioning of the hereditary material may be modified following either a natural or an experimentally-induced diapause. The result is a more or less pronounced return to an initial, juvenile morphogenetic mode of expression.

Miniaturization of meristems

In addition to what we have just mentioned concerning the stolon of *Nephrolepis biserrata*, certain other processes are quite instructive (Espagnac, 1973). Upon separation from the mother plant, cell division ceases in the apical meristem. The meristem eventually reduces to the initial apical cell, plus a few surrounding cells. At this point, juvenile-type morphogenesis resumes. In this sequence of events an important point, in our opinion, must be clarified: what do we mean by 'cessation' of activity? Although the end of the stolon does not exhibit any visible external morphological modifications, clearly very important internal changes are taking place there. Equally, in all the cases of diapause which we have mentioned above, similar reorganizations are also taking place; here, too, there are reductions in the number of cells which are later involved in morphogenesis.

In vitro cultivation affords an opportunity to shed additional light on these processes. For example, a comparison of individuals of *Dahlia* raised under either normal or *in vitro* conditions shows that in the latter situation the apical meristem is smaller in both volume and cell number (Watelot-Gonod & Favre, 1981). We are thus tempted to relate the return to an ontogenetically early type of activity with the decrease in the number of meristematic cells involved. Moreover, a single somatic cell in culture may undergo a morphogenesis that mirrors that of the unicellular zygote (Steward, Mapes & Mears, 1958; Reinert, 1958).

Similar processes have been found in fungi. Thus, in *Podospora anserina*, Marcou (1961) was able to show that although a mycelium evolved towards senescence, explanting a very small fragment of it could lead to a rejuvenated culture. Chevaugeon (1968) and Chevaugeon & Lefort (1960) have performed comparable experiments with several other fungal species grown *in vitro* and have obtained a mycelium whose form is 'modified' when compared to the usual 'initial' mode; this presumably corresponds to a new pattern of genetic activity. Return to the initial mode itself takes place from one or a few cells.

Does this mean that any isolated cell, or any miniaturized meristem, will give rise to a juvenile-phase individual? We do not believe so, and two studies support our position. The first deals with a fungus, *Ustilago cynodontis* in which Chevalier (1967) demonstrated that *under all the experimental conditions he tried* cells isolated from the modified sector of the mycelium (similar to that described by Chevaugeon in other fungi and referred to above) always transmitted the modified functioning mode. The second has to do with experiments conducted in phanerogams whose meristems were microsurgically reduced in size This was achieved in *Phyllanthus* with dimorphic branches by Bancilhon (1969). In *P. distichus*, reduction of the orthotropic axis meristem leads to an apex with juvenile characteristics; these are not exhibited if a similar operation is conducted on the meristem of a plagiotropic axis.

The available data indicate that juvenile modes of functioning require a reduction in the number of cells involved. Is cell number the sole condition? One of the possible interpretations of the role of miniaturization in this phenomenon is that it may – but does not always – induce physiological conditions, particularly nutritional deficiencies, which evoke certain types of gene activity. The diapause that frequently accompanies this phenomenon may in addition provide the time necessary for the cytophysiological state to revert to an earlier mode of functioning.

Directing interactions in the morphogenetic progression

Indirect and direct actions may be distinguished among the interactions intervening in the orientation of the morphogenetic progression.

Indirect correlative effects

The correlations which influence certain intrinsic phenomena which we have just mentioned may be classified in this category. We refer particularly to those which may be collectively called 'cessations of functioning', though we must recall the reservations already formulated concerning the validity of such a term. In the aerial part, it is frequently the leaves that are involved. Their activity is at the origin of inhibitions and dormancy, of senescence and flowering.

Inhibitions and dormancy

In certain cases, the distinction between inhibition and dormancy is quite clear-cut, but in others less so. We use the term 'diapause' to designate these latter cases.

It has been demonstrated that leaves play an essential role in the onset of diapause and that their influence may extend over quite large distances. Let us take as an example the case of a species of the Loganiaceae from intertropical Africa, *Anthocleista nobilis* (Nozeran, 1956). The trunk of this 10–15 m tree bears in the axils of the foliar scars, latent buds enveloped by two scale leaves transformed into hard spines. These buds usually develop into branches only when flowering occurs at the terminal meristem. However, cutting the trunk causes outgrowth of these buds near the cut at any time irrespective of flowering. This shows that the terminal vegetative buds of the principal axis normally inhibit the functioning, several metres away, of the secondary axis meristems which it has produced.

The source of the stimuli may, on the other hand, be quite close to where the induction will take place: an example is the imposition of diapause on a bud located in its axil. Such inhibitory inductions may be transmitted either basipetally or acropetally. Basipetal action is quite easy to detect by incising the axis below the terminal bud. Acropetal transmission has also been demonstrated in *Syringa vulgaris* (Champagnat, 1951) and *Phyllanthus* (Bancilhon, 1969). In many species, the correlative inhibition is due mainly to the young leaves of the terminal bud (Snow, 1929; Champagnat, 1965); in *Gleditsia triacanthos* these leaves are almost the only ones involved (Nozeran, Bancilhon & Neville, 1971).

The role played by leaves in dormancy may depend on the stage of development they have reached. In certain varieties of *Vitis vinifera*, Nigond (1966) has shown that the young leaves and the growing vegetative apices exert a strongly *inhibitory* action on the initiation of dormancy in the buds. After a certain stage in their development, however, the leaves appear to *promote* dormancy; later still, they again exert an *inhibitory* effect (Fig. 13.3a).

Similarly, the 'flushing' in shoots of *Theobroma cacao* is influenced by leaves since it is eliminated by removal of the young leaves (Vogel, 1975). Comparable phenomena have been noted by Hallé & Martin (1968) in *Hevea brasiliensis* and by Bancilhon (personal communication) in explanted plagiotropic branches of *Phyllanthus distichus*. It should be added that such rhythmic phenomena may be the result of the action not only of leaves that have reached a certain stage of development near the shoot tip but also of leaf primordia newly formed by it.

The correlative action of green leaves that have reached a certain size may also explain the scale-like structure of the first leaves of diapaused buds; this has been demonstrated in *Gleditsia triacanthos* (Nozeran *et al.*, 1971). The inhibition-inducing mechanisms, whether in the presence of dormancy or not, seem to be related, probably in a determinate way, to the action of the leaves and also to the latter's stage of development. Activities such as these are part of a wider spectrum of phenomena. Leaves, products of active shoot meristems, can exert an inhibitory feedback to such meristems. This has been demonstrated in various *Dipsacus* spp. (Cutter, 1964) and in *G. triacanthos* (Bancilhon & Neville, 1966). Removal of the leaves at a sufficiently early stage results in a

Fig. 13.3. Schematic representation of processes intervening in relation to age of the leaf on (a) bud dormancy in *Vitis vinifera*, and (b) flowering in *Scrophularia arguta* and *Phyllanthus amarus*.

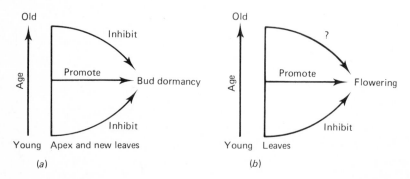

significant increase (almost a doubling) of the organogenic activity of the meristem. Other types of lateral 'phyllomorphic' formations, e.g. the plagiotropic branches of *Phyllanthus distichus* (Bancilhon & Neville, 1966) (Fig. 13.4a) show the same sort of activity. Even more remarkable is the observation reported by Ducreux (1975) that young-stage pleuridia in *Chara* exert an inhibitory effect on the rate of division of the apical cell (Fig. 13.4b).

Senescence and flowering

Until now we have dealt with the reversible cessation of activity. The phenomena to which we will now turn foreshadow the irreversible cessation, or death, of a meristem.

Certain senescent processes (already mentioned on p. 379) affect, under natural conditions, the distal ends of vegetative branches in certain species (*Ulmus, Vitis,* etc.). They are, in part, related to a particular type of functioning of the leaves. Thus, in *Gleditsia triacanthos*, Neville (1969) has shown that removal of the young leaves delays the cessation of growth and the onset of senescence. We may ask ourselves if comparable processes are responsible for the development of short branches (brachyblasts) in *Pinus*, spines in *G. triacanthos*, and also for the determinate stems of Euphorbiaceae (such as *Anthostema*), Apocynaceae (such as

Fig. 13.4. Schematic representation of correlations occurring in the terminal portion of (a) *Gleditsia triacanthos* and *Phyllanthus distichus*, and (b) *Chara*.

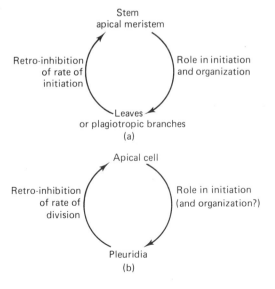

Alstonia), and Sterculiaceae (such as *Theobroma*) where senescence of the meristem inevitably follows a certain period of functioning. Significantly, an early breakdown of correlations in *Pinus* and in *Gleditsia* can cause such determinate branches to become vegetatively indeterminate.

Let us now turn to flowering which constitutes an end-point for the meristem which gives rise to the flower. It has been shown in several plants that leaves of a certain morphogenetic stage may either activate or inhibit flowering; in *Scrophularia arguta* and in *Phyllanthus amarus* Chouard & Lourtioux (1959) and Bancilhon (1969), respectively, have demonstrated that young leaves inhibit the flowering process, whereas when older they activate it (Fig. 13.3b). Thus, the occurrence of flowering implies the presence of leaves which have acquired a particular functional state by their exposure to specific ecological conditions (e.g. a particular photoperiod). A single leaf rendered inductive is sufficient, in some cases, to trigger flowering, and the process thereafter 'infects' other meristems of the individual, and even meristems which have been grafted onto the plant.

Direct correlative effects

We will classify under this heading actions which lead to new morphological structures. Such structures are secondary in relation to the organ on which they are induced, though they are obviously still part of the overall morphogenetic plan. We believe that we may compare this mode of action with the one which animal embryologists have termed 'organizer'.

We will select a few typical examples demonstrating the correlative determinism of various parts of the life-cycle. A whole assortment of different situations has been found in relation to the determination of plagiotropic development of certain stems. In *Ajuga reptans*, Pfirsch (1962) has shown that horizontal growth of the stolons is influenced by the presence of open flowers. After their death, which coincides with seed formation, or their removal at an earlier stage, the stolon gives rise to an erect, rosette axis. In *Stachys sylvatica*, Pfirsch (1962, 1965) has also shown that the creeping plagiotropic stolons of this plant are, at first, induced by a correlative action of the erect axis. It is the nodes below the terminal apex which determine the plagiotropic orientation of the buds. Each of the latter then progressively conveys information to the next during growth of the stolon. The original plagiotropic nature of the meristem is therefore due to the foliar axis from which it emerged, while the continuity of this type of morphogenesis results from an autocorrelative process as the plagiotropic stolon itself branches.

In *Phyllanthus*, with its dimorphic branches, Bancilhon (1965) has demonstrated that the plagiotropic orientation of a lateral bud meristem depends on the meristem of the orthotropic axis which formed it (Fig. 13.4a). The parallel between this process and that of an organizing centre in an animal embryo is shown by the fact that if a lateral (branch) meristem formed by an orthotropic meristem is removed, after a sufficient period, from the latter's influence, it continues to function indefinitely in a plagiotropic mode.

The programming of the morphogenesis of lateral branches in *Phyllanthus urinaria* by messages originating in the orthotropic meristem not only determines their plagiotropism but also restricts their growth (Bancilhon, 1965). Such a situation produces secondary axes exhibiting mirror-image characteristics of a 'phyllomorphic', compound leaf (Nozeran, Ducreux, Neville & Rossignol-Bancilhon, 1980).

Such evidence tends to prove that the terminal meristem of the shoot axis in vascular plants also determines the secondary meristems towards foliar morphogenesis. This hypothesis is supported by various findings. Wardlaw (1949) was able to demonstrate the organizing influence of the main axis meristem of the fern *Dryopteris aristata* on the meristematic primordia it produced which were consequently directed towards leaf morphogenesis; on the other hand, their early isolation from the main meristem induced them to develop into typical axes. Although more difficult to perform with phanerogams, similar experiments were conducted on *Phaseolus vulgaris* by Pellegrini (1961), yielding a similar result: depending on whether or not it remains connected with its parent apical meristem, a secondary meristem will grow as either a leaf or an axis.

Neville (1964) has demonstrated comparable phenomena within a leaf. In the unipinnate compound leaf of the young individual of *Gleditsia triacanthos*, early removal of the terminal meristems of the rachis induces a modification of lateral meristems which, instead of giving rise to leaflets, become secondary rachides.

Analysis of the branching in certain algae, for instance in *Sphacelaria* (Ducreux, 1977) or in Rhodophyceae (L'Hardy-Halos, 1975), has yielded comparable results.

All these examples illustrate that during the course of morphogenesis, meristems can be produced which qualitatively differ from the meristems which gave rise to them. They may express new potentialities unrevealed at the start of the life-cycle. Some of them either exhibit self-maintaining processes, or show evidence of a gradual programming that leads to a specific type of growth. An instance of the latter is the plagiotropic determination of the dimorphic branches in *Phyllanthus*. Bancilhon (1965) has demonstrated that acquisition of the organizer property in the orthotropic meristem after seed germination is progressive. The first plagiotropic branches formed by the individual, instead of being indefinitely set in that mode of growth, are capable of reverting to an orthotropic mode. This reversal to what is evidently a juvenile mode occurs even after a period of plagiotropic growth. This shows clearly that although these lateral meristems have been exposed to a 'plagiotropizing' influence, of itself this is insufficient to induce the self-maintenance of the plagiotropic mode of growth. Dyanat-Nejad (1970) has shown a similar phenomenon in the roots of *Theobroma cacao*. By removing the tap root meristem at a level distal to developing secondary root primordia the latter develop as either secondary tap roots, plagiotropic roots, or roots of intermediate character which become tap roots after a more or less lengthy plagiotropic phase. The outcome depends on the period for which the primordia were exposed to the influence of the tap root.

These examples imply the existence of gradients, which also have an expression at the cellular level. The observations of Aghion-Prat (1965) on *Nicotiana tabacum* (cv. Wisconsin 38) are instructive in this respect. If fragments of stem, taken from different levels of the flowering plant, are explanted *in vitro* they give rise to callus, and thence to neoformed buds. The buds are vegetative if the stem explant is taken immediately below the inflorescence. But explants taken from the subapical region of the inflorescence form callus which directly bears newly-formed flowers. Explants from the intermediate region of the floral axis give rise to callus which produces branches ending in a flower but which possess a vegetative region that is larger the nearer the explant was to the base of the inflorescence. A gradient is also evident in cells from superficial layers of leaves which, when placed in an adequate culture environment, produce either vegetative buds or flowers, depending on the location of the explanted cells in relation to the inflorescence (Tran Thanh Van, 1973*a,b*). (See also analogous examples given by McDaniel, Chapter 11, and Carr, Chapter 12.)

Gradients also correspond to physiological states characterized by other properties. For instance, Meins, Lutz & Foster (1980) have shown in tobacco pith tissue the effect of explant position in the stem on the capacity of such tissue to grow with or without exogenous cytokinin. Also, Stoutemyer & Britt (1965) and Banks (1979) have shown that different types of tissue cultures can be obtained depending on whether the explant originates from a juvenile or a mature portion of *Hedera helix* (ivy), callus from juvenile stems regenerates shoots, while callus from mature stems regenerates only embryos (Banks, 1979).

Other studies have identified physiological variations in relation to the season. For example, Meins & Lutz (1980) found this for the induction of cytokinin habituation in primary pith explants of tobacco.

Age has been studied in relation to differentiation by Phillips (1981) using tubers of *Helianthus tuberosus*. He has shown that their capacity for xylogenesis varies as a function of their maturity. Similarly, Wielgat, Wechselberger & Kahl (1979) have demonstrated in *Solanum tuberosum* an age-dependent variation in RNA transcription in response to wounding and gibberellic acid treatment.

Another principle may be drawn from the examples we have just discussed. They clearly reveal the diverse nature of the 'organizer' part in plants, a region which varies greatly in size, and may be reduced in certain cases to a meristem, and probably even to a very small number of cells in some algae. It seems, moreover, that induction is much more coercive, and will lead to a more stable situation, if the organizer centre is small.

It is also important to recall that there are the contradictory aspects to correlative phenomena, such as activation or inhibition within the same leaf (which we have already mentioned on p. 390) that operate at different stages of its life-cycle. In certain cases, these contradictory aspects may be exhibited simultaneously. Favre (1973) had demonstrated this through the study of cuttings of *Vitis* grown *in vitro*. Their buds are *at the same time* activators and inhibitors of root neoformation. The aspect that is expressed depends on where the bud is located. If near the base it will exert a negative effect on rhizogenesis, if near the tip, a positive effect.

Similarly, should we not also consider youth and senescence among the expression of these contradictory modes of functioning? We have in mind the example of the shoot derived experimentally from the short branch, or brachyblast, meristem of *Pinus*. We mentioned earlier that it may exhibit all the morphological characteristics of the young seedling. But its destiny when taken as a cutting, depends on the site of its formation on the tree (Franclet, 1979). It thus 'remembers' certain characteristics that appeared more or less early in the development of the plant part which gave rise to it.

Memory also seems to be a part of development. In young individuals of *Phyllanthus distichus* flowering usually takes place only when the plants have reached a certain size. But Bancilhon (1965) has shown that very young seedlings, as soon as they form their first plagiotropic secondary branches, show a tendency to flower. Later, with further development, this tendency decreases. Thus, we have individuals which, although very young, obviously exhibit characteristics typical of the end of the plant's life-cycle. The young seedling appears to carry a 'memory' of a certain type of functioning of the flower from which it was formed.

CONCLUSION

We have deliberately made choices from among all the interesting available investigations on the morphogenetic progression of plants. It is noteworthy that they were mostly performed in the last few decades. They far from exhaust the possibilities of further advances in our knowledge of morphological phenomena occurring in the plant life-cycle, but they allow us to formulate a few conclusions.

One of them is prompted by the complexity of the interconnected systems that control plant morphogenesis. These systems have meaning only when viewed within the framework of the whole organism. Naturally, changes in ecological conditions must also be taken into account in the general morphogenetic progression; their integration occurs through their local action on particular parts of the plant which are 'competent' to respond.

In spite of this complexity, whose very magnitude offers a mighty challenge to analysis, there remains a profound impression of *order* permeating the construction of the individual. If the integrity of the organism is at stake, things rapidly go sour, as happens in tissue cultures. The culture becomes anarchic, and genetic aberrations arise in the cells which comprise it. For example, Banks-Izen & Polito (1980) have shown in *Hedera helix* that the proportion of non-diploid cells varies with time in culture according to whether calluses originate from juvenile or adult stems.

If one is successful in promoting from a single cell a new integrated ontogeny, e.g. the neoformation of a new individual, the latter may be quite different from the one from which the culture originated. It is possible, using this procedure, to obtain a polymorphic progeny from a non-chimaeral plant. This has been described in various publications and reviewed by Skirvin (1978). New prospects thus arise for the agronomical improvement of various plants, for example sugar cane (Heinz, Krishnamurthi, Nickell & Maretzki, 1977), or *Solanum tuberosum* (Nozeran & Rossignol-Bancilhon, 1979; Rossignol-Bancilhon, Nozeran, Quraishi & Darpas, 1980).

Lower plants, e.g. the thallophytes, offer the advantage of providing valuable information at the cellular level. Recent studies on *Sphacelaria cirrosa* by Ducreux (personal communication) are quite promising in this respect. In this species, although an isolated apical cell is non-functional, such is not the case with the subapical cell. Following the destruction of the apical cell, the sub-apical cell gives rise to various morphogenetic pathways, the outcome of which depends on how long the cell has remained connected with the rest of the plant (Fig. 13.5).

Fig. 13.5. Experiments conducted on the apex of *Sphacelaria cirrhosa*. (a) Organization of the apex. (Scale bar, 100 μm). (b–f) Role of correlative influences in the expression of morphogenetic potentialities in the apical cell and in the sub-apical cell. Ap, apical cell; SAp, sub-apical cell; N_1, N_2, N_3, nodes; IN_1, IN_2, IN_3, internodes (destroyed portions are marked by X and by letters in parentheses); R, branch; RH, rhizoid. (After G. Ducreux, personal communication.)

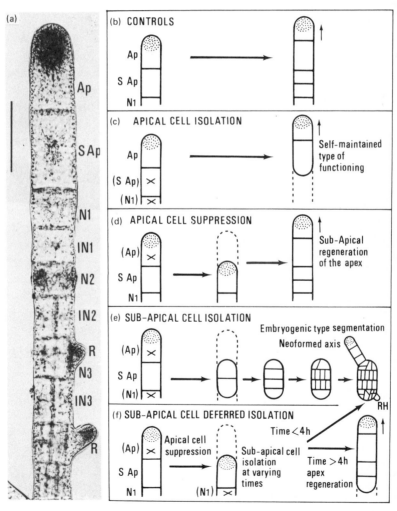

We must not omit the biochemical aspects of morphogenesis. Until now we must admit that, with the exception of results concerning the regulation of growth processes, little real progress has been made. It might be worthwhile to reconsider developmental biochemistry in terms of the new relationships which have been demonstrated in the morphogenetic progression. Some avenues are being opened: recently, Cabanne, Dalebroux, Martin-Tanguy & Martin (1981) have shown that in *Nicotiana tabacum* var. *xanthi* the presence or absence of various hydroxycinnamic acid amides depends on the morphogenetic stage reached by the plant. These amides have not been detected in the juvenile plants. Similarly, Moncousin (1982) has shown in *Cynara scolymus* an indirect relation between rejuvenation, measured in terms of rooting capabilities, and peroxidase content.

Other researchers have attempted to link various morphogenetic phenomena directly with changes in gene expression. A study by Wechselberger, Wielgat & Kahl (1979) illustrates rhythmic changes of the activity of chromatin-bound, DNA-dependent RNA polymerase, chromatin-template accessibility, and the rRNA and chromosomal protein content in the course of the formation, dormancy, and sprouting of potato tubers (see also p. 393). Furthermore, Domoney & Timmis (1980) have shown that in *Hedera helix* the increase in RNA amounts associated with the phase change from juvenile to mature forms correlates with a high degree of 'elasticity' of rRNA gene redundancy in the various cell populations. With regard to the DNA itself, Kessler & Reches (1977) and Nagl (1979) claim to have found evidence that its differential replication in leaves of *Hedera helix* is also associated with phase change. Moreover, Wardell & Skoog (1973) have shown that the DNA content of stem tissue from a flowering tobacco plant is correlated with that tissue's capacity to form flowers *in vitro*.

Further genetical studies will help develop our understanding in this domain. A prime example is furnished by individuals from *Phyllanthus urinaria* which arrest growth at various steps in their morphogenetic cycle according to the genotype of their parents (Haicour, 1982).

In another field, recent advances have been made, originating with studies by Desbiez (1975), in the analysis of 'memory' processes (already alluded to on p. 393) in correlative phenomena. In *Bidens pilosus* certain stimuli that affect bud development can be memorized, without any apparent loss of information, for 15 days or more (Thellier, Desbiez, Kergosien & Champagnat, 1981).

All these new break-throughs provide an added incentive to intensify research on the events characterizing the morphogenetic progression – nurtured by sometimes contradictory processes – as an integrated correlative complex. The task is not easy, but such is the price of progress in our knowledge of these phenomena. Studies conducted in this spirit have, in just a few years, already opened up many new perspectives and have helped us make much headway in our understanding, for instance, of vegetative multiplication. Such advances but foreshadow the many more of tomorrow.

References

Aghion-Prat, D. (1965). Néoformation de fleurs *in vitro* chez *Nicotiana tabacum* L. *Physiologie Végétale*, **3**, 229–303.

Bancilhon, L. (1965). Sur la mise en évidence d'un rôle 'organisateur' du méristème apical de l'axe orthotrope de *Phyllanthus*. *Comptes rendus hebdomadaires des Séances de l'Académie des Sciences, Paris, Série D*, **260**, 5327–9.

– (1969). Étude expérimentale de la morphogénèse et plus spécialement de la floraison d'un groupe de *Phyllanthus* (Euphorbiacées) à rameaux dimorphes. *Annales des Sciences Naturelles, Botanique, 12ème Série*, **10**, 127–224.

Bancilhon, L. & Neville, P. (1966). Action régulatrice des jeunes organes latéraux à rôle assimilateur sur l'activité du méristème de la tige principale chez *Phyllanthus distichus* Müll. Arg. et *Gleditsia triacanthos* L. *Comptes rendus hebdomadaires des Séances de l'Académie des Sciences, Paris, Série D*, **263**, 1830–3.

Banks, M. S. (1979). Plant regeneration from callus from two growth phases of English ivy, *Hedera helix* L. *Zeitschrift für Pflanzenphysiologie*, **92**, 349–53.

Banks-Izen, M. S. & Polito, V. S. (1980). Changes in ploidy level in calluses derived from two growth phases of *Hedera helix* L., the English ivy. *Plant Science Letters*, **18**, 161–7.

Billard, J. P., Binet, P., Boucaud, J., Coudret, A. & Le Saos, J. (1976). Halophilie et résistance au sel. Réflexion sur l'halophilie et quelques uns de ses aspects physiologiques. *Études de Biologie Végétale. Hommage au Professeur P. Chouard.* ed. R. Jacques, pp. 39–55. Paris.

Borchert, R. (1976). The concept of juvenility in woody plants. *Acta Horticulturae*, **56**, 21–33.

Boucaud, J. & Ungar, I. A. (1978). Halophilie et résistance au sel dans le genre *Suaeda* Forsk. *Bulletin de la Société Botanique de France*, **125**, Actualités Botaniques, no. 3–4, 23–35.

Bouzid, S. (1975). Quelques traits du comportement de boutures de *Citrus* en culture *in vitro*. *Comptes rendus hebdomadaires des Séances de l'Académie des Sciences, Paris, Série D*, **280**, 1689–92.

Cabanne, F., Dalebroux, M. A., Martin-Tanguy, J. & Martin, C. (1981). Hydroxycinnamic acid amides and ripening to flower of *Nicotiana tabacum* var. *xanthi* n.c. *Physiologia Plantarum*, **53**, 399–404.

Carvalho, A., Krug, C. A. & Mendes, J. E. T. (1950). O dimorfismo dos ramos em *Coffea arabica* L. *Bragantia*, **10**, 151–9.

Champagnat, P. (1951). Les corrélations d'inhibition sur la pousse herbacée du Lilas: II. Un curieux exemple d'inhibition acropète. *Bulletin de l'Association Philomatique Alsace et Lorraine*, **9**, 54–6.

– (1965). Physiologie de la croissance et de l'inhibition des bourgeons: dominance apicale et phénomènes analogues. In *Encyclopaedia of Plant Physiology*, vol. 15/1, ed. W. Ruhland, pp. 1106–64. Berlin, Heidelberg & New York: Springer-Verlag.

Chaperon, M. (1979). Maturation et bouturage des arbres forestiers. *Micropropagation d'Arbres Forestiers, AFOCEL, Paris. Études et Recherches*, **12**, 19–31.

Charrier, A. (1969). Contribution à l'étude de la morphogénèse et de la multiplication végétative du cacaoyer (*Theobroma cacao*). *Café, Cacao, Thé*, **13**, 97–115.

Chevalier, S. (1967). Analyse d'une différenciation contagieuse des thalles de l'*Ustilago cynodontis* (Pass.) P. Henn., cultivé *in vitro*. *Thèse Doctorat d'État, Université de Paris*, 101 pp.

Chevaugeon, J. (1968). Étude expérimentale d'une étape du développement du *Pestalozzia annulata* B. et C. *Annales des Sciences Naturelles, Botanique, 12ème Série*, **9**, 417–32.

Chevaugeon, J. & Lefort, C. (1960). Sur l'apparition régulière d'un 'mutant' infectant chez un champignon du genre *Pestalozzia*. *Comptes rendus hebdomadaires des Séances de l'Académie des Sciences, Paris, Série D*, **250**, 2247–9.

Chouard, P. (1977). Conclusion du Colloque sur la multiplication végétative chez les végétaux (Orsay, 1974). *Bulletin de la Société Botanique de France*, **124**, 191–3.

Chouard, P. & Lourtioux, A. (1959). Corrélations et réversions de croissance et de mise à fleurs chez la plante amphicarpique *Scrofularia arguta* Sol. *Comptes rendus hebdomadaires des Séances de l'Académie des Sciences, Paris, Série D*, **249**, 889–91.

Cutter, E. G. (1964). Phyllotaxis and apical growth. *The New Phytologist*, **63**, 39–46.

Darwin, C. (1979). *The Origin of Species by Means of Natural Selection*. New York: Avenal Books (reprint of original edition of 1859).

David, A., David, H., Faye, M. & Isemukali, K. (1979). Culture *in vitro* et micropropagation du pin maritime (*Pinus pinaster* Sol.), *Micropropagation d'Arbres Forestiers, AFOCEL, Paris. Études et Recherches*, **12**, 33–40.

Desbiez, M-O. (1975). Base expérimentale d'une interprètation des corrélations entre le cotylédon et son bourgeon axillaire. *Thèse Doctorat d'État, Université de Clermont-Ferrand*, 174 pp.

Domoney, C. & Timmis, J. N. (1980). Ribosomal RNA gene redundancy in juvenile and mature ivy (*Hedera helix*). *Journal of Experimental Botany*, **31**, 1093–110.

Dormling, I. (1973). Photoperiodic control of growth and cessation in Norway spruce seedlings. *Growth Process Symposium on Dormancy in Trees, (IUFRO Division 2, Work part 2.01.4)*, 16 pp.

Ducreux, G. (1975). Corrélations et morphogénèse chez le *Chara vulgaris* L. cultivé *in vitro*. *Revue Générale de Botanique*, **82**, 215–357.

– (1977). Étude expérimentale des corrélations et des possibilités de régénération au niveau de l'apex de *Sphacelaria cirrhosa* Agardh. *Annales des Sciences Naturelles, Botanique, 12ème Série*, **18**, 163–84.

Dyanat-Nejad, H. (1970). Contrôle de la plagiotropie des racines latérales chez *Theobroma cacao* L. *Bulletin de la Société Botanique de France, Mémoires*, **117**, 183–92.

– (1971). Corrélations intervenant dans la morphogénèse de l'appareil souterrain du cacaoyer (*Theobroma cacao* L.). *Café, Cacao, Thé*, **15**, 105–14.

Espagnac, H. (1973). Les axes polymorphes de *Nephrolepis biserrata*. Analyse expérimentale du déterminisme de leurs structures. *Annales des Sciences Naturelles, Botanique, 12ème Série*, **14**, 223–86.

Favre, J. M. (1973). Effets corrélatifs de facteurs internes et externes sur la rhizogénèse d'un clone de vigne (*Vitis riparia* × *Vitis rupestris*) cultivé *in vitro*. *Revue Générale de Botanique*, **80**, 279–361.

Favre, J. M. & Grenan, S. (1979). Sur la production de vrilles, de fleurs et de baies chez la vigne cultivée *in vitro*. *Annales d'Amélioration des Plantes*, **29**, 247–52.

Franclet, A. (1969). Vers une production en masse de cyprès de forme contrôlée. *Proceedings 2nd FAO/IUFRO World Consultation on Forest Tree Breeding, Washington*, **69**, 11–8.

– (1979). Rajeunissement des arbres adultes en vue de leur propagation végétative. *Micropropagation d'Arbres Forestiers. AFOCEL, Paris. Études et Recherches*, **12**, 3–18.

Goublaye de Nantois, T. de la (1980). Rajeunissement chez le Douglas (*Pseudotsuga menziesii*) en vue de la propagation végétative. *Diplôme d'Études Approfondies Physiologie Végétale, Université de Paris VI*.

Gregory, L. E. (1951). Una nota sobre el enraizamiento des clones de *Hevea. Turrialba*, **1**, 201–3.

Hallé, F. & Martin, R. (1968). Étude de la croissance rythmique chez l'hévéa (*Hevea brasiliensis* Müll. Arg. Euphorbiacée Crotonoïdae). *Adansonia, Série 2*, **8**, 475–503.

Haicour, R. (1982). Eléments d'analyse de la structure et de l'évolution d'une espèce rudérale pantropicale *Phyllanthus urinaria* L. (Euphorbiacées). *Thèse Doctorat d'État, Université de Paris Sud, Centre d'Orsay*, 24 pp.

Heinz, D. J., Krishnamurthi, M., Nickell, L. G. & Maretzki, A. (1977). Cell, tissue and organ culture in sugarcane improvement. In *Applied and Fundamental Aspects of Plant Cell, Tissue and Organ Culture*, ed. J. Reinert & Y. P. S. Bajaj, pp. 4–17. Berlin, Heidelberg & New York: Springer-Verlag.

Kazandjian. B. (1977). La multiplication du chêne par bouturage. *Mémoire de 3ème année, Ecole Nationale des Ingénieurs des Travaux des Eaux et Forêts, Domaine des Barres, Nogent-sur-Vernisson*.

Kechaou, M. (1971). La régénération des oliviers âgés dans la région de Sfax. *INRAT Tunisie, Conférence Prononcée au Centre d'Amélioration et de Démonstration de la Technique Oléicole*, Cordoue, 25 Février 1971, 28 pp.

Kessler, B. & Reches, S. (1977). Structural and functional changes of chromosomal DNA during aging and phase change in plants. *Chromosomes Today*, **6**, 237–46.

Lahondère, C. & Neville, P. (1959). Sur le rameau court et les aiguilles de quelques pins. *Naturalia Monspeliensia, Botanique*, **11**, 29–47.

L'Hardy-Halos, M. T. (1975). A propos de corrélations morphogènes contrôlant l'initiation des ramifications latérales chez les algues à structures cladomiennes typiques. *Bulletin de la Société Phycologique de France*, **20**, 1–6.

Marcou, D. (1961). Notion de longévité et nature cytoplasmique du déterminant de la sénescence chez quelques champignons. *Annales des Sciences Naturelles, Botanique, 12ème Série*, **2**, 653–764.

Meins, F. Jr. & Lutz, J. (1980). The induction of cytokinin in primary pith explants of tobacco. *Planta*, **149**, 402–7.

Meins, F. Jr., Lutz, J. & Foster, R. (1980). Factors influencing the incidence of habituation for cytokinin of tobacco pith tissus in culture. *Planta*, **150**, 264–8.

Moncousin, C. (1982). Contribution à la caractérisation biochimique et physiologique de la phase juvénile de l'artichaut (*Cynara scolymus* L.) au cours de sa multiplication végétative conforme et accélérée en culture *in vitro*. *Thèse Docteur Ingénieur, Université de Paris Sud, Centre d'Orsay*, 209 pp.

Nagl, W. (1979). Search for the molecular basis of diversification in phylogenesis and ontogenesis. *Plant Systematics and Evolution, Supplement*, **2**, 3–25.

Nancy-David, C. (1977). Eléments d'analyse du déterminisme de la formation de différents types de tiges, y compris de tubercules, chez la pomme de terre (*Solanum tuberosum* L. var. BF15) cultivée *in vitro*. *Thèse Doctorat 3ème Cycle, Université de Paris Sud, Centre d'Orsay*, 82 pp.

Neville, P. (1964). Corrélations morphogènes entre les différentes parties de la feuille de *Gleditsia triacanthos* L. *Annales des Sciences Naturelles, Botanique, 12ème Série*, **5**, 785–98.

– (1969). Morphogénèse chez *Gleditsia triacanthos* L. III. Étude histologique et expérimentale de la sénescence chez des bourgeons. *Annales des Sciences Naturelles, Botanique, 12ème série*, **10**, 419–28.

Nigond, J. (1966). Recherches sur la dormance des bourgeons de la vigne. *Thèse Doctorat d'État, Université de Paris Sud, Centre d'Orsay*, 170 pp.

Nozeran, R. (1956). Sur la structure des épines et des bourgeons dormants d'*Anthocleista nobilis* C. Don. *Naturalia Monspeliensia, Botanique*, **8**, 167–75.

– (1968). Intérêt de la connaissance de la morphogénèse des plantes supérieures pour la conduite de leur multiplication végétative. *Revue Horticole Suisse*, **41**, 247–58.

– (1978). Reflexions sur les enchaînements de fonctionnements an cours du cycle des végétaux supérieurs. *Bulletin de la Société Botanique de France*, **125**, 263–80.

Nozeran, R. & Bancilhon, L. (1972). Les cultures *in vitro* en tant que technique pour l'approche de problèmes posés par l'amélioration des plantes. *Annales d'Amélioration des Plantes*, **22**, 167–85.

Nozeran, R., Bancilhon, L. & Neville, P. (1971). Intervention of internal correlations in the morphogenesis of higher plants. *Advances in Morphogenesis*, **9**, 1–66.

Nozeran, R., Ducreux, G., Neville, P. & Rossignol-Bancilhon, L. (1980). Ramifications et corrélations morphogènes. *Bulletin de la Société Botanique de France*, **127**, *Actualités Botaniques*, no. 2, 59–69.

Nozeran, R. & Rossignol-Bancilhon, L. (1977). La multiplication végétative chez les végétaux vasculaires. *Bulletin de la Société Botanique de France* (*Colloque 'La multiplication végétative chez les végétaux'*), **124**, 59–96.

– (1979). Les cultures *in vitro* et la production de 'semences' de pomme de terre. *Séminaire International de la Pomme de Terre, Institut de Développement des Cultures Maraîchères, Staoueli-Alger*, **1**, 133–46.

Pellegrini, O. (1961). Modificazione delle prospettive morfogenetiche del germoglio. *Delpinoa, N.S.*, **3**, 1–12.

Pfirsch, E. (1962). Recherches sur le conditionnement plagiotropique chez quelques plantes à stolons. *Thèse Doctorat d'État, Université de Strasbourg*, 121 pp.

– (1965). Déterminisme de la croissance plagiotropique chez les stolons épigés de *Stachys silvatica* L. *Annales des Sciences Naturelles, Botanique, 12ème Série*, **6**, 339–60.

Phillips, R. (1981). Direct differentiation of tracheary elements in cultured explants of gamma-irradiated tubers of *Helianthus tuberosus*. *Planta*, **153**, 262–6.

Privat, G. (1958). La culture in vitro des tissus de lierre (*Hedera helix* L.). *Naturalia Monspeliensia, Botanique*, **10**, 91–6.

Reinert, J. (1958). Morphogenese und ihre kontrolle an Gewebekulturen aus Karotten. *Die Naturwissenschaften*, **45**, 344–5.

Rossignol-Bancilhon, L., Nozeran, R., Quraishi, A. & Darpas, A. (1980). Début d'exploitation d'un des moyens d'extension de la variabilité chez la pomme de terre: les néoformations sur cals. *Colloque Eucarpia 'Application de la Culture in vitro à l'Amélioration des Plantes Potagères'*, pp. 192–200. Versailles: CNRA.

Schaffalitzky de Muckadell, M. (1954). Juvenile stages in woody plants. *Physiologia Plantarum*, **7**, 782–96.

Skirvin, R. M. (1978). Natural and induced variation in tissue culture. *Euphytica*, **27**, 241–66.

Snow, R. (1929). The young leaf as an inhibiting organ. *The New Phytologist*, **28**, 345–58.

Steward, F. C., Mapes, M. O. & Mears, K. (1958). Growth and organized development of cultured cells. II. Organization in culture grown from freely suspended cells. *American Journal of Botany*, **45**, 705–8.

Stoutemyer, V. T. & Britt, O. K. (1965). The behavior of tissue culture from English and Algerian ivy in different growth phases. *American Journal of Botany*, **52**, 805–10.

Thellier, M., Desbiez, M-O., Kergosien, Y. & Champagnat, P. (1981). Mise en mémoire de signaux morphogènes chez *Bidens pilosus* L. *Comptes rendus hebdomadaires des Séances de l'Académie des Sciences, Paris, Série III*, **292**, 1187–90.

Tran Thanh Van, M. (1973a). *In vitro* control of *de novo* flower, bud, root, and callus differentiation from excised epidermal tissues. *Nature*, **246**, 44–5.

– (1973b). Direct flower neoformation from superficial tissue of small explants of *Nicotiana tabacum* L. *Planta*, **115**, 87–92.

Vogel, M. (1975). Recherche du déterminisme du rythme de croissance du cacaoyer. *Café, Cacao, Thé*, **19**, 265–90.

Wardlaw, C. W. (1949). Experiments on organogenesis in ferns. *Growth*, **9** (supplement), 93–113.

Wardell, W. L. & Skoog, F. (1973). Flower formation in excised tobacco stem segments III. Deoxyribonucleic acid content in stem tissue of vegetative and flowering tobacco plants. *Plant Physiology*, **52**, 215–20.

Watelet-Gonod, M. C. & Favre, J. M. (1980–81). Miniaturisation et rajeunissement chez *Dahlia variabilis* (variété Télévision) cultivé *in vitro*. *Annales des Sciences Naturelles, Botanique, 13ème Série*, **2 & 3**, 51–67.

Wechselberger, M., Wielgat, B. & Kahl, G. (1979). Rhythmic changes in transcriptional activity during the development of potato tubers. *Planta*, **147**, 199–204.

Wielgat, B., Wechselberger, M. & Kahl, G. (1979). Age-dependent variations in transcriptional response to wounding and gibberellic acid in a higher plant. *Planta*, **147**, 205–9.

14

Phyllotaxis

W. W. SCHWABE

The term *phyllotaxis* is concerned with the description of patterns in which not only leaves but other organs are arranged on the plant stem or, perhaps more importantly, on the growing point of the shoot. According to the age of the organ under consideration, these descriptions may refer to cylindrical structures (as on stems) or conical surfaces (as on growing points), although the latter are often considered in the form of plane projections, especially in mathematical analyses.

Types of pattern

There are a number of visually very distinct types of pattern and these may be arranged in what is a convenient and probably also a logical order going from whorled to spiral systems: some examples of these types are seen in Fig. 14.1 to 14.6

Where leaves and other organs are inserted on the axis so that there are several at virtually the same level at each node, the resulting patterns are described as 'whorled'. This phyllotactic system is well exemplified by the Equisetales, and in higher plants by species of *Peperomia*, *Galium* and others, in which the numbers of organs per whorl may be large or as few as four. Whorls with fewer primordia at the same level (i.e., three) are often referred to as 'tricussate' and in descending order these lead to whorls with paired insertions at each node, the 'opposite and decussate' system. In each of these whorled systems the members of the next higher whorl tend to be positioned exactly above the gaps of the preceding whorl, as long as the numbers remain constant and with equiangular spacing within the whorl, the divergence depends on the number of leaves at each level. The next step in the reduction of numbers is the system where only a single leaf is inserted at each node. Where these are positioned on opposite sides of the stem, i.e., at an angular separation or 'divergence' of 180°, the pattern is referred to as 'alternate', or 'distichous' phyllotaxis. The same sequence of single leaf insertion also leads to 'spiral' phyllotaxis, where the divergence angle between successive primordia (in age) is not an approximate 180°, but, taking the smaller angle of the full circle, the divergence tends to the Fibonacci angle of 137°30′.*

It should be made clear here that in the whorled arrangements it is not necessary to assume that the formation of all organ initials at one level needs to occur absolutely simultaneously. In the older literature this point has been the subject of dispute ('false whorls'), but clearly the vertical distances between members of one whorl and those of the next are of subtantially greater magnitude. When looked on in this sequence, the various patterns found clearly grade from one to another and differences would seem to be quantitative rather than qualitative.

When the contacts made by leaves, scales, florets, etc., or by the young leaf primordia at the apex or on the stem, are considered, projections of whorled systems usually reveal straight lines radiating from the centre of the system, e.g., a cross with 90° angles for opposite and decussate, or 120° for tricussate, systems (Fig. 14.7). The spiral leaf arrangement in turn makes visible the contacts between adjoining primordia in such a manner that two sets of spiral lines become obvious which are referred to as the 'contact parastichies'. These contact spirals are recognizable as running in opposite directions, and counts of the numbers of spirals in the right- and left-handed directions nearly always yield numbers which are successive terms of the main Fibonacci series, though rather rarely examples have been recorded where such numbers are taken from accessory Fibonacci series, or have been recorded as anomalous. The visible or 'contact parastichies' are, however, not the only ones which may be drawn, and Richards (1951) showed that with different angles of intersection of the parastichy lines, other pairs of spirals from the Fibonacci series may be drawn through the same set of points representing primordial centres. Clearly, the visible pair is not the only one but characterizes the shape of the organ (see Fig. 14.8).

In addition to these patterns, there are a number, often given specific terms, in which a whorled system may itself be twisted spirally ('spired') displacing the vertical arrangement. In another, somewhat uncommon

---

*This angle is derived from the Fibonacci series, in which each term is the sum of the preceding two terms, i.e., where $F_n = F_{(n-1)} + F_{(n-2)}...(F_0 = F_1 = 1)$. The series can also be derived from the recurrent fraction of $1/1 + 1/1 + ...$ which converges to 1.618. When applied to the circle this divides 360° into two angles of 222.5° and 137.5°, the latter and smaller being the 'Fibonacci angle'. The value of 1.618 is, of course, the equivalent of the 'golden mean' section in geometry.

Fig. 14.1. *Equisetum arvense*
– multiple whorl.

Fig. 14.2. *Cruciata laevipes*
– 4-membered whorl.

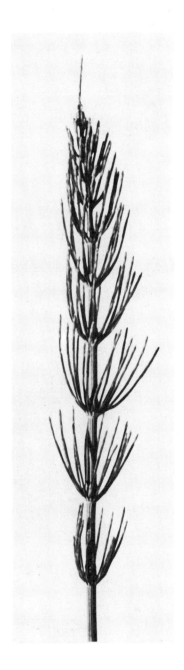

Fig. 14.3. *Epilobium adenocaulon* Fig. 14.4. *Kalanchoe* sp.
– transition to tricussate system.   – opposite and decussate.

Fig. 14.5. *Ravenala madagascariensis* – alternate or distichous system.

Fig. 14.6. *Helianthus annuus* – spiral phyllotaxis (34 : 55).

pattern, two-membered whorls are found in which the normal spiral system may be said to obtain if one member of each pair is disregarded; as Richards (1951) pointed out, these may be treated as two interlocked Fibonacci spiral systems. These so-called bijugate systems (as found in *Cephalaria* or *Dipsacus*) are regarded as a special case of the much rarer systems of multijugy.

Fig. 14.7. *Bryophyllum tubiflorum* – tricussate plant seen from above, six radii.

The transition between patterns

Phyllotactic patterns are not fixed for any one plant or species, though they are often characteristic of a species. They may nevertheless change

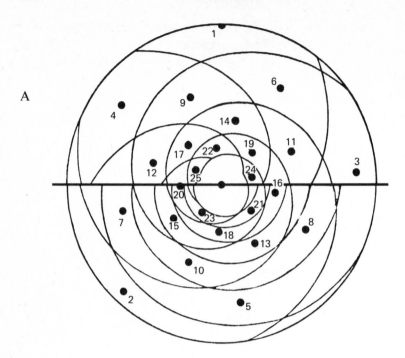

A

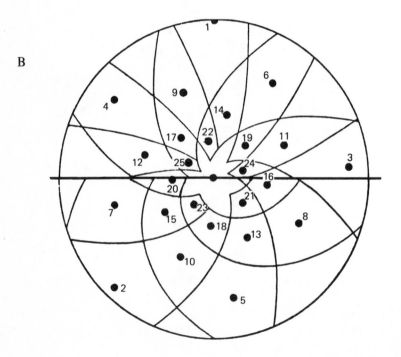

B

drastically, even in the same plant according to age, development, or after artificial interference. This common observation also points to the fact that the underlying processes giving rise to the various patterns do not differ qualitatively.

Most dicotyledonous plants start with oppositely-placed cotyledons, and for one or two nodes the seedling may continue the opposite and decussate pattern. Frequently the members of a pair may get slightly out of step and a transition occurs which may lead to the spiral system with obvious single leaf insertion per node, e.g., in the flax seedling (Fig. 14.9). In many other species, the opposite and decussate pattern is maintained throughout the vegetative growth period and transition to spiral phyllotaxis occurs only at the onset of flowering, e.g., in *Epilobium* spp. In other species, often monocotyledonous, single leaf insertion is the normal, vegetative pattern, from the first leaf onwards, resulting in a distichous leaf arrangement; this is then followed by a more or less rapid transition to the spiral arrangement when the plant becomes floral – a common occurrence in members of the Orchidaceae (e.g., *Epidendrum* spp.). Nor is the reverse transition from a spiral to a distichous system an uncommon occurrence; it is often found in the Bromeliaceae (e.g., *Vriesia* spp.) (Fig. 14.10), again on the transition to flowering.

While these transitions between apparently different patterns are perhaps more visibly striking, the changes within the spiral system itself from lower to higher orders of phyllotaxis (i.e., higher terms of the Fibonacci series of visible parastichies) or vice versa, are very common and no less marked. Where flowering can be photoperiodically induced, changes in phyllotaxis are often associated with this, even in consistently spiral systems (e.g., Erickson & Meicenheimer, 1977). The significance of referring briefly to these transitions will be clearer when the theories on the origins of phyllotaxis are discussed below.

Fig. 14.8. Identical positions of primordial centres can give rise to very different parastichy patterns according to the angle of intersection of parastichies. (A), top half 3 : 5 system; bottom half, 2 : 3. (B), top half 8 : 13 system; bottom half, 5 : 8. In (A) the 3-parastichies are common to both halves, in B the 8-parastichies are common.

Control of pattern formation

Basically all the phyllotactic patterns must originate at the growing points, though later twisting of petioles or other changes may impose secondary changes which may superficially become more apparent to the eye. The changes at the growing point in turn are due to 'growth' of the apical tissue and the rate and pattern of organ formation. The growth of the apical tissue and organogenesis have been reviewed by Romberger (1963), Cutter (1965), Richards (1948, 1956), Gifford & Corson (1971), Halperin (1978), and others.

From the organizational point of view it is important to remember that the system is a dynamic one with continuous or discontinuous growth of the apical tissue as such, whose average doubling times range, for example, from 20 days in a fern (*Dryopteris aristata*) to two days in *Chrysanthemum*.

The so-called 'bare' apex is constant neither in its size nor growth rate, but nevertheless represents a tightly controlled system. All cells within this apical 'cap' are so regulated as to maintain it as an integrated tissue – something which is perhaps even more obvious in the root. Such rigid control can be maintained, even where there is only one apical cell as in ferns (e.g., Wardlaw, 1956), and this includes the prothallus (cf., Albaum, 1938). However, some distance below the bare apex other meristematic centres can arise; that is, after they have reached a certain distance from, or below, the growing point, cells enter what has been termed the region of the 'anneau initial' (Buvat, 1955). It is in this zone, around and below

Fig. 14.9. Transition from opposite and decussate system in seedling to spiral system in *Linum usitatissimum*, seen in transverse section (COT, cotyledons).

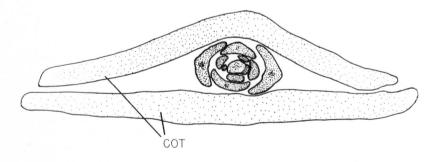

COT

Fig. 14.10. *Vriesia* sp. showing transition from spiral to distichous phyllotaxis upon flowering.

the terminal growing point, where cells or groups of cells can assume new activities by way of establishing new centres of meristematic activity. While meristematic cells can divide and expand at different rates and to different extents and ultimately mature and differentiate by assuming different specialized metabolic functions, perhaps the most important aspect at this level is that the plane of cell division may be differentially regulated (see also Lyndon, 1968, 1970 *a*, *b*). As long as the cells are under the regulating influence of the terminal growing point, the plane of cell division is such as to maintain the integrity of the terminal growing point itself. When new primordia arise, cells or groups of cells seem to 'reorient' the plane of cell division for one or more cycles in a periclinal direction; this then gives rise to elevated mounds of cells developing independently into new organs.

Whereas the terminal growing point, and of course all lateral stem growing points giving rise to cauline structures, have a more-or-less radial symmetry, the organs initiated on the flank of the growing points are foliar in nature and display bilateral symmetry. Moreover, except in the case where the terminal growing point on transition to flowering is converted into a single flower or terminal inflorescence, the growth of the terminal growing point itself is normally indeterminate, whereas growth of the leaf primordia is determinate.

The basic quest of studies into the origin of phyllotaxis is, of course, concerned not so much with the fact that groups of cells can start to form new organs, but with the determination of the positions in which these new organs arise. The first of the controlling factors is, therefore, the role of the bare apex itself, which prevents the formation of independent organ centres for some distance below it (and beside it, in very flat apices) and thus maintains its own integrity and indeterminacy of growth. Very little is known of the underlying mechanism, which resembles in some manner that of 'apical dominance', but which acts over only very short distances and affects foliar primordia, rather than cauline meristems, as in true apical dominance.

Interaction between primordia

It has long been agreed that the positioning of the new primordia arising in the 'anneau initial' is very largely controlled by the presence of existing (leaf) primordia especially the most recently formed ones. This notion was already expressed in the theories of Hofmeister (1868), who characterized the site of the new primordium as 'die grösste Lücke' ('biggest space'), and Schwendener (1878) in the mutual pressure theory. However, there have been other theories for this positioning, such as that the vascular system may determine the position of a new primordium from below, or perhaps more mysteriously, that there are generating centres or spirals (hélices foliaires) which at intervals 'bud off', if that is the right term, positions for new leaf primordia. These theories and the support for them will be briefly discussed later; see also Table 14.1, p. 420 et seq.

However, there is good evidence that in plants in general, there is a mutual influence of meristematic centres on one another. This is exemplified not only at the terminal growing point, but also within developing organs where the spacing of such centres leads to clearly non-random arrangements. Examples of this are found in glands, multicellular hairs, stomata forming from stomatal mother cells, and so on (e.g., Bünning, 1948, 1965). The resulting distribution of such morphogenetic centres is then such that minimum distances between them are maintained. Even in such examples as surface warts in fruits, e.g., the fleshy spines of the jackfruit (*Artocarpus heterophyllus*), an available area is divided up into more-or-less equal-sized areas with the boundaries between them being formed by the midlines between centres, which ultimately leads to the formation of similarly sized excrescences (Fig. 14.11).

The mechanisms of such mutual interaction of meristematic centres is thus a crucial problem which may also be directly relevant to phyllotaxy. Clearly, messages must be interchanged at least between adjoining centres and maybe beyond, possibly by competition for substances required for cell multiplication and/or expansion by each of the incipient centres, or by, what seems more likely, mutual inhibition, i.e. zones surrounding each centre in which independent cell division is prevented – a situation reminiscent of the behaviour of the bare apex itself.

There is so far no firm evidence that the mechanism of positioning control exerted by the terminal apex is the same as the 'mutual' interaction between primordia – it may be similar, as has been assumed in several theories, but it could also be different.

Theories of phyllotaxis

The theoretical treatment of the causes of leaf positioning has always been based on two approaches: (a) a purely mathematical and descriptive approach, and this group includes the majority of the more recent papers, and (b) an experimental approach attempting to draw conclusions from changes in phyllotaxis brought about by deliberate action. In this chapter the latter aspect will be given greater emphasis, since mathematical approaches can only show that hypothetical mechanisms are possible, but cannot prove their actual operation.

The older theories of phyllotaxis may generally be said to confound the method of description of the patterns and hypotheses of the forces or mechanisms bringing about the origin of phyllotaxis. As Richards (1951) in his classic paper stressed, measures are required for the description of phyllotaxis, which are independent of any theory of the origin, i.e., 'measures whose validity of usefulness do not depend on the assumptions of some particular phyllotaxis theory', but allow a faithful reconstruction of the pattern.

For such a description (especially in the spiral systems) two pieces of information are required: the radial distances of successive primordia from the centre of the system, and their angular divergence. Assuming in spiral phyllotaxis that this divergence in practice will, with random errors, approximate to the Fibonacci angle, the radial distances must be measured to give the plastochrone ratio 'r', which then yields the Phyllotaxis Index (P.I. $= 0.38 - 2.39 \log_{10} \log_{10} r$), which has been so designed as to give integral values for consecutive pairs of parastichy numbers, i.e., the 1 : 2 system being approximately 1, the 2 : 3 system being 2, the 3 : 5 being 3, and so on.

For the detailed method of the derivation of the P.I. and the 'Equivalent Phyllotaxis Index' on conical surfaces (E.P.I. $=$ P.I. $+ 2.3925 \log_{10} \sin\theta$, where θ is the half angle of the apical cone), reference should be made to Richards (1951).

However, while the plastochrone ratio is mainly a measure of the apical growth made between successive plastochrones, the divergence angle is the result of the mutual interaction of the primordia.

Non-experimental approaches

In the course of time a series of theories has been proposed on a variety of possible mechanisms, and in the recent spate of papers on this topic several of the older ideas have been revived. Most are concerned to account for the mutual interaction of primordia on one another and to give reasons for the Fibonacci angle being approached as the limiting divergence. Many of these have been discussed in the literature, often in considerable detail, and only some of these will be briefly described here, brief comments being included in the listing in Table 14.1.

The minimum available space theory was enunciated in detail by Snow & Snow (1933, 1962), who made considerable efforts to prove the validity of ideas first formulated by G. van Iterson (1907). Basically, it postulates that a new primordium must arise when, in the apical region, a 'large enough

Fig. 14.11. Jackfruit, *Artocarpus heterophyllus*. The surface of the fruit shows subdivision of area between fleshy spines along midline between centres.

gap' appears above and between the young and actively growing primordia. Generally, this would fit in with most observations, but the concept is not sufficiently distinct to allow it to be distinguished from so-called 'field theories', as has been pointed out by several authors. To distinguish this concept from a field theory a *specific* size for the 'minimum required space' would need to be defined for any species and shown to apply in reality. This, however, has never been shown or postulated and certainly in many species the minimum area occupied by a new primordium is not constant in time and varies relative to developmental state.

One of the earliest theories was based on the idea of mutual pressures of older organs on younger ones (Schwendener, 1878). Recently this idea has been revived by Adler (1974, 1975) and by Williams (1974); see also Roberts (1977) who applied the concept to semi-decussate phyllotaxis. It is difficult to demonstrate that such pressures exist on the soft apical tissues of the terminal meristem and leaf primordia. Williams has concluded that the relatively dense packing must be taken as evidence for pressures. Before mutual pressure on older or younger organs can be accepted as causal, it is essential to prove their existence, and no incontrovertible proof exists. In any case, a mutual pressure theory cannot be applied to fern apices where the individual primordia are well separated from one another at initiation (cf. Church, 1904). In apices from which older leaf primordia were removed no change of phyllotaxis was recorded; and in the many surgical experiments carried out by Snow & Snow (1931, et seq.) only direct surgical interference at the growing point itself caused modification of phyllotaxis. Adler (1974) has been concerned only with the theoretical consequence of such pressures and suggests the possibility of positional changes after initiation. In fact, the faster growth rate of the primordia themselves compared with the apical tissues usually leads to expansion in an outward direction and sections cut through apical buds usually show distinct gaps between primordia. Recent results of experiments (Wrigley & Schwabe, unpublished) with bilateral pressures of 3–7 kg/cm^2 applied laterally on young *Chrysanthemum* buds have shown that it is possible to get large distortions in the position of the outer young leaves, in the bud, yet no modification of the normal spiral phyllotaxis at the apex itself resulted (Fig. 14.12).

Another kind of theory is based on the proposition that primordial initiation at the apex is determined from below by the vascular system and has been suggested a number of times by several authors (e.g., Esau, 1942; Larson, 1977). However, here too the evidence is not cogent and the positioning is not specifically related to the vascular pattern. Moreover, as

Wardlaw (1943) showed, there are many instances of adventitious shoots arising from a variety of tissues where the new shoot meristem and primordial initiation has proceeded normally, well before any new connections are made with the existing vascular system; in fact, the new xylem may often develop basipetally to link up with the older tissues. Finally, the experiments by Snow & Snow (1947), in which incipient vascular tissue was severed below the initiating region of the apex, confirmed that the exclusion of vascular influences from below did not lead to any changes in the initiating pattern.

Another type of theory is based on the postulate that in the shoot apex internal spirals (hélices foliaires) are operating which 'cut off' new primordia at regular intervals. This, as Richards (1951) pointed out, amounts to a multiple equivalent of the 'genetic spiral' of the Schimper-Braun system where instead one set of parastichies is selected. The idea was first proposed by Plantefol (1948) and later elaborated by Loiseau (1969), and is the basis for most publications in the French literature. A mathematical basis for the theory has not been established.

Fig. 14.12. Transverse section through terminal bud of vegetative *Chrysanthemum morifolium* subjected to lateral pressure between two plates (approx. 7 kg cm^{-2}) causing severe distortion of outer primordia. The innermost primordia are still inserted at approximately the Fibonacci angle. (Successive divergences from outside inwards: 147°, 155°, 130°, 106°, 134°, 137°.) Direction of pressure is arrowed.

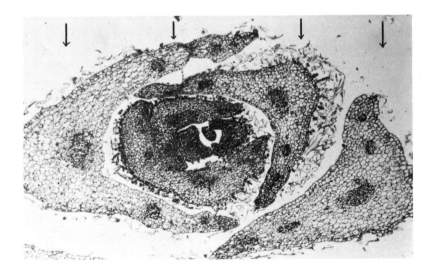

Table 14.1. *List of 'Theories of Phyllotaxis' with some salient features*

Authors	Basic description or mechanism postulated	Some important assumptions made and additional comments
(A) *Largely descriptive systems*		
Braun, A. (1831) Schimper, C. F. (1835, 1836)	All leaves inserted on single genetic spiral	With a rational divergence angle exact orthostichies can be found. System described by ratio of complete revolutions made over number of leaves inserted. Common ratios given 1/2, 1/3, 2/5, 3/8, 5/13 ...
Bravais, L. & Bravais, A. (1837)	Irrational divergence angle, converging to 137.5°	No single genetic spiral found in multijugate systems. Leads on to 'hélices foliaires multiple' of Plantefol.
Hofmeister, W. (1868)	New leaf formed in largest gap 'grösste Lücke'	Greatest elasticity furthest away from last primordium.
Schwendener, S. (1878)	Mutual pressure of primordia	Pressure from older primordia inhibits organogenic potential. Pressure may also give some displacement of primordia after formation.
Church, A. H. (1904)	Rejecting the existence of true orthostichies, systems are described in terms of pairs of visible parastichies	Orthogonal intersection of logarithmic spirals is generally assumed. He also defines 'bulk ratio' of young primordia.
Schüepp, O. (1938, 1966)	Tension of apical surface by excessive growth of tunica. Supposed to have same division rate as corpus	See also Snow & Snow (1951) and Lyndon (1971). No experimental evidence.

Plantefol, L. (1948) Loiseau, J.-E. (1969)	Operation of multiple generative spirals (hélices foliaires multiples et centres générateurs foliaires)	Assumption that (a) leaves are arranged along the most common helices, and (b) that along the same helix the leaves are contiguous. Foliar helices seem thus identifiable with one set of visible parastichies.
Williams, R. F. (1974)	Mechanical theory of physical constraint exerted by older primordia	Conclusion drawn from close packing of primordial leaves in bud, e.g. in *Ficus* spp. Pressure of older primordia must affect sites of initiation.
(B) *Mathematical systems not involving experimental approaches* Iterson, G. van (1907)	Initiation of new primordium in large gap between existing older primordia; not identical with Hofmeister's concept of the largest gap (p. 220). Generally regards outlines of new primordia as circular	
Schoute, J. C. (1913, 1922, 1925, 1936)	Fields of inhibition emanating from primordial centres and terminal apex, determining sites of origin of new primordia. Theory complicated by postulate of pseudoconchoid curve	
Richards, F. J. (1948, 1951)	Field theory of inhibition. Description of system by Fibonacci angle and plastochrone ratio (ratio of radial distances of two successive primordial centres from centre of apex). Defines	Reasons for convergence to ideal angle not fully developed, but indicated by defining site of new primordium as placed between the two youngest existing ones and nearer to the older one of these.

Table 14.1. (*continued*)

Authors	Basic description or mechanism postulated	Some important assumptions made and additional comments
	Phyllotaxis Index on centric system and Equivalent Phyllotaxis Index on cone. (See text for detail)	
Adler, I. (1974, 1977)	Contact pressure between adjacent leaves, with subsequent positioning of primordia even after initiation	Considers normalized cylindrical model. The assumption is made that contact pressure maximizes the minimum distance between leaf primordia (Maxmin principle).
Adler, I. (1975)	Space filling model	Can explain equal distribution of leaves on helix. Does not explain why divergence tends towards ideal angle.
Thornley, J. H. M. (1975*a*, *b*)	Inhibition by morphogen production, diffusion and degradation	Steady-state model with circular diffusion round meristematic ring, using polar coordinates for parastichy intersection. Morphogen degraded by first-order reaction. Assumes spatial competence = sufficient amount of tissue for organization.
Gierer, A. & Meinhardt, H. (1972)	Activation and inhibition in a diffusion theory	
Mitchison, G. J. (1977)	Inhibitor theory with short-range action but long-living inhibition	Theory applies to spiral phyllotaxis only. Gradually increasing circumference of anneau

		initial subsumed. Steady state; not clear how initial placement made according to inhibition theory.
Veen, A. H. & Lindenmayer, A. (1977)	Computer model for an inhibitor system with surface diffusion and inhibitor breakdown. On single cell basis. Threshold for each cell to initiate new leaf	Cylindrical surface chosen. Uniform concentration in cell, but gradient across cell walls. Central apex and primordia both produce same inhibitor.
Hellendoorn, P. H. & Lindenmayer, A. (1974)	Computer model applied to *Bryophyllum tubiflorum*	Modified theories for opposite and decussate, and tricussate systems. Decay of inhibition proportional to its concentration.
Roberts, D. W. (1977, 1978)	Contact pressure and 'chemical contact pressure'	Chemical inhibitor with preferential longitudinal movement in both directions. The contours of chemical contact pressure believed to attenuate.
Richter, P. H. & Schranner, R. (1978)	Inhibitor theory relating to spatial and temporal influences	Golden section of angle applied to genetic spiral. Genetic spiral as primary pattern of Gierer & Meinhardt (1972) system with inhibition as long-range interaction. Consider expansion in discoidal and helical systems. Argument given which makes divergence angle of 137.5° plausible but no mathematical proof given.
Young, D. A. (1978)	Inhibitor diffusion computer model for steady state in cylindrical shoot apex system	Two-dimensional diffusion of inhibitor, destruction by first-order kinetic process. Threshold level. For cylinder, parastichies higher than (3,5) not obtained.

Table 14.1. (*continued*)

Authors	Basic description or mechanism postulated	Some important assumptions made and additional comments
Thomas, R. L. (1975) Thomas, R. L. & Cannell, M. G. R. (1980)	Defines system using Braunian divergences and the generative spiral with a constant angle	By use of mathematical identities, Richards' system of plastochrone ratio and 'ideal' angle are linked to Braunian orthostichies and generative spiral.
Jean, R. V. (1976)	'Relational "tree" of normal asymmetric phyllotaxis'	An asymmetric growth factor and growth matrices are proposed. The growth operators are then subjected to spectral analysis of matrices.
Jean, R. V. (1982)	Hierarchical control	Possibly via vascular system cf. Bolle (1939).
Dixon, R. (1981)	Genetic selection of golden section of angle as 'ideal spacing'	Fibonacci angle assumed for computer modelling.
(C) *Theories based largely on experimental approaches*		
Snow, M. & Snow, R. (1931, 1933, 1935, 1952, 1955, 1959, 1962)	First available space (following on van Iterson)	Microsurgical. Experimental studies giving induced changes in phyllotaxis. 'Available' space not defined, nor constant for any species. See also text for summary of results.
Wardlaw, C. W. (1950)	Surgical experiments on apices of fern species and others, have shown that vascular differentiation controlled from above (apex)	See also Section A of this Table.

Wardlaw, C. W. (1956, 1957)	Application of auxins and other substances directly to bare apex, led to various disruptions, or left pattern of leaf initiation intact. No systematic changes of phyllotaxis were obtained
Girolami, G. (1953)	Suggestion of direct influence from primary vascular system on phyllotaxis in *Linum*
Schwabe, W. W. (1971)	Inhibition with polar basipetal transport of the inhibitor, instead of attenuation or destruction
Larson, P. R. (1975, 1977)	Vascular traces and phyllotaxis. Influence must move in upward direction
Green, P. B. (1980)	Hoop-reinforcement of cellulose in tunica cells

	Based on change of phyllotaxis by application of auxin transport inhibitors and natural changes.
	Experimental evidence for upward transport of labelled carbon compounds along vascular traces. No evidence for primordial sites being predetermined by upward flow, and downward movement of a vascular tissue determinant not excluded.
	Based on surface observations on apices with opposite and decussate phyllotaxis. Not extended to spiral phyllotaxis.

A few more recent publications have resurrected concepts of earlier work. For example, Thomas (1975) includes in his theoretical treatment the orthostichies postulated by Braun (1831) and Schimper (1835, 1836). This concept was rejected by Bravais & Bravais (1837) and later dismissed by Church (1904) and others, one of the problems being that it is virtually impossible to determine in practice with sufficient accuracy that a true orthostichy exists and can be distinguished from Fibonacci divergences. This was further developed by Thomas & Cannell (1980), using the ratio of the distance of two adjacent primordia from the centre of the system and the constant angle between any tangent to a logarithmic generative spiral.

The majority of theories of phyllotaxis are based on the concept of 'fields' emanating from young existing primordia positioning the site of formation of the next one. Generally, this has been in terms of inhibition, and the concept goes back to Schoute (1913). It was accepted by Richards (1951), Richards & Schwabe (1969) and Schwabe (1971), and has formed the basis of the mathematical treatments by Thornley (1975*a*, *b*), Veen & Lindenmayer (1977), Mitchison (1977), Gierer & Meinhardt (1972), and others.

It is difficult to do justice to the various mathematical theories in an attempt to summarize their content and the underlying assumptions in the available space of this paper; they need to be read in full. Hence only some of the salient aspects are summarized in Table 14.1.

Perhaps it will be useful at this stage to list the aspects which need to be explained by any mathematical theory to fit the major facts:

(a) the influence of the *bare apex* and failure of primordia to form on its surface (at least until such time as when it is 'consumed' in inflorescence formation, etc.);
(b) the total failure of primordia ever to form below existing, older primordia;
(c) the pattern formation in whorled systems ranging from multi-membered whorls to the limiting system of distichous phyllotaxy;
(d) the spiral systems of varying degrees of complexity, with divergences closely approximating the Fibonacci angle;
(e) the transitions between systems (e.g. whorled to spiral and *vice versa*) and changes in complexity of spiral systems;
(f) the occurrence of multijugate and similar patterns.

Moreover, some explanation would ultimately be required to account for the occurrence of unusual divergences and also fasciations.

Most theories based on the concept of morphogens acting to form fields (of inhibition) are based on the pre-condition of continuing an existing system and thus do not require the intervention of processes discussed by Turing (1952) in which inequalities develop from an initial uniform situation. *De facto* this must be the right approach since in higher plants, even in the embryo, patterns are already extant which must affect subsequent patterning, and a uniform, unpolarized situation probably never exists.

Experimental approaches

Many fewer approaches have been made to discover the underlying mechanisms of phyllotactic positioning by actual experimentation. The first substantive work on this aspect is due to the husband and wife team of the Snows (1931, 1933, 1935). By using microsurgical techniques they succeeded in modifying phyllotaxis of living apices, and showed that transitions between systems could be brought about by changing the strengths of the mutual influences of primordia upon one another. Much of their work was done with *Epilobium hirsutum* and *Lupinus albus*, and the main conclusions which were derived from this work may be summarized as showing that:

(a) primordia do influence one anothers' position, so that destruction of an initial will cause re-positioning of the next one to arise, closer to the site of the destroyed one;

(b) destruction of one member of a paired system (opposite and decussate) will cause at least temporary transition to a spiral system;

(c) microsurgical cuts on the bare apical surface suggest that, if anything, the tunica is under tension and that the suggestion by Schüepp (1916) that organ formation is due to wrinkling by excessive tunica growth is not tenable (see also Snow, 1951);

(d) undercutting of the site of the presumptive primordium which excludes vascular influences from below, did not modify its positioning, suggesting that vascular influences, if any, are not decisive.

Wardlaw (1949, 1950) and Wardlaw & Cutter (1954) also carried out surgical experiments on apices with similar results. Wardlaw (1955) seems to have been the first to apply growth regulatory substances directly to the growing point surface to modify phyllotaxis, but succeeded only to the extent of demonstrating that such treatment can cause disruption of existing patterns. Since then Schwabe (1971) has shown that growth regulators such as tri-iodobenzoic acid (TIBA) can cause substantial changes in phyllotaxis if applied through the vascular system, i.e., a change from spiral to distichous phyllotaxis in *Chrysanthemum* (cf. below).

Cutter (1964) has shown that application of gibberellin to apices of *Dipsacus* can change (increase) the divergence angle to some extent – possibly by elongating early internodes. Meicenheimer (1981) obtained a transition in *Epilobium hirsutum* from the opposite and decussate to the spiral system by treatment with naphthylphthalamic acid (NPA), a change which occurs naturally in this species on flowering and is reflected by a change in the divergence to 137.5°. Charlton (1974) obtained some changes in a water plant by use of dwarfing substances.

The importance of vertical spacing

While Richards' postulate of plastochrone ratio and divergence yields all the essential data for a description and reconstruction of the pattern either in plane projection or on a cone (using the PI and EPI, respectively), the third dimension appears to have been somewhat neglected in theories on the origin of phyllotaxis. This applies to theories such as Thornley's (1975*a*), where the controlling events are viewed only in terms of a morphogen operating in a ring, or that of Adler (1974, 1977), which is concerned with the two dimensions of a cylinder and the pressure packing around it.

   However, Schwabe (1971) has shown that chemical modification of the phyllotactic pattern is possible. In experiments with *Chrysanthemum*, the most striking effect obtained by TIBA treatment fed into the vascular system from below was an elongation of the incipient first (youngest) and second internodes by relatively very substantial amounts, i.e. in the apical region where the determining primordia of the existing complement must act on the new ones to arise ('presumptive sites' according to Snow). These experiments revealed that an extra elongation of some 15–20 μm above the normal, equivalent to the vertical diameter of two cells (which are virtually isodiametric at this level), is sufficient to change the phyllotaxis from a spiral to a distichous pattern (i.e. a change in the divergence from 137.5° to one of 180°). If it is assumed that the primordia influence each other by mutual repulsion, then this would mean that the second-youngest primordium (*below* this elongated, incipient internode) is deprived of all influence to the extent that the new one can arise directly above it, i.e., it is unaffected by the presence of the second-youngest existing primordium. Schwabe had pointed out that this was likely to be due to an effect of TIBA on the elongation of this internode, rather than to a direct effect on the new primordium. Other substances such as NPA and morphactins (unpublished data) have since been shown to have similar effects. There is good reason to believe that the influence emanating from the developing primordia is more probably inhibitory than a competition for a growth promoter diffusing upwards from below, since the older primordia would in fact be in a stronger position to intercept this than the younger ones above them.

Since then a detailed study on this 'vertical distance effect' has been done by McDonald (1975) and Schwabe & McDonald (unpublished). These investigations were concerned with the correlation between the phyllotactic patterns and the internode lengths between the youngest primordia at the apex, especially where there were transitions from one system to another, either occurring naturally, or brought about artificially.

These studies revealed a remarkable consistency. Wherever the internodes concerned, i.e. the youngest and second youngest, were elongated, the divergence increased and usually reached 180°. Where these internodes shortened, phyllotaxis became spiral. This was so regardless whether such changes in the internode lengths were brought about by natural changes or by the onset of flowering or by chemical treatment. Table 14.2 gives some of these data. The associated changes in shape at the apex are shown by the examples of *Epidendrum vitellinum* and *Epilobium adenocaulon* before and after transition to flowering, *Hedera helix* in the juvenile and mature condition, as well as *Chrysanthemum* treated with NPA (Figs. 14.13–14.16); also given in the figures are the vertical distances between the youngest primordia.

How is this correlation to be interpreted in the context of mutual inhibitory effects between primordia? Clearly the effects of any pattern of inhibitors must allow for threshold values, below which the inhibition fails and a new primordium may become initiated. This requirement has been met in several theoretical treatments by the assumption that the inhibitory morphogen acts only over short distances and is then *somehow* inactivated (see Table 14.1, e.g. Richter & Schranner (1978)). This assumption of inactivation or inactivators of the inhibitory substances is very important, and it could be said that the pattern formation may be due to the *inactivators of the inhibitors*, though the simplicity of a field system must be lessened by the introduction of a second morphogen activity, albeit negative. But is it necessary to assume this second morphogen activity? In the case of the positional control of phyllotaxis it would seem a simpler assumption that only one species of morphogen is needed and not two, but that the movement of the single substance is not random but polar. Polar transport of substances in plants is not a novel phenomenon but thoroughly well established for the oldest-known type of growth

W. W. Schwabe

Fig. 14.13. Apical profile of vegetative and flowering *Epidendrum vitellinum*. Horizontal bars indicate vertical distances between the youngest primordia.

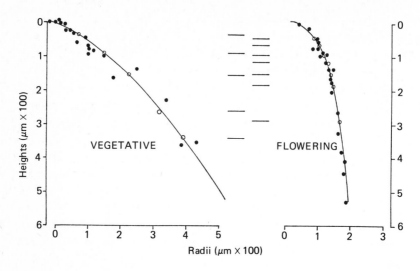

Fig. 14.14. Apical profile of vegetative and flowering *Epilobium adenocaulon*. Horizontal bars indicate vertical distances between the youngest primordia.

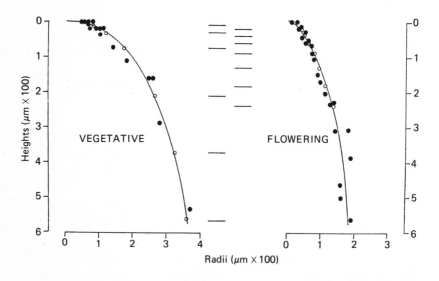

regulator, the auxins; recent discoveries of changed polarity (Wright, 1982) are highly exceptional. If it is assumed that the transport of the inhibitors is polar, the loss of effect of primordia at a greater vertical downward displacement from the apex is at once explained (Fig. 14.17). The degree of inhibitor production could then even increase with increasing size and growth of the primordium, which would not be at all implausible, but nevertheless the effect would 'diffuse' or move away in a downward direction. The total failure of ever finding new primordia between or below older ones is at once explained as the results of the very high level of inhibition which would be reached below the anneau initial itself.

The rapid loss of any effect with downward displacement by growth of any one primordium would be explained, and *per contra*, the increasing accuracy of positioning of primordia with very flat apices would equally be explained. While this accuracy reaches the level of minutes of arc in such inflorescences as the sunflower head, the effects of a large number of participating older primordia would be involved.

Considering the situation at the moment of initiation at an apex of 'normal' shape (conical), the youngest existing primordium has the major inhibitory effect, being still in the zone of the anneau itself, the next youngest primordium having less influence. Yet clearly the inhibition must traverse the whole tissue of the meristem, unless surface diffusion is postulated, as has been done by Thornley (1975a), though biologically there is no good reason for assuming such a rather unusual behaviour. The inhibition from the youngest primordium is thus likely to be at a minimum 180° away, and if there is only one determining primordium present at any one time, distichous phyllotaxis results, e.g. in the cereals. Where the second youngest primordium or others are also still capable of adding to the sum total of inhibition at the apex, a less-symmetrical pattern of inhibition must result and lead to a lateral shift which will tend to approximate the 137.5°. Here, however, it is obviously necessary to

Fig. 14.15. Apical profile of juvenile and mature *Hedera helix*. Horizontal bars indicate vertical distances between the youngest primordia.

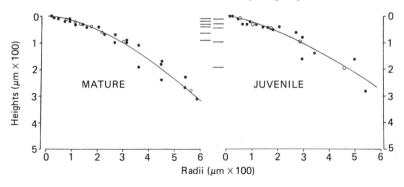

Table 14.2. *Correlation between natural or induced changes in phyllotaxis altering the divergence angle and the length of the first incipient internode between primordia P1 and P2*

Species	Condition	Incipient internode length between primordia P1 and P2 (μm)	Mean divergence angle	Least significant difference
Chrysanthemum	control	18.7	136° ⎱	11°
	50 ppm TIBA-treated	29.5	162° ⎰	
	control	10.8	133° ⎱	11°36'
	100 ppm NPA-treated	24.0	158° ⎰	
Epilobium adenocaulon	vegetative	22.5	179°[a] ⎱	15°
	flowering[b]	12.1	138° ⎰	
Epidendrum vitellinum	vegetative	56.5	179° ⎱	11°
	flowering	22.3	135° ⎰	
Hedera helix	juvenile	17.0	134°31' ⎱	5°53'
	adult	9.5	179°36' ⎰	
Zea mays	normal	32.5	180° ⎱	16°
	ABPHYL genotype	15.0	142° ⎰	

[a] between sets of paired leaves.
[b] 13 days after start of long-day induction.

postulate *some* non-polar movement in at least an obliquely upward direction. How could this be accounted for on a polar transport theory? The answer could be sought along the following line: Polarity of transport of any substance would seem to depend on transport through living material, and in the plant tissue this would be represented by the symplast. (See also Mitchison (1981) for the theory of increasing permeability of the plasmalemma with increasing flux.) On the other hand, some non-polar (non-directional) diffusion could occur through the apoplast. If one estimates the cross-sectional area of the meristematic apical tissue which is occupied by cell walls (in the absence of any air spaces) this ranges from about 1 to 5% of the total. It is this apoplast area through which non-polar diffusion of the inhibitor may be envisaged to proceed, and this would approximately supply just the right amount of inhibitor from the second-youngest primordium under 'normal' circumstances. McDonald & Schwabe have calculated the minimum diffusion rates required to vary from 1 to 3 μm h^{-1}, which is many times smaller than acropetal rates recorded (Jacobs, 1954; McCready, 1968).

Comparison of the known diffusion rates obtained for auxins, the actual, very short distances involved at the apex, and the time for diffusion

Fig. 14.16. Vertical distances between successive leaf primordia (incipient internodes) of *Chrysanthemum* plants treated with naphthylphthalamic acid (NPA) and controls, sampled at 50 days. The control shows spiral phyllotaxis, the treated plants show alternate phyllotaxis.

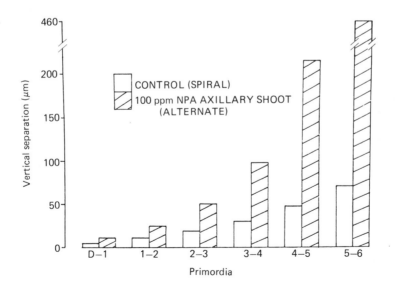

available during the plastochrone, are all fully adequate for such pattern control by a polarly-transported inhibitor.

The question arises at once whether this could be a known or unknown substance, and auxins with their established polarity movement would be the obvious candidates. Moreover, the fact that most, if not all, of the auxin emanating from the apical bud is believed to come from the leaf primordia would fit the pattern well, while also agreeing with the suggestion that at least at the level near the shoot apex, no destruction of auxin in known to occur (as was postulated for other non-polar morphogens).

The problem of establishing that auxin is involved, or alternatively isolating and identifying another morphogen, is as yet barely possible because of the minute amounts of tissue involved and the even more minute quantities of morphogen involved. However, the fact that auxin-transport inhibitors such as TIBA, NPA and some morphactins can cause

Fig. 14.17. Diagrammatic pattern of inhibitor flow from two primordia at different distances from apex.

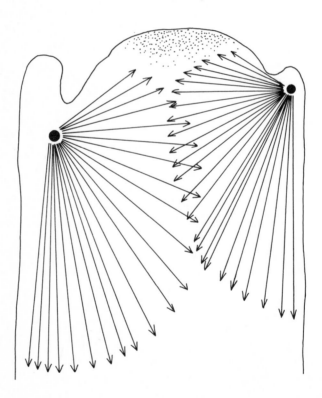

changes of phyllotaxis – seemingly by causing elongation of the youngest internode – suggest that this in turn is brought about by damming up the auxin itself. Interestingly, gibberellin treatment, which causes often very large elongation of internodes, usually affects only the older internodes, but where it lengthens internodes 1 or 2, as in the ivy (McDonald, 1975), there is a corresponding increment in the divergence angle.

The question whether the bare apex is itself a source of the known inhibitor is still wide open. Possibly, one could make further assumptions in which the bare apex itself is the source of a quite separate inhibitor, or a whole range of other so-far unsupported speculations might be made. Moreover, it would seem equally likely to seek an explanation in terms of the surface (tunica) or deeper (corpus) cells in the apical dome lacking the capacity, or competence, to divide in a plane which would initiate a primordium until it is some distance removed from the tip itself. Here too there is room for an experimental approach.

In conclusion, the more experimental approach discussed here suggests that the positioning of new primordia can be explained sufficiently by a single inhibitor emanating from new growth centres, with a polar (downward) directional transport. In many ways the plant shoot apex is thus a much simpler system than any animal system of morphogenesis where one must account not only for the differentiation of cells, but also for their active movement, for which more than one gradient of morphogen concentration is needed.

In the shoot apex only one gradient system of inhibitor for each centre and down the young stem needs to be postulated.

Thus one might, following Wolpert's analogy (1981, 1982), postulate that each new primordium was aware of its position merely due to finding itself at a point of minimum inhibition at the time the cells had reached the competence to become independent of the terminal meristem.

The theoretical, mathematical treatments lead to the same conclusion in as much as the divergence is found to reach the limiting angle of 137.5°, but, of course, the underlying assumptions are different in most cases. A different computer-based calculation using the above assumptions is now in the process of being completed, which should meet the requirements outlined earlier.

Apical size and type of organ produced

A final word may be said here on the type of organ produced at the apex. Usually, these are leaf primordia, and their production is associated with the continued vegetative growth of the bare apex. On transition to flowering, whether by the formation of a large receptacle with numerous florets or by way of a single flower being produced, the bare apex is usually 'consumed' in the process, and one or more floral apices are formed. On this transition, the foliar element of the primordial initiation frequently diminishes and may become so vestigial as to disappear virtually completely, e.g. composite inflorescences where the individual florets represent cauline structures, although it is possible by appropriate environmental conditions such as adjusted photoperiod to strengthen the foliar development to some extent (Schwabe, 1951). Another aspect of great interest is the question why floral organs are produced on such cauline structures. In many instances such floral apices are smaller than the vegetative ones (i.e., the floret apex rather than the inflorescence receptacle which is often many times as large as the vegetative apex). Is it possible that under such conditions of a generally reduced-size apex the equivalents of leaf primordia become sepals and petals, etc.? This aspect is worthy of some quantitative study and possibly some experimental approach. Although the foliar system of phyllotaxis is often continued right into the inflorescence, the actual arrangements in the flower or floret are commonly whorled and would indicate that the apical changes conducive to this type of pattern have occurred.

Note added in proof: An interesting analysis of actual apices of a range of phyllotactic types is described by Rutishauser (1982) who utilizes four parameters for the characterization: divergence angle, the arc of leaf insertion, the contact parastichy pattern and the plastochrone ratio; these were determined from cross-sections and scanning electron micrographs of apices.

References

Adler, I. (1974). A model of contact pressure in phyllotaxis. *Journal of Theoretical Biology*, **45**, 1–79.
— (1975). A model of space filling in phyllotaxis. *Journal of Theoretical Biology*, **53**, 435–44.
— (1977). The consequences of contact pressure in phyllotaxis. *Journal of Theoretical Biology*, **65**, 29–77.
Albaum, H. G. (1938). Inhibition due to growth hormones in fern prothallia and sporophytes. *American Journal of Botany*, **25**, 124–33.
Bolle, F. (1939). Theorie der Blattstellung. *Verhandlungen des botanischen Vereins der Provinz Brandenburg*, **79**, 152–92.
Braun, A. (1831). Vergleichende Untersuchung über die Ordnung der Schuppen an den Tannenzapfen als Einleitung zur Untersuchung der Blattstellung Überhaupt. *Verhandlungen der Kaiserlichen Leopoldinisch-Carolinischen deutschen Akademie der Naturforscher*, **15**, 197–401.

Bravais, L. & Bravais, A. (1837). Essai sur la disposition des feuilles curvisériées. *Annales des Sciences Naturelles, 2ème Série*, **7**, 42–110.

Bünning, E. (1948). *Entwicklungs- und Bewegungsphysiologie der Pflanzen*. Berlin: Springer-Verlag.

— (1965). Die Entstehung von Mustern in der Entwicklung von Pflanzen. In *Encyclopedia of Plant Physiology*, vol. XV/1, ed. W. Ruhland, pp. 383–408. Berlin, Heidelberg & New York: Springer-Verlag.

Buvat, R. (1955). Le méristème apical de la tige. *Année Biologique, 3ème Série*, **31**, 595–656.

Charlton, W. A. (1974). Studies in the *Alismataceae*. V. Experimental modification of phyllotaxis in pseudostolons of *Echinodorus tenellus* by means of growth inhibitors. *Canadian Journal of Botany*, **52**, 1131–42.

Church, A. H. (1904). *On the Relation of Phyllotaxis to Mechanical Laws*. London: Williams & Norgate.

Cutter, E. G. (1964). Phyllotaxis and apical growth. *The New Phytologist*, **63**, 39–46.

— (1965). Recent experimental studies of the shoot apex and shoot morphogenesis. *The Botanical Review*, **31**, 8–113.

Dixon, R. (1981). The mathematical daisy. *New Scientist*, **92**, 792–5.

Erickson, R. O. & Meicenheimer, R. D. (1977). Photoperiod induced change in phyllotaxis in *Xanthium*. *American Journal of Botany*, **64**, 981–8.

Esau, K. (1942). Vascular differentiation in the vegetative shoot of *Linum*. I. The procambium. *American Journal of Botany*, **29**, 728–47.

Gierer, A. & Meinhardt, H. (1972). A theory of biological pattern formation. *Kybernetik*, **12**, 30–9.

Gifford, E. M. & Corson, G. E. (1971). The shoot apex in seed plants. *The Botanical Review*, **37**, 143–229.

Girolami, G. (1953). Relation between phyllotaxis and primary vascular organisation in *Linum*. *American Journal of Botany*, **40**, 618–25.

Green, P. B. (1980). Organogenesis – a biophysical view. *Annual Review of Plant Physiology*, **31**, 51–82.

Halperin, W. (1978). Organogenesis at the shoot apex. *Annual Review of Plant Physiology*, **29**, 239–62.

Hellendoorn, P. H. & Lindenmayer, A. (1974). Phyllotaxis in *Bryophyllum tubiflorum*: morphogenetic studies and computer simulations. *Acta Botanica Neerlandica*, **23**, 473–92.

Hofmeister, W. (1868). *Allgemeine Morphologie der Gewächse*. Leipzig: Engelmann.

Iterson, G. van, Jr. (1907). *Mathematische und mikroskopische-anatomische Studien über Blattstellungen*. Jena: Fischer Verlag.

Jacobs, W. P. (1954). Acropetal auxin transport and xylem regeneration. *American Naturalist*, **88**, 327–37.

Jean, R. V. (1976). Growth matrices in phyllotaxis. *Mathematical Biosciences*, **32**, 165–76.

— (1982). The hierarchical control of phyllotaxis. *Annals of Botany*, **49**, 747–60.

Larson, P. R. (1975). Development and organization of the primary vascular system in *Populus deltoides* according to phyllotaxy. *American Journal of Botany*, **62**, 1084–99.

— (1977). Phyllotactic transitions in the vascular system of *Populus deltoides* Bartr. as determined by ^{14}C labeling. *Planta*, **134**, 241–9.

Loiseau, J.-E. (1969). *La Phyllotaxie*. Paris: Masson & Cie.

Lyndon, R. F. (1968). Changes in volume and cell number in the different regions of the shoot apex of *Pisum* during a single plastochron. *Annals of Botany*, **32**, 371–90.

— (1970a). Rates of cell division in the shoot apical meristem of *Pisum*. *Annals of Botany*, **34**, 1–17.

— (1970b). Planes of cell division and growth in the shoot apex of *Pisum*. *Annals of Botany*, **34**, 19–28.

— (1971). Growth of the surface and inner parts of the pea shoot apical meristem during leaf initiation. *Annals of Botany*, **35**, 263–70.

McCready, C. C. (1968). The acropetal movement of auxin through segments excised from petioles of *Phaseolus vulgaris*. In *The Transport of Plant Hormones*, ed. Y. Vardar, pp. 108–23. Amsterdam: North-Holland Publishing Co.

McDonald, E. A. (1975). Effects of Primordial Spacing on Phyllotaxis. *Ph.D. Thesis, University of London*.

Meicenheimer, R. D. (1981). Changes in *Epilobium* phyllotaxy induced by *N*-1-naphthylphthalamic acid and α-4-chlorophenoxyisobutyric acid. *American Journal of Botany*, **68**, 1139–54.

Mitchison, G. J. (1977). Phyllotaxis and the Fibonacci Series. *Science*, **196**, 270–5.

— (1981). The polar transport of auxin and vein patterns in plants. *Philosophical Transactions of the Royal Society of London*, B, **295**, 461–71.

Plantefol, L. (1948). *La Théorie des Hélices Foliaires Multiples*. Paris: Masson & Cie.

Richards, F. J. (1948). The geometry of phyllotaxis and its origin. *Symposia of the Society for Experimental Biology*, **2**, 217–45.

— (1951). Phyllotaxis: its quantitative expression and relation to growth in the apex. *Philosophical Transactions of the Royal Society of London*, B, **235**, 509–64.

— (1956). Spatial and temporal correlations involved in leaf pattern production at the apex. In *The Growth of Leaves*, ed. F. L. Milthorpe, pp. 66–76. London: Butterworth Scientific Publications.

Richards, F. J. & Schwabe, W. W. (1969). Phyllotaxis: a problem of growth and form. In *Plant Physiology, a Treatise*, vol. VA, ed. F. C. Steward, pp. 79–116. New York & London: Academic Press.

Richter, P. H. & Schranner, R. (1978). Leaf arrangement: geometry, morphogenesis, and classification. *Die Naturwissenschaften*, **65**, 319–27.

Roberts, D. W. (1977). Contact pressure model for semi-decussate related phyllotaxis. *Journal of Theoretical Biology*, **68**, 583–95.

— (1978). The origin of Fibonacci phyllotaxis – an analysis of Adler's contact pressure model and Mitchison's expanding apex model. *Journal of Theoretical Biology*, **74**, 217–33.

Romberger, J. A. (1963). Meristems, Growth and Development in Woody Plants. *United States Department of Agriculture Technical Bulletin*, no. 1293, 1–214.

Rutishauser, R. (1982). Der Plastochronquotient als Teil einer quantitativen Blattstellungsanalyse bei Samenpflanzen. *Beiträge zur Biologie der Pflanzen*, **57**, 323–57.

Schimper, C. F. (1835). Vorträge über die Möglichkeit einer wissenschaftlichen Verständnisses der Blattstellung, nebst Andeutung der hauptsächlichen Blattstellungsgesetze und insbesondere der neuentdeckten Gesetzt der Aneinanderreihung von Cyclen verschiedener Masse. *Flora, oder allgemeine botanische Zeitung*, **18**, 145–92 and 737–56.

— (1836). Geometrische Anordnung der um eine Achse peripherischen Blattgebilde.

Verhandlungen der Schweizerischen Naturforschungs Gesellschaft, **21**, 113–17.

Schoute, J. C. (1913). Beiträge zur Blattstellungslehre. I. Die Theorie. *Recueil des Travaux Botaniques Néerlandais*, **10**, 153–325.

— (1922). On whorled phyllotaxis. I. Growth whorls. *Recueil des Travaux Botanique Néerlandais*, **19**, 184–218.

— (1925). On whorled phyllotaxis. II. Late binding whorls of *Peperomia*. *Recueil des Travaux Botaniques Néerlandais*, **22**, 128–72.

— (1936). On whorled phyllotaxis. III. True and false whorls. *Recueil des Travaux Botanique Néerlandais*, **33**, 670–87.

Schüepp, O. (1916). Untersuchungen über Wachstum und Formwechsel von Vegetationspunkten. *Jahrbucher für wissenschaftliche Botanik*, **57**, 17–79.

— (1938). Über periodische Formbildung bei Pflanzen. *Biological Reviews*, **13**, 59–92.

— (1966). *Meristeme. Wachstum und Formbildung in den Teilungsgeweben höherer Pflanzen.* Basel: Birkhäuser Verlag.

Schwabe, W. W. (1951). Factors controlling flowering in the *Chrysanthemum*. II. Daylength effects on the further development of inflorescence buds and their experimental reversal and modification. *Journal of Experimental Botany*, **2**, 223–37.

— (1971). Chemical modification of phyllotaxis and its implications. *Symposia of the Society for Experimental Biology*, **25**, 301–22.

Schwendener, S. (1878). *Mechanische Theorie der Blattstellungen.* Leipzig: Engelman.

Snow, M. (1951). Experiments on spirodistichous shoot apices. I. *Philosophical Transactions of the Royal Society of London*, B, **235**, 131–62.

Snow, M. & Snow, R. (1931). Experiments on phyllotaxis. I. The effect of isolating a primordium. *Philosophical Transactions of the Royal Society of London*, B, **221**, 1–43.

— (1933). Experiments on phyllotaxis. II. The effect of displacing a primordium. *Philosophical Transactions of the Royal Society of London*, B, **222**, 353–400.

— (1935). Experiments on phyllotaxis. III. Diagonal splits through decussate apices. *Philosophical Transactions of the Royal Society of London*, B, **225**, 63–94.

— (1947). On the determination of leaves. *The New Phytologist*, **46**, 5–19.

— (1951). On the question of tissue tensions in stem apices. *The New Phytologist*, **50**, 184–5.

— (1952). Minimum areas and leaf determination. *Proceedings of the Royal Society of London*, B, **139**, 545–66.

— (1959). Regulation of sizes of leaf primordia by older leaves. *Proceedings of the Royal Society of London*, B, **151**, 39–47.

— (1962). A theory of the regulation of phyllotaxis based on *Lupinus albus*. *Philosophical Transactions of the Royal Society of London*, B, **244**, 483–514.

Snow, R. (1955). Problems of phyllotaxis and leaf determination. *Endeavour*, **14**, 190–9.

Thomas, R. L. (1975). Orthostichy, parastichy and plastochrone ratio in a central theory of phyllotaxis. *Annals of Botany*, **39**, 455–89.

Thomas, R. L. & Cannell, M. G. R. (1980). The generative spiral in phyllotaxis theory. *Annals of Botany*, **45**, 237–49.

Thornley, J. H. M. (1975*a*). Phyllotaxis. I. A mechanistic model. *Annals of Botany*, **39**, 491–507.

— (1975*b*). Phyllotaxis. II. A description in terms of intersecting logarithmic spirals. *Annals of Botany*, **39**, 509–24.

Turing, A. M. (1952). The chemical basis of morphogenesis. *Philosophical Transactions of the Royal Society of London*, B, **237**, 37–72.

Veen, A. H. & Lindenmayer, A. (1977). Diffusion mechanism for phyllotaxis. Theoretical, physico-chemical and computer study. *Plant Physiology*, **60**, 127–39.

Wardlaw, C. W. (1943). Experimental and analytical studies of pteridophytes. I. Preliminary observations on the development of buds on the rhizome of the ostrich fern (*Matteuccia struthiopteris* Tod.) *Annals of Botany*, **7**, 171–84.

— (1949). Experimental and analytical studies of pteridophytes. XIV. Leaf formation and phyllotaxis in *Dryopteris aristata* Druce. *Annals of Botany*, **13**, 163–98.

— (1950). The comparative investigation of apices of vascular plants by experimental methods. *Philosophical Transactions of the Royal Society of London*, B, **234**, 583–604.

— (1955). Responses of a fern apex to direct chemical treatments. *Nature*, **176**, 1098–100.

— (1956). The inception of leaf primordia. In *The Growth of Leaves*, ed. F. L. Milthorpe, pp. 53–65. London: Butterworth Scientific Publications.

— (1957). Experimental and analytical studies of pteridophytes. XXXV. The effects of direct applications of various substances to the shoot apex of *Dryopteris austriaca* (*D. aristata*). *Annals of Botany*, **21**, 85–120.

Wardlaw, C. W. & Cutter, E. G. (1954). Effect of deep and shallow incisions on organogenesis at the fern apex. *Nature*, **174**, 734–5.

Williams, R. F. (1974). *The Shoot Apex and Leaf Growth*. Cambridge: Cambridge University Press.

Wolpert, L. (1981). Positional information and pattern formation. *Philosophical Transactions of the Royal Society of London*, B, **295**, 441–50.

— (1982) Differences between cell differentiation *in vitro* and *in vivo*. In *Differentiation in vitro*, ed. M. M. Yeoman & D. E. S. Truman, pp. 267–79. Cambridge: Cambridge University Press.

Wright, M. (1982). The polarity of movement of endogenously produced IAA in relation to a gravity perception mechanism. *Journal of Experimental Botany*, **33**, 929–34.

Young, D. A. (1978). On the diffusion theory of phyllotaxis. *Journal of Theoretical Biology*, **71**, 421–32.

15

Positional information in the specification of leaf, flower and branch arrangement

D. J. CARR

In the preceding chapter Schwabe has provided an introduction to the very large literature on phyllotaxis and an excellent summary of the current theories which have been put forward to explain the origin of the patterns in which leaf or flower primordia originate at the shoot apex. Phyllotaxis is a very broad topic, which has not been monographed except for the book by Loiseau (1969) which propounds chiefly the views of the French School. A large segment of plant morphology is derived from or concerns the vast range of patterns of arrangements of leaves and branches, bringing order into what would otherwise be a bewildering array of phenomena in both higher and lower plants. This patterning is the result of control systems specifying not only where and when the lateral organs developing from the shoot apex will be produced, but also their subsequent rearrangement or change of form through growth. Positional information is deployed in this patterning and is communicated over distances that may range from the diameter of the cell or a few cells, to metres.

Differential growth and cell death as form-shaping factors

Differential growth

Before turning to examine some of the morphological consequences that relate to phyllotactic patterns it will not be out of place to deal briefly with two form-shaping factors, not previously given much prominence in this book. The more important is that of *differential growth*. Differential growth is involved in a wide variety of morphogenetic phenomena in plants, as for instance, post-genital fusion of organs (especially of the flower, references in Chapter 10), and torsions of various kinds affecting organs or whole shoots, as well as stem curvatures. As a result, the initial form, orientation or mutual disposition of organs becomes changed during ontogeny. A famous example, first brought to attention by F. O. Bower in his studies of ferns, is that of the fern sorus and its sporangia. These reproductive structures develop from the marginal meristem of the frond. Their final position, whether they remain marginal or 'shift' to the underside of the frond, depends on the degree of further activity of the marginal meristem and growth of the subjacent tissue (see Wardlaw, 1968).

The characteristic morphology of leaves of higher plants is also the result of particular patterns of differential growth. Unusual activity of the adaxial (and/or abaxial) meristem and relative inactivity of the marginal and plate meristems of the leaf primordium can result in a leaf shape which departs radically from the usual form in which the lamina is flattened dorsiventrally. The resultant leaf may be rounded in section, as in *Acorus calamus* (Kaplan, 1970), or flattened laterally, as in *Acacia* phyllodes (Carr & Burdon, 1975). The various wavy forms of the margins of the leaves of different species of oak (*Quercus* spp.) are thought of by Melville (1960) as derived from patterns of the concentration of morphogens, which arise in the manner suggested by Turing (see Chapter 1 by Meinhardt). At the time the patterns first become evident the leaf primordia are small enough to be in the theoretical size range for diffusion-reaction patterning.

The phyllotactic pattern initiated at the apex may also be altered by differential growth. Subsequent to their initiation as primordia at the shoot apex, leaves may 'shift' with respect to each other and to the axis on which they are borne, in some instances so as to quite obscure the original pattern. For instance, a group of leaves which arise successively at the apex may appear, when mature, as constituents of a 'false whorl' on the stem, due to lack of growth of the internodes between them (e.g., many species of *Lilium*, *Euphorbia*, *Acacia verticillata* (Velenovsky, 1907), *Elodea canadensis*, *Stylidium* spp.). Adult leaves of eucalypts arise as opposite, initially synchronous and equal primordia in a decussate arrangement. The leaves of a pair become separated by an intercalated region of stem – an 'intranode' (D. J. & S. G. M. Carr, 1959), resulting in a leaf arrangement described by taxonomists as 'alternate'.

Differential growth may affect the direction of growth of the shoot. In many trees, the lateral branches are more-or-less plagiotropic, i.e., they are horizontal or inclined at an angle to the trunk. Yet the growing tips of these branches may be turned upwards (as in pines and ash) or downwards (*Araucaria excelsa* and cedars). The more-or-less horizontal orientation of the branch is brought about by differential growth of the upper and lower parts of the shoot at some distance behind the tip (see Brown, 1971).

These ontogenetical changes are just as much a part of the positional mechanism guiding morphogenesis as that which determines the initial positioning of the primordia of the organs themselves.

Cell death in morphogenesis

In animals a good deal of the shaping of organs takes place as a result of programmed cell death (Saunders & Fallon, 1966); cells at certain locations in the developing embryo are eliminated to yield, for example, the shape of the hand, much as a sculptor would remove stone or plaster to the same end. Cell death in plants (Barlow, 1982) contributes relatively little, in general, to the shaping or repositioning of plant organs, in contrast with its importance in their anatomical development (e.g., the hyaline cells of *Sphagnum* – see Chapter 7, and the outer cells of the root cap – see Chapter 10). An excellent instance, that of cell death in shaping the palm leaf, is described by Eames (1953) and Corner (1966). As part of its development, marginal strips which join the tip of the leaf primordium and the tip itself are set off early in development as tracts of tissue which are destined to die and to be shed during final leaf expansion. During early stages, the tip and marginal strips constitute a very large part of the primordium and are important in providing a fairly rigid framework within which the developing lamina becomes variously and intricately folded. When the leaf finally expands out of the bud, the marginal strips break away from the leaflets and, in some instances, remain attached to the tip, hanging from it as two thin, dried 'reins'. The development of the leaves and in particular the origin of the folds, which give them the appearance of a closed fan, has been elegantly documented in studies using scanning electron microscopy and light microscopy by Kaplan and his colleagues (Kaplan, Dengler & Dengler, 1982; Kaplan, 1983). The locations of these processes in the developing leaf differ in different species and result in taxonomically-identifiable leaf shapes; they constitute a further instance of positional information in morphogenesis.

Other examples in which cell death, as well as differential growth, are involved in leaf morphogenesis are to be found in species of *Monstera*. The peculiar and instantly recognizable fenestrations in the mature leaf result from the death of groups of cells in the leaf primordium (Melville & Wrigley, 1969; see also Kaplan, 1983). The necrotic islands of tissue are spaced at regular distances from each other and from the veins; the final shape of the fenestrations results from regionally different degrees of expansion of the growing lamina. Melville & Wrigley (1969) suggest that a diffusion-reaction (Turing-type) mechanism is responsible for the regular patterning of veins and necrotic areas. They also suggest a similar mechanism as determining the patterning of fenestrations in leaves of the water plant, *Aponogeton fenestralis*, as well as the spacing of differently-coloured areas in variegated leaves, such as *Calathea*, in which pale and dark green areas alternate along the length of the lamina.

Morphological consequences of positional effects at the shoot apex

The theory of multiple foliar helices

Because I shall wish to use its descriptive power in what comes later, I propose to enlarge a little on the French theory of phyllotaxis mentioned in the preceding chapter by Schwabe. The theory of 'multiple foliar helices' has not received a 'good press' outside France. This has been due partly to the suspicion of British authors of 'abstract' theories (one remembers the scepticism with which the highly abstract, largely German, concept of endogenous rhythm was, and in some instances, still is, received by British plant physiologists, despite its successes in explaining a great variety of phenomena), partly to differences in interpretation and definitions on the part of its adherents and their early tendency to derive phyllotactic data from an examination of the positions of mature leaves on the stem – which, as explained above, can lead to gross errors – instead of the positions of primordia at the apex (Cutter, 1959). One of the French concepts, that of the *anneau initial*, has certainly been accepted and used by other authors (see Chapter 14). According to Plantefol's (1948) theory, certain alignments of the leaves on a plant are conspicuous in being the only unchanging – or the least changing – ones during development. These unchanging alignments are the foliar helices. Although they are often multiple, each helix 'produces' the same number of primordia; the leaves along a helix are usually equal in all respects and the helix extends into the *anneau initial* zone of the shoot apex. Dicotyledons clearly start with two helices and these are retained in decussate or bijugate systems, but in others the numbers of helices may increase during development. Decussate systems were treated by the Bravais brothers (Bravais & Bravais, 1837) as a special – although rather common – case of bijugy; Plantefol's theory sustains this view. Some species of Monocotyledons have only one foliar helix. From a causal point of view, each helix terminates in the apex in a leaf-generating centre, conceived of as a site of physiologically-specialized activity, provoking, by a sort of homeogenetic induction, the formation of the next primordium. Plantefol supposes that an organizer property of the apex itself regulates synchrony, rhythm and harmony of the foliar helices. Tacitly, this organizer is made responsible for the differences between different types of phyllotaxis (decussate, bijugate, spirodistichous) which depend on two foliar helices (Cutter, 1959). Some French writers (Camefort, 1956; Loiseau, 1969) have endeavoured to reconcile the views of the 'available space' theorists with those of the French School.

Direct observation of certain plants strongly supports the theory: in the dwarf Andean alpine, *Phyllactis rigida* (Valerianaceae), a rosette plant of hemispherical form which attains a diameter of 60 cm, each of the opposite decussate leaves is united by a thin membrane with the leaf at 90° to it at the next node; thus the leaves are arranged in two continuous ribbons around the stem (Benoist, 1932). Plants with spirodistichous leaf arrangements (e.g., species of Cucurbitaceae) in which the leaves are inserted in two twisted rows around the stem, also lend support to the concept of two foliar helices.

The results of many surgical experiments on shoot apices – for instance those of Snow (1935) – are easily explained by the theory. Dividing a decussate apex diagonally (Snow, 1935) yields two shoots of like spirality due to a separation of the two foliar helices. The same explanation can be given for the similar results obtained by Cutter & Voeller (1959) repeating Snow's experiment on bijugate *Dryopteris* apices. Loiseau (1969) and others have provided a considerable body of supporting experimental evidence of this kind, although no doubt the results can be accommodated by other theories. One of the most convincing demonstrations is the change in the phyllotaxis of *Stellaria media*, normally decussate, to a single spiral by treatment with phenylboric acid (Roche, 1971; Fig. 15.1); this is explained as due to suppression of one of the foliar helices.

Plantefol's theory explains very simply the most diverse phyllotactic arrangements, as well as aberrations from them and developmental transitions between them. It thus possesses considerable advantages over other theories, despite its speculative nature. Certainly, from a morphological point of view, there is merit in being able to describe in a uniform way the very large variety of phyllotactic patterns which exist in nature, many of which must be treated as exceptions to the rules of other phyllotactic theories. However, so far it has not been applied to the most complex systems, such as the sunflower (*Helianthus annuum*) capitulum.

Fig. 15.1. *Stellaria media* (chickweed) plant with modified phyllotaxis, raised from a germinating seed treated with phenylboric acid. Alternate leaves numbered 1–6, with axillary buds (ax). (From Roche, 1971.)

The sunflower capitulum

'When a bricklayer builds a factory chimney, he lays his bricks in a certain, steady orderly way, with no thought of the spiral patterns to which his orderly sequence inevitably leads' (D'Arcy Thompson, 1942, p. 921). A great deal of phyllotactic theory has concerned itself with the fully developed sunflower head (e.g., Church, 1904). Marc & Palmer (1981) have shown that the complexity of spirals of flowers begins to be laid down at the edge of a bare meristematic disc, up to one millimetre in diameter (Fig. 15.2a), and that these spirals progress inwards from the edge over the disc (Fig. 15.2b). The first primordia are laid along the edge of the disc, much in the manner of cherry stones laid on the edge of a plate, but simultaneously. However, at their initiation, the primordia are not touching but are uniformly distant from each other. Imagine circular rows of cherry stones being laid on the plate within the outer row; evidently, as the circumference available diminishes, the number of stones which can be laid in each circle diminishes inwards towards the centre of the plate. Using the chimney analogy, the number of bricks in each course diminishes upwards, to make a tapering column. These simple models are, however, biologically unrealistic since not only are the individual primordia growing but also the meristematic disc itself is expanding during the process of construction. This means that the circumferential space available for the second and successive 'courses' of primordia increases,

Fig. 15.2. Development of the sunflower head. (a) Formation of a disc-like receptacle; (b) appearance of flower primordia at its rim. Scale bars: in (a)100 μm, in (b) 200 μm. (Scanning electron micrographs courtesy of Dr Jan Marc.)

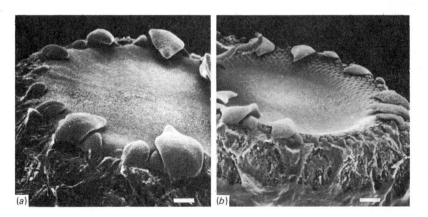

so that the same number can be laid down as in the first 'course'. If the ratio of the two rates, that of disc expansion and that of floret initiation, remains constant, the phyllotaxis will continue to be regular; also its degree of complexity will have been determined at the outset by the positioning of the first 'course' of primordia at the rim. The circumference available at that stage must somehow be assessed for the number of primordia which can be fitted into it, together with the inter-primordial spaces. One possible model for this would be a Turing-type diffusion-reaction system (see Chapter 1). Initial sizes of primordia and inter-primordial distances would presumably remain uniform throughout. As Palmer & Marc (1982) put it, 'once floral initiation has commenced, the process becomes self-perpetuating and continues as long as free uncommitted space is available on the surface of the receptacle'.

Under optimal growing conditions one disc floret is laid down every 10 min or so for 15 days, on the basis of the data of Palmer & Marc (1982). Using data and illustrations in Church (1904) of large sunflower heads and assuming optimum conditions and the same timing of overall growth and development, 4300 florets would be laid down at a rate of one every five minutes. This is in stark contrast to the mere 14–16 leaf pairs laid down by the apex during the course of some months growth prior to flowering (see Chapter 11). Even if one supposes that the positions of all the primordia in a 'course' (up to about 150 in the largest heads – Church, 1904) are determined simultaneously (not, as in bricklaying, one by one), the processes of positional determination of each primordial centre and its 'acquisition' of 100–120 cells as its region of operation or 'capital' for growth, are to be considered as comparatively rapid events for positional determination. Assuming the rates calculated above apply to the largest heads, one obtains a figure of between 12 and 25 h per 'course' of primordia.

The initial size of the bare disc, before any flowers are initiated, is probably determined nutritionally. Flower heads on lateral branches are smaller than the terminal head; and if nutrition fails (e.g., due to boron deficiency, see p. 363) during development, the regular progression and regular phyllotaxis are interrupted, the expansion of the disc fails to keep up with the ingrowing files of flower primordia, and their arrangement becomes chaotic – as it frequently does in late season (Church, 1904, p. 26; Palmer & Marc, 1982, and unpublished). If, during the mid-phase of capitulum development, the meristematic disc expands faster than the rate of initiation of floret primordia, the phyllotaxis will rise in complexity, i.e., the number of parastichies or intersecting helices will increase; if the opposite occurs, the complexity of the phyllotaxis will fall.

Right- and left-handedness

Rutishauser (1981) has provided evidence that even in the simplest phyllotactic pattern of Dicotyledons, the opposite, decussate arrangement, the members of a pair of leaf primordia are rarely exactly equal or synchronous in origin. In addition there are many examples in the literature of irregularities and anomalies in phyllotaxis during development. In particular, the position of a primordium is often not an exact one but fluctuates statistically around what may be thought of as a mathematically 'ideal' position. All these uncertainties suggest uncertainties in the operation of position-determining mechanisms such as one would expect of biological rather than purely mechanical systems. Rutishauser (1981, 1982) has developed a case for more exact descriptions to be made of the phyllotactic systems exhibited by plants as an increasingly necessary basis for developmental theories. This is to some extent a reaction to the recent spate of mathematical theories, sometimes elegant but always highly idealistic.

One of the consequences of the frequent inequality of even the opposite primordia in a decussate system is the tendency to the development of spirality which may be left-handed or right-handed. In most flowering plants, there is little tendency for one of these directions to prevail, either between plants or between branches of a single plant. Nevertheless, the proportions of right-handed and left-handed plants in some populations have been found to be unequal, suggesting, perhaps, some genetic control of spirality. In the strawberry, Sironval (1949) found 65% right-handed to 35% left-handed plants. In certain populations, Davis (1963) found a preponderance of left-handed coconut palms (*Cocos nucifera*); surprisingly, the left-handed palms were more vigorous and yielded a larger crop than right-handed palms. Analysing 400 cones of *Pinus laricio*, Church (1904, p. 92) found proportions of 71 left-handed to 39 right-handed; Church also cites other instances of preponderance of one direction of spirality over the other in pine cones (op. cit., p. 351). In *Crataegus monogyna*, the hawthorn, the spirality of the lateral buds is opposite to that of the parent shoot (i.e., antidromous) with a frequency of 94%. Maize grains occur in paired rows on the cob; grains from a right-hand row yield right-handed seedlings much more frequently than left-handed seedlings, and *vice versa* (Compton, 1912).

In lower plants there is often a consistent adherence to one direction or the other (Schmucker, 1925). In the red algae, the 'leaves' are arranged in left-handed spirals; in the green alga, *Chara*, the direction of the spiral thickenings of the oospore wall is always left-handed, whereas the inter-nodal cells are right-handed. Cell division of the apical cell of roots on the right-hand side of the water-fern, *Azolla*, is clockwise (right-handed); similarly, in roots on the left-hand side of the plant, division of the apical cell proceeds anticlockwise (left-handed) (Gunning, Hughes & Hardham, 1978). The spirality extends to subsequent divisions which demarcate the first phloem and xylem elements (see Fig. 10.2, p. 288).

These positionally-determined differences must depend on a slight pre-ferment of one primordium in a sequence (or, in the case of *Azolla* roots, of one of two walls of the three-sided apical cell) over its alternate, thus shifting the whole of the rest of the development into one direction or the other. Handedness in higher plants may derive from a particular bias in the pattern (or 'handedness') of cell division. Lindenmayer & Rozenberg (1979) have shown how a small change in the specification of planes of division can lead to a change from, say, a right-handed to a left-handed sequence of new cell walls, or *vice versa*. In the stem cambium of trees the angles of pseudo-transverse divisions of the fusiform initials determine the direction of spiral grain of the wood (Bannan, 1966). In some trees this may change seasonally, or from one year or period of years to another.

Ontogenetical changes in phyllotaxis

Among the most striking features of phyllotaxis are the changes which take place during ontogeny. Dicotyledonous seedlings begin with (usu-ally) opposite cotyledons and the first few pairs of leaves may also be opposite and decussate; in some families (e.g., Labiatae) this pattern per-sists throughout the life of the plant. However, in a large number of species the phyllotaxis changes early in seedling development and its com-plexity may increase with age or size of the plant. It has been stated that this ontogenetical change in phyllotaxis is dependent on an increase in the size of the apex (Allsopp, 1965; Cutter, 1965). Evidently a larger apical meristem would allow for a greater degree of phyllotactic complexity if that depends on 'available space' or the interaction of 'fields' of either inhibition or stimulation. Although the number of foliar helices is ex-pected by the French theory to increase as the apex increases in diameter (as it does in *Linaria* (Loiseau, 1969)) there are many exceptions (e.g., members of persistently decussate families of flowering plants). The avail-able evidence certainly allows doubts to be raised on the universality of the assumption of an ontogenetical increase in both apical size and phyllo-tactic complexity.

One way in which phyllotaxis may remain unchanged while the apex enlarges in diameter is by increase in the size of the leaf primordia, so that their insertions take up a larger proportion of the 'available space' (Williams, 1974). Seedlings of certain eucalypts (e.g., *Eucalyptus longicornis* and *E. socialis*) have narrow juvenile leaves arranged in complex spirals and produced at relatively broad apices. The adult leaves of these species are arranged in the usual sub-decussate pattern and are produced in pairs at an apex which is *smaller* than that of the seedling (D. J. & S. G. M. Carr, unpublished).

While there can be no doubt that, in most instances, the size of the apical meristem increases in the immediate post-germination phase, there is no reason to suppose that it continues to do so or that it is necessarily accompanied by an increase in phyllotactic complexity. Elm seedlings, for instance, begin with a short sequence of leaves arranged spirally, but soon revert to a distichous arrangement (i.e., the leaves are arranged in two ranks) and this then persists throughout the life of the tree. The same is true of a number of European trees (e.g., beech, lime). In other deciduous trees, the plagiotropic branches have a less complex phyllotaxis than erect shoots; the persimmon (*Diospyros lotus*) has spirally arranged leaves on erect shoots, while those on the plagiotropic branches are distichous. In *Cupressus macrocarpa* the size of the apex appears to become fixed early in seedling growth and remains at that size despite a number of changes in phyllotaxis; this is not true of other conifers, however (Camefort, 1953).

Thus there may not even be a consistent relationship between apical size and phyllotactic complexity. According to Bierhorst (1959) the number of leaf teeth in each whorl in *Equisetum* 'is a function of the size of the shoot apex at the time the leaves are initiated'. The number of leaves in a whorl in *Hippuris* increases from as few as 2–6 in seedlings to as many as 12 in strong shoots, and such variations have been associated with changes in size of the apical meristem (McCully & Dale, 1961). The similar study by Loiseau & Grangeon (1963) of *Ceratophyllum* and *Hippuris* also included measurements of apical height and diameter, in relation to leaf numbers in the whorls. From their data one can calculate a correlation coefficient $r = 0.17$ between apical size (height) and leaf numbers in whorls, indicating only a very loose correlation.

There is some experimental evidence for a relationship between leaf size and shape and apical size. Repeated defoliation of a rhizome of the fern *Onoclea sensibilis* led to a reduction in the size of its apical meristem and the production by it of small, simple juvenile leaves (Wardlaw, 1968). Similar treatments of the fern, *Nephrolepis biserrata*, leading to similar results, are described in Nozeran's chapter. Moreover, surgery resulting in reduction in the size of the apex of orthotropic (but not of plagiotropic) shoots of the flowering plant, *Phyllanthus*, led to the production of juvenile leaves (for references see the chapter by Nozeran). Cutter (1964) has shown a relationship between experimentally changed size of apical meristem and phyllotactic complexity in the fern *Dryopteris aristata* but was unable to find a difference in either apical size or rate of leaf inception between plants with spiral or bijugate phyllotaxis (Cutter & Voeller, 1959; Voeller & Cutter, 1959).

The reproductive apex

The shoot apex may enlarge during the change to the reproductive state. This is most obvious in plants with terminal, relatively large inflorescences, such as those of members of the Compositae. It is shown in the familiar diagrams of the early stages in the inception of the terminal male inflorescence of *Xanthium* (Salisbury, 1963). Following the induction of flowering, the diameter of the apex of *Chrysanthemum* increases four-fold during the course of a few hours (Schwabe, 1959). Associated with this increase in size is a sudden and marked increase in phyllotaxis, leading to complex arrangements such as that of the sunflower head. As Loiseau (1969) points out, there is, in many cases, no change in phyllotaxis in plants with racemose inflorescences (e.g., tobacco, members of the family Labiatae, etc.). In other cases, the number of foliar helices does increase from the vegetative to the flowering phase (e.g., in *Cleome spinosa, Reseda luteola, Carex, Plantago*). Whether or not this is due to increased size of the apex or diminished size of the primordia is not clear. The induction of flowering may result in a spurt of growth and an increase in the longitudinal component of apical growth; this has been held responsible for a change from decussate to spiral phyllotaxis in *Epilobium adenocaulon* (Schwabe, 1963; see also Chapter 14).

It seems to be generally assumed that phyllotaxis will increase in complexity with the onset of flowering and decrease thereafter. An obvious candidate for examination of this hypothesis is the bottlebrush, *Callistemon speciosus*, in which the same apex produces first a suite of leaves, then a closely-set suite of flowers and subsequently reverts to leaf production. Examination shows that the phyllotaxis remains throughout as $(3+5)$. Finally, the floral apices themselves, which give rise to individual flowers, are often much smaller than vegetative apices or inflorescence apices, which produce leaves or bracts respectively. Yet the parts of the flower may be arranged in complex spirals (as they are in water-lilies or magnolias) or, more usually, in whorls. Evidently, the informational controls regulating the positioning of floral parts at the apex differ, at least quantitatively, from those which regulate the positioning of leaf primordia.

Associated with ontogenetical changes in phyllotaxis are changes in the potentialities of the primordia produced. Although these may show no overt differences at inception, the leaves produced may develop as bracts, prophylls, cataphylls, foliage leaves (differing in form according to whether they are juvenile or adult), floral organs, etc. Some, at least, of these striking differences in performance of lateral primordia depend on the positional relationships of the apex to the rest of the plant, or to positional relationships at the apex itself; this will be emphasized in the next section.

Differential development of leaves and branches associated with phyllotaxis: anisophylly and anisoclady

An extreme of the inequality of leaf primordia of a pair in a decussate system is the subsequent unequal development of the leaves themselves, or *anisophylly*, which is a feature of some flowering plant genera. The topic is treated *in extenso* by Troll (1937, pp. 353–408). Anisophylly may be gravitationally determined, as in *Acer saccharum*, in which the lower of two primordia develops into a larger leaf than the upper. However, anisophylly is frequently genetically determined and a characteristic of species or even of whole families (Gesneriaceae, Melastomataceae, etc. – Loiseau, 1969). In lower plants, an extreme example is the water fern, *Salvinia natans*, in which the lower leaf of each whorl of three develops as a set of root-like processes that take the place of roots, which the plant lacks. Strangely, a similar transformation of some of the leaves into root-like organs is true of a number of rootless Australian species of *Drosera*, in which each of the scale leaves on the stem is prolonged below into three root-like processes which bear absorptive hairs (Velenovsky, 1909). Anisophylly is often associated with dorsiventrality, whether overtly that of a horizontally-growing stem or covertly that of an erect shoot which may nevertheless betray its dorsiventrality in this and other morphological and anatomical features.

Anisophylly has been produced experimentally by apical surgery in the decussate, normally isophyllous plant, *Phlox drummondii* (Tort, 1969); the result was that the leaves of one foliar helix were smaller than those of the other. In *Tococa guyanensis* (Melastomataceae) the petiole of one leaf of each decussate pair develops a structure called a *domatium*, or ant-housing. The domatium-bearing leaves are situated on one of the two foliar helices and thus are arranged helicoidally on the plant. The two halves of a leaf may be differentially developed, one growing larger than the other. The larger half may be that directed, on its foliar helix, away from the shoot apex (i.e., *catadromic*, as in beech; see also Fig. 15.4) or that which lies towards the shoot apex (*anadromic*, as in elm and *Begonia* spp.).

Buds occupying certain positions on the shoot may be capable of a special performance denied to buds in other positions (*anisoclady*). Unequal development of the auxillary buds of plants of otherwise normal, isophyllous appearance is widespread (members of Caryophyllaceae, Gentianaceae, Asclepiadaceae, Rubiaceae, etc.). The leaves subtending the preferred axillary buds, or (+) buds, are commonly situated on a single foliar helix (Fig. 15.3a) – thus leading to *helicoidal anisoclady* – or they may be dispersed in a more complex way, usually occupying a sector of the shoot (Fig. 15.3b) (*sectorial anisoclady*, e.g. *Ceratophyllum demersum* – Loiseau & Grangeon, 1963). This topic has been admirably reviewed by Cutter (1966) and by Dormer (1972). The (+) buds may develop as flowers (as in *Vinca* and *Cuphea*). Plants of *Stellaria media* (chickweed) usually have opposite, decussate leaves and show the usual Caryophyllaceous helicoidal anisoclady. Plants occur which have unusual

Fig. 15.3. Diagrams of phyllotaxis, to illustrate anisoclady. The preferred buds are indicated as (+). (a) Helicoidal anisoclady in Caryophyllaceae, the direction of the spiral indicated by a dotted line. (b) Sectorial anisoclady in Acanthaceae, showing the distribution of (+) buds on adjacent orthostichies. The sequence of leaf pairs is given by the numbers. (After Goebel, 1913.)

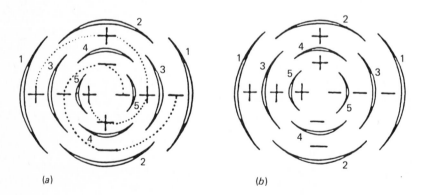

trimerous whorls; these also show anisoclady, and the 'preferred' leaves lie along one or two of the foliar helices (Roche, 1970). Helicoidal anisoclady is found in red algae: the long and short branches of *Asparagoides armata* are arranged helicoidally (Connolly, 1911).

In the Acanthaceae, which have sectorial anisoclady (Fig. 15.3b), the preferred buds may develop as the sole flower-bearing shoots (Danert, 1953), while shoots developed from (−) buds remain vegetative, or even become spiny. The distribution of the (+) buds on certain orthostichies betrays a concealed dorsiventrality, even of isophyllous vertical shoots; horizontal shoots, however, are clearly dorsiventral and may be anisophyllous (Fig. 15.4). The asymmetry may also affect the root system: the primary root bears lateral roots preferentially on radii corresponding to the orthostichies of (+) buds of the plumule. Also, the (+) side of the shoot may have a greater capacity for regeneration than the (−) side (Fig. 15.5).

Fig. 15.4. Horizontal shoot of *Goldfussia glomerata* (Acanthaceae) from the upper side, showing both anisophylly and anisoclady. The preferred leaf in each pair *aa*, *bb*, etc., is indicated as (+), the other as (−). The leaves of a pair are unequal in size, the larger ones being inserted on the dorsal side; the halves of each leaf are also unequally developed, the catadromic half (also turned towards the dorsal side) being the larger. The preferred buds are in the axils of the *smaller* leaves. (After Troll, 1937.)

Fig. 15.5. Stem cutting of *Blechum brownii* (Acanthaceae). The first roots (arrowed) appear below the leaf which subtends an axillary shoot, i.e., a (+) bud. (After Goebel, 1913.)

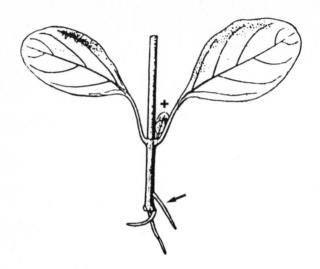

In her review of the topic, Cutter wrote

'In genera with spiral phyllotaxis, e.g. *Limnobium* and *Hydrocharis*, the leaves with buds can be considered to lie along one of the two helices which can be drawn through the leaves. In *Halophila*, on the other hand, which belongs to the same family (Hydrocharitaceae) buds are formed in the axil of every second leaf, but the leaves are arranged in opposite pairs, each pair making a slight angle with the preceding pair, so that all the buds lie more or less on one side of the stem' [Fig. 15.6]. 'In *Naias*, the leaves occur in not quite opposite pairs, and two spirals can be drawn through them; the leaves along one spiral subtend axillary buds' (Cutter, 1966, with omission of the references).

These positional differences in performance, with their profound and persistent effects on the general morphology of the plant, arise at the inception of each axillary meristem and represent differences in positional information allied to the phyllotactic system of the plant. So far, little has been done to explore them experimentally. However, in the water plant, *Hygrophila*, which has decussate phyllotaxis and helicoidal anisoclady, Cutter (1972) has shown that, subsequent to the initiation of the unequally-sized buds at the newest node near the apex, the larger or

Fig. 15.6. *Halophila ovalis*, to show unusual sectorial anisoclady. (a) Portion of the creeping stem with two nodes, each bearing a pair of scale leaves (1,2; 3,4). Scale leaves 2 and 4 each subtend a shoot bearing two oval, opposite foliage leaves (L_1 and L_2). Leaves 1 and 3 subtend no axillary shoots, but a root (W) has developed from each of the associated nodes. Leaves L_1 and L_2 each subtend either a shoot (S) or a female flower (B). (b) The phyllotactic diagram shows that, between the leaf pair (1,2) the shoot lies to the left, between the leaf pair (3,4) it lies to the right, and so on, so that the buds lie on a zig-zag line along the shoot. (After Troll, 1937, modified from Bayley Balfour.)

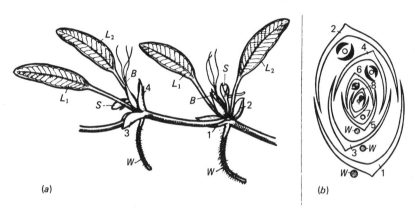

(a)   (b)

preferred bud inhibits the smaller ones. In grafting experiments, Cutter & Chiu (1972) have shown that this inhibition (interpreted in terms of the outgrowth of the buds on isolated nodes) is transmitted across a graft union and is not related to competition for nutrients. If longitudinally-sliced half nodes are bound together but prevented from grafting, both buds may grow out. These experiments reveal nothing of the causes which underly the *distribution* of buds of two different sizes along the two foliar helices.

References

Allsopp, A. (1965). Heteroblastic development in cormophytes. In *Encyclopaedia of Plant Physiology*, vol. XV/1, ed. W. Ruhland, pp. 1172–235. Berlin, Heidelberg & New York: Springer-Verlag.

Bannan, M. W. (1966). Spiral grain and anticlinal divisions in the cambium of conifers. *Canadian Journal of Botany*, **44**, 1515–38.

Barlow, P. W. (1982). Cell death – an integral part of plant development. In *Growth Regulators in Plant Senescence. Monograph 8 of the British Plant Growth Regulator Group*, ed. M. B. Jackson, B. Grout & I. A. MacKenzie, pp. 27–45. Wantage: British Plant Growth Regulator Group.

D. J. Carr

Benoist, R. (1932). La phyllotaxie du *Phyllactis rigida* Pers. *Bulletin de la Société Botanique de France*, **70**, 490–5.

Bierhorst, D. W. (1959). Symmetry in *Equisetum*. *American Journal of Botany*, **46**, 170–9.

Bravais, L. & Bravais, A. (1837). Essais sur la disposition des feuilles curvisériées. *Annales des Sciences Naturelles, Botanique, 2ème Serie*, **7**, 42–110.

Brown, C. L. (1971). Growth and form. In *Trees: Structure and Function*, ed. M. H. Zimmerman & C. L. Brown, pp. 125–67. Berlin, Heidelberg & New York: Springer-Verlag.

Camefort, H. (1953). Évolution de la phyllotaxie et de la taille du point végétative de *Cupressus macrocarpa* Hartw. pendant les trois premières années de son developpement. *Comptes rendus hebdomadaires des Séances de l'Académie des Sciences de Paris, Série D*, **236**, 847–9.

– (1956). Étude de la structure du point végétatif et des variations phyllotaxiques chez quelques Gymnospermes. *Annales des Sciences Naturelles, Botanique, 11ème Serie*, **17**, 1–185.

Carr, D. J. & Burdon, J. J. (1975). Temperature and leaf shape in seedlings of *Acacia aneura*. *Biochemie und Physiologie der Pflanzen*, **168**, 307–18.

Carr, D. J. & Carr, S. G. M. (1959). Developmental morphology of the floral organs of eucalypts. I. The inflorescence. *Australian Journal of Botany*, **7**, 109–41.

Church, A. H. (1904). *On the Relation of Phyllotaxis to Mechanical Laws*. London: Williams & Norgate.

Compton, R. H. (1912). A further contribution to the study of right- and left-handedness. *Journal of Genetics*, **2**, 53–70.

Connolly, C. J. (1911). Beiträge zur Kenntnis einiger Florideen. *Flora*, **103**, 125–70.

Corner, E. J. H. (1966). *The Natural History of Palms*. London: Weidenfeld & Nicholson.

Cutter, E. G. (1959). On a theory of phyllotaxis and histogenesis. *Biological Reviews of the Cambridge Philosophical Society*, **34**, 243–63.

– (1964). Phyllotaxis and apical growth. *The New Phytologist*, **63**, 39–46.

– (1965). Recent experimental studies of the shoot apex and shoot morphogenesis. *The Botanical Review*, **31**, 7–113.

– (1966). Patterns of organogenesis in the shoot. In *Trends in Plant Morphogenesis*, ed. E. G. Cutter, pp. 220–34. London: Longmans.

– (1972). Regulation of branching in decussate species with unequal lateral buds. *Annals of Botany*, **36**, 207–20.

Cutter, E. G. & Chiu, H.-W. (1972). Grafting experiments on correlative effects between lateral buds. *Annals of Botany*, **36**, 221–8.

Cutter, E. G. & Voeller, B. R. (1959). Changes in leaf arrangement in individual fern apices. *Journal of the Linnean Society (Botany)*, **56**, 225–36.

Danert, S. (1953). Über die Symmetrieverhältnisse der Acanthaceen. *Flora*, **140**, 307–25.

Davis, T. A. (1963). The dependence of yield on asymmetry in coconut palms. *Journal of Genetics*, **58**, 186–215.

Dormer, K. J. (1972). *Shoot Organization in Vascular Plants*. London: Chapman & Hall.

Eames, A. J. (1953). Neglected morphology of the palm-leaf. *Phytomorphology*, **3**, 172–89.

Goebel, K. (1913). *Organographie der Pflanzen*. Vol. 1: *Allgemeine Organographie*. 2nd edn. Jena: Fischer.

Gunning, B. E. S., Hughes, J. E. & Hardham, A. R. (1978). Formative and proliferative cell divisions, cell differentiation, and developmental changes in the meristem of *Azolla* roots. *Planta*, **143**, 121–44.

Kaplan, D. R. (1970). Comparative foliar histogenesis in *Acorus calamus* and its bearing on the phyllode theory of monocotyledonous leaves. *American Journal of Botany*, **57**, 331–61.

– (1983). The development of palm leaves. *Scientific American*, **249**, no. 1, 88–95.

Kaplan, D. R., Dengler, N. G. & Dengler, R. E. (1982). The mechanism of plication inception in palm leaves: problem and developmental morphology. *Idem*: histogenetic observations on the pinnate leaf of *Chrysalidocarpus lutescens*. *Idem*: histogenetic observations on the palmate leaf of *Rhapis excelsa*. *Canadian Journal of Botany*, **60**, 2939–75; 2976–98; 2999–3016.

Lindenmayer, A. & Rozenberg, G. (1979). Parallel generation of maps: developmental systems for cell layers. *Lecture Notes in Computer Science*, **73**, 301–16.

Loiseau, J.E. (1969). *La Phyllotaxie*. Paris: Masson & Cie.

Loiseau, J.-E. & Grangeon, D. (1963). Variations phyllotaxiques chez *Ceratophyllum demersum* et *Hippuris vulgaris*. *Bulletin de la Société Botanique de France, Mémoires*, **110**, 76–91.

Marc, J. & Palmer, J. H. (1981). Photoperiodic sensitivity of inflorescence initiation and development in sunflower. *Field Crop Research*, **4**, 155–64.

McCully, M. E. & Dale, H. M. (1961). Variations in leaf number in *Hippuris*. A study of whorled phyllotaxis. *Canadian Journal of Botany*, **39**, 611–25.

Melville, R. (1960). A metrical study of leaf-shape in hybrids: II Some hybrid oaks and their bearing on Turing's theory of morphogenesis. *Kew Bulletin*, **14**, 161–77.

Melville, R. & Wrigley, F. A. (1969). Fenestration in the leaves of *Monstera* and its bearing on the morphogenesis and colour patterns of leaves. *Botanical Journal of the Linnean Society*, **62**, 1–16.

Palmer, J. H. & Marc, J. (1982). Wound-induced initiation of involucral bracts and florets in the developing sunflower inflorescence. *Plant & Cell Physiology*, **23**, 1401–9.

Plantefol, L. (1948). *La Théorie des Hélices Foliaires Multiples. Fondements d'une Théorie Phyllotaxique Nouvelle*. Paris: Masson & Cie.

Roche, D. (1970). Manifestations de l'anisocladie hélicoidale après une variation phyllotaxique provoquée chez la stellaire (*Stellaria media* [L.] Vill.). *Bulletin de la Société Botanique de France, Mémoires*, **117**, 205–14.

– (1971). Suppression d'une hélice foliaire par action de l'acide phenylborique sur la stellaire [*Stellaria media* (L). Vill.]. *Comptes rendus hebdomadaires des Séances de l'Académie des Sciences de Paris, Série D*, **273**, 2522–5.

Rutishauser, R. (1981). *Blattstellung und Sprossentwicklung. Dissertationes Botanicae*, **62**, pp. 127 + 18 plates. Vaduz: Cramer.

– (1982). Der Plastochronquotient als Teil einer quantitativen Blattstellungsanalyse bei Samenpflanzen. *Beiträge zur Biologie der Pflanzen*, **57**, 323–57.

Salisbury, F. B. (1963). *The Flowering Process*. New York & Oxford: Pergamon.

Saunders, J. W. & Fallon, J. F. (1966). Cell death in morphogenesis. In *Major Problems in Developmental Biology*, ed. M. Locke, pp. 289–315. New York & London: Academic Press.

Schmucker, T. (1925). Rechts- und Linkstendenz bei Pflanzen. *Botanisches Centralblatt*, **41**, 51–85.

Schwabe, W. W. (1959). Some effects of environment and hormone treatment on reproductive morphogenesis in *Chrysanthemum*. *Journal of the Linnean Society (Botany)*, **56**, 254–61.

– (1963). Morphogenetic responses to climate. In *Environmental Control of Plant Growth*, ed. L. T. Evans, pp. 311–36. New York & London: Academic Press.

Sironval, C. (1949). Recherches organographiques et physiologiques sur le developpement du fraisier des quatre-saisons à fruits rouges. *Mémoires de l'Académie Royale Belgique, Classe Sciences*, **26**, 3–184.

Snow, R. (1935). Experiments on phyllotaxis. III. Diagonal splits through decussate apices. *Philosophical Transactions of the Royal Society of London*, B, **225**, 63–94.

Thompson, D'Arcy, W. (1942). *On Growth and Form*. Cambridge: Cambridge University Press.

Tort, M. (1969). Anisophyllie hélicoidale provoquée et ontogenie foliare chez le *Phlox drummondii*. *Comptes rendus hebdomadaires des Séances de l'Académie des Sciences de Paris, Série D*, **269**, 2522–5.

Troll, W. (1937). *Vergleichende Morphologie der höheren Pflanzen*. Vol. 1: *Vegetationsorgane*, Part 1. Berlin: Gebrüder Bornträger.

Velenovsky, J. (1907 & 1909). *Vergleichende Morphologie der höheren Pflanzen*. Vol. 2 (1907); Vol. 4 (Supplement) (1909). Prague: Fr. Řivnáč.

Voeller, B. R. & Cutter, E. G. (1959). Experimental and analytical studies of pteridophytes XXXVIII. Some observations on spiral and bijugate phyllotaxis in *Dryopteris aristata* Druce. *Annals of Botany*, **23**, 391–6.

Wardlaw, C. W. (1968). *Essays on Form in Plants*. Manchester: Manchester University Press.

Williams, R. F. (1974). *The Shoot Apex and Leaf Growth*. Cambridge: Cambridge University Press.

16

Positional and temporal control mechanisms in inflorescence development

ARISTID LINDENMAYER

Inflorescences are compound flowering structures. They are branching systems of stem segments on which flowers, leaves, scales and apices are borne. While they have been extensively studied by morphologists as static structures, there are not many observations that treat inflorescences as dynamic, developing organs and there have been few attempts by physiologists to find the control factors responsible for their developmental processes and integrative mechanisms. These systems are, on the other hand, easily observable and their development is very reproducible. For these reasons, and because of their great variety, they are also eminently suited for theoretical investigations of positioning and timing phenomena in development. The purpose of this paper is to present such theoretical studies with the hope that they can serve as a framework within which descriptions and experimental observations on inflorescence development can be placed and evaluated.

When we are faced with the problem of describing the development of an inflorescence as a whole the following observational techniques can be immediately utilized. First of all, we have to obtain an accurate record of the branching development. There are two measures that can be employed: (1) finding the plastochron intervals for the individual branches of the inflorescence, and (2) finding the paracladial relationships among the various mother-daughter branch-pairs. The plastochron intervals are the time periods between the production of consecutive appendages on a given branch; they are the inverse of the branching frequencies (Erickson & Michelini, 1957). The paracladial relationships are the ratios of plastochron intervals between a branch and the part of its mother branch which extends apically above it (Frijters & Lindenmayer, 1976; Lindenmayer, 1977). After the vegetative apices turn into flowering apices on a branching system we can also obtain (3) the positions of flowers on the branches, and (4) the sequence of flower opening (or some other developmental stage) along the different branches (Schüepp, 1942).

To explain the observed sequence of flowering on an extended shoot system, we can consider two main control factors. The first and most obvious is the distribution of the flowering hormone (florigen) throughout the plant. This hormone, whose existence is generally accepted but whose chemical nature eludes us, is produced in leaves under certain photo-periodic regimes and is transported from the leaves to the vegetative apices of the plant. After the arrival of the hormone, the apex is rapidly (within hours) transformed from a vegetative to a flowering condition (this has been documented extensively, particularly for various members of the Compositae such as *Chrysanthemum* and *Xanthium*). Furthermore, in some of the plants we have studied (e.g., *Mycelis muralis* and *Hieracium murorum*, which are also in the Compositae) the leaves are confined mainly to a rosette at the base of the shoot system; thus the distribution of the flowering hormone forms a relatively simple pattern, consisting solely of the acropetal transport of the hormone in each branch. In this way, time sequences can be obtained for the arrival of the hormone at the vegetative apices; this, in turn, can explain certain flowering sequences.

A second control factor we may consider is the role of apical dominance in developing inflorescences. This manifests itself when parts of the inflorescence are removed, giving rise to an apparent lifting of inhibitory (or an increase of stimulatory) activity in the lateral buds and branches below the removed portion (Jauffret, 1970; Favard, 1970; Sell, 1967, 1980). This phenomenon has been extensively described and various mechanisms have been proposed for it (cf. Vardar & Kaldewey, 1972; Tucker & Mansfield, 1973). The most likely mechanism is the production of an inhibitory hormone (auxin) by the main apices and its basipetal transport in the branches. The production of this hormone only takes place in vegetative apices and stops when they are transformed to flowering apices. This factor provides us with yet another timing and positioning mechanism for the observed flowering sequences.

In one of our models only the flowering hormone distribution is used, while in a second model both flowering hormone and apical-inhibitory hormone distributions are computed. The latter model provides more flexible and satisfactory results. In the last section the place of these models among other hypothetical morphogenetic mechanisms is discussed, in particular with respect to the achievement of timing and positioning processes in developing and growing structures.

Descriptions of inflorescences

We shall make use of Troll's (1964) definitions of inflorescence types. There are two main types, monotelic and polytelic; the former is defined by Weberling (1965) as follows: 'In the *monotelic* inflorescence the apex of the inflorescence axis ends with a terminal flower. This also applies to all the floral branches below the terminal flower of the main axis. All of them, whether branched or not, proved to be homologous elements, and they are all referred to by the term *paracladia* because these branches repeat the structure of the main axis of the flowering system... The paracladia may be called *enriching branches* because their presence enables the number of flowers in the flowering system to be increased. Consequently the whole area which produces enriching branches may be designated as an *enriching field*. In the lower part of the flowering shoot this zone is commonly preceded by a *field of inhibition* within which the formation of paracladia is prevented. The transition between these two areas may be more or less abrupt. The same areas can be recognized in the individual paracladia if these are not reduced to their terminal flower or restricted to the terminal flower and their enriching field only.' Schematic drawings of monotelic inflorescences are shown in Fig. 16.1.

In the other main type of inflorescence, the *polytelic*, there are no terminal flowers on the main axis or on the paracladia. These inflorescences are developmentally more complex than the monotelic ones since the flowers are borne laterally in structures called 'florescences' which can then be repeated on side branches further below (these can also be considered as paracladia). Schemes for polytelic inflorescences are also shown in Fig. 16.1.

The Compositae is one of those families of flowering plants which are characterized exclusively by polytelic inflorescences. The capitula (heads) typical of this family are florescences which occur at the ends of the main axis and many of the side branches. The arrangement of the capitula on the whole plant, however, resembles the monotelic inflorescences, with paracladia of various orders on which capitula substitute for individual flowers.

The two most important aspects of inflorescence patterns are: (1) the spatial distribution of capitula-bearing branches in the entire structure, and (2) the temporal sequence of flowering of the various capitula (each capitulum considered a unit). As to the first, spatial, aspect, morphologists consider that on fully developed inflorescences there are three types of spatial distributions of flowers or florescences: *basitonic* for the case

Fig. 16.1 Diagrams of monotelic (I–IV) and polytelic (V, VI) inflorescences. T, terminal flower; Pc, paracladium; Pc′, Pc″, paracladia of second and third order; HF, main florescence (Hauptfloreszenz); CoF, coflorescence; BZ, field of enrichment (Bereicherungszone); HZ, field of inhibition (Hemmungszone); PF, partial florescence. Terminology according to Troll (1964). (Reproduced from Weberling (1965) with permission.)

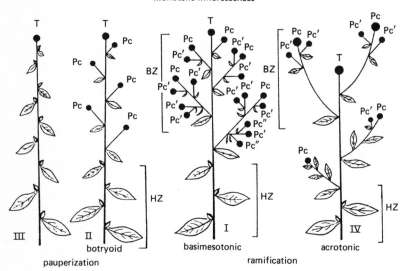

Monotelic inflorescences

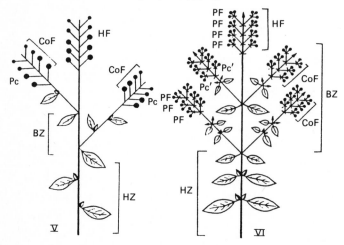

Polytelic inflorescences

where there are greater numbers of flowering structures on the lower part of an axis of the inflorescence (but above the zone of inhibition); *acrotonic* for the case where there are greater numbers on the upper part of an axis of the inflorescence; and *mesotonic* when most of the flowering structures appear on the middle part of an axis (cf. Fig. 16.1). As to the second, temporal aspect, they speak of *acropetal* flowering when flowers open in an upward sequence along an axis, of *basipetal* when they open in a downward sequence, and of *divergent* flowering when the first flowers open at the middle of an axis and opening then proceeds from there both upwards and downwards. We should note that the terminal flower of each axis usually opens before the commencement of the flowering sequence of the lateral flowers on this axis.

These spatial and temporal aspects of flower distribution are independent of each other. Thus one can observe any combination of the three spatial types with the three temporal types among the species. It is these combinations that make inflorescences in general so interesting from a developmental point of view, since presumably a few, fundamentally simple, developmental mechanisms are responsible for the large number of different structures. In the Compositae only a few of the possible patterns are realized, the spatial distribution of capitula is mostly basitonic and the flowering sequence is either strictly basipetal or divergent. The reason that their inflorescences are particularly intriguing lies precisely in this combination of basitonic and basipetal distributions, to account for which one needs more complicated mechanisms than for some of the other spatio-temporal flowering patterns. Clearly, other families with other patterns should be investigated before generalizations can be carried any further.

Finally, we should point out that the theoretical considerations which are presented here should be considered as motivation to further experimental work on inflorescence development. There have been surprisingly few physiological attempts at this problem. Parts of inflorescences have been removed in some cases, and there have also been observations on flower development under hormonal treatment. Apart from these meagre data, there is a large literature on flower induction and the florigen. Unfortunately, it has no direct bearing on the complex integration mechanisms required for the development of large-scale flowering structures.

Observation of inflorescence development

We consider first of all how the spatial and temporal distribution of flowering structures can be observed. We take as an example the common wild lettuce, *Mycelis muralis* (L.) Dumort., in the family Compositae.

An average plant produces several hundred capitula in the course of a two-month (approx.) blooming period, on branches with up to five orders of branching. Such plants were observed during a six-week period, from the end of May until the beginning of July. A sample of serial observations made on a single plant is shown in Fig. 16.2 (from Janssen, 1983).

At the beginning of the observations, the plant was about 0.5 m tall and had above the basal rosette of leaves an unbranched main axis with about 25 leafy nodes. Since the apex of the main axis has turned into a flowering apex, no further nodes are produced on this axis. The only capitulum bud which could be seen was at the tip of the main axis. On day four there were capitulum buds on the three top-most nodes below the terminal bud

Fig. 16.2. Developmental record of the inflorescence on a single plant of *Mycelis muralis*. The black dots indicate vegetative buds or unopened capitula, the open circles capitula which have opened, the circles with crosses the capitula which have withered.

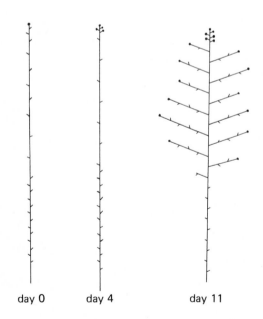

day 0   day 4   day 11

(b)

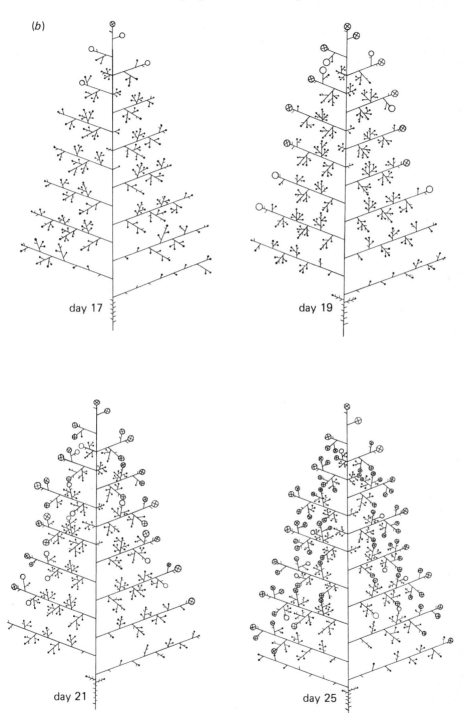

day 17

day 19

day 21

day 25

of the main axis. On day 11 there were buds at the ends of 14 more first-order branches.

By day 17 the terminal capitulum of the main axis has opened and withered (capitula are open for one or two days only, after which the ray florets wither). There were open capitula on the ends of the four top-most branches, and buds at the ends of an additional 11 branches. Furthermore, buds could be seen at the ends of a large number of second-, third-, and even of some fourth-order branches. Accessory branches appeared opposite a number of first-order branches, and these were also bearing buds. On day 19, the terminal capitula on the nine top-most branches were withered, on the following three branches downward they were open, and there were four open capitula on second-order branches on the topmost-but-one first-order branch. By day 21, the terminal capitula of many second-order branches were in bloom, and there was one capitulum of a third-order branch open (on the fourth first-order branch). On day 25, flowering of second-order, and some third-order branches progressed down to the fourteenth first-order branch, and below this level of the flowering branch system some new first-order branches showed buds. On later days increasingly higher-order branches came into bloom, basipetally along each branch.

In order to investigate the spatial distribution of branching and flowering development (also called 'vigour' by some authors) we can make use of various measures: e.g., the number of branches of order $n+1$ along a branch of order n as a function of the position of this branch, the lengths of entire branches or internodes, the total number of apices borne on a branch, etc. Among these measures the first mentioned has the property that once the terminal capitulum of a branch of order n has been initiated the number of $(n+1)$-order branches on it remains constant. Thus, these are the easiest data to collect, particularly for the case of $n=1$.

In Fig. 16.3 we plot the number of second-order branches on first-order ones as a function of the serial number of the first-order branches on the main axis. We see that the points lie approximately on a line of slope 1/2. Below the 11th branch the numbers fall off rapidly and irregularly (zone of inhibition).

The same kind of plot can be obtained for each sequence of branches of order n along a branch of order $n-1$ counting the number of $(n+1)$-order branches. For wild lettuce plants the slopes for $n=1$ lie usually between 1/3 and 1; for garden lettuce (*Lactuca sativa*) the slope for $n=1$ is mostly close to 1. Other members of the Compositae exhibit widely varying slopes for $n=1$, from shallow (1/5) to very steep slopes (>2).

The functions represented by these plots resemble the vegetative para-cladial functions, but are also distinct from them in important ways. For these reasons we call them 'floral paracladial functions' (Janssen & Lindenmayer, in preparation). These functions (and the specific values of their slopes) can be derived mathematically from the values of branching frequencies and signal-propagating rates on the basis of the models presented in the following section.

The temporal sequences of flowering can be dealt with in various ways. One of these is to plot the date of opening of the terminal capitula on branches of order n against the serial numbers of these branches along branches of order $n-1$. Doing this for the main axis (of order 0) we obtain the plot shown in Fig. 16.4. Blooming progresses rapidly in the first two days after the opening of the terminal capitulum of the main axis and is followed by an asymptotic slowing of flowering down the axis. Similar plots can also be obtained for each first-order branch; this shows the flowering sequence of second-order branches. In principle, the same plots could be done for higher-order branches as well.

The most interesting question concerning these sequences is the correlation between opening times of the capitula of subsequent branching orders. An attempt at showing such correlation was made in the following way. We plotted in Fig. 16.5 the serial numbers of first-order branches on

Fig. 16.3. Plot of the number of second-order branches against the serial number of the first-order branch on which they occur (a floral paracladial function). The slope is approximately 1/2. Data obtained from Fig. 16.2, day 25.

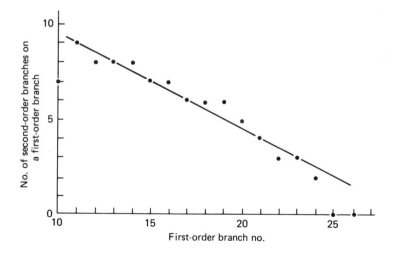

A. Lindenmayer

one coordinate axis (numbering the branches from base to apex), the serial numbers of second-order branches on another axis (also numbered from base to apex), and the number of third-order branches on the given second-order branches on a third axis. The times of flowering are shown at each position by a letter of the alphabet.

This kind of plot involving three orders of branching gives us information about both the branching vigour and the flowering sequences as functions of first- and second-order positions in the whole structure. The flowering times of the terminal capitula of first-order branches can be seen on the left at the ends of short projecting lines. By following the more or less horizontal lines for each first-order branch, we can follow the flowering sequence for that branch. For instance, the first-order branch at position 20 (on the vertical axis) has a terminal capitulum flowering on day 18 and has four second-order branches (as can be read on the diagonally-placed axis), the two most apical of which flower on day 20 and the two basal ones flower on day 21. These last two branches have one third-order branch each, while the others have none (this can be read with reference to the horizontal axis).

The main conclusions from this plot can be summarized in a schematic drawing (Fig. 16.6) of an imaginary plane intersecting the three axes on

Fig. 16.4. Plot of the serial number of the first-order branches on which the terminal capitula have opened against the day of opening. Same plant as in Fig. 16.5.

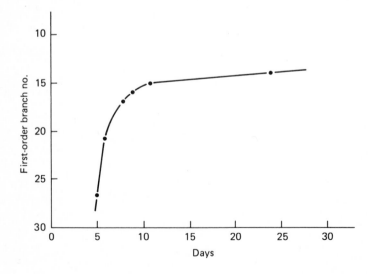

Fig. 16.5. Plot of serial numbers of first-order branches versus those of second-order branches and versus the number of third-order branches which occur on the second-order ones. The opening of the capitula at the apices of the different first- and second-order branches are indicated by letters. The capitulum at the apex of the main axis is shown at the very top of the figure. These data refer to a different plant from the one shown in Fig. 16.2.

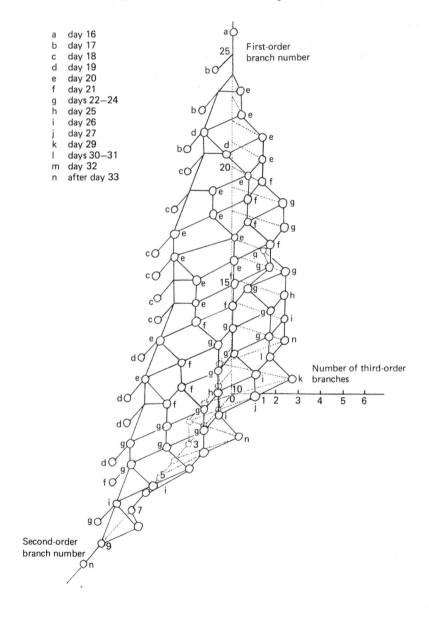

a	day 16
b	day 17
c	day 18
d	day 19
e	day 20
f	day 21
g	days 22–24
h	day 25
i	day 26
j	day 27
k	day 29
l	days 30–31
m	day 32
n	after day 33

which the flowering times are represented by alternating white and dark areas. We see a diagonal striped pattern appearing with the earliest open flowers occurring along the left margin, and flowering progressing from there towards the lower right corner of the plane. This representation thus gives us an overall wave-like pattern in which the maturing of capitula occurs throughout the plant on the first- and second-order branches. This pattern confirms the statement that flowering is basipetal on both of these branch orders, and it gives a tentative time scale relationship between the two orders. Propagation measures obtained this way for all branching orders are essential for finding quantitative mechanisms for the flowering process in the entire inflorescence.

Developmental mechanisms

What kind of mechanisms, within the framework of developmental physiology, could be responsible for the morphogenesis of such inflorescences?

Fig. 16.6. Schematic representation of the flowering order seen in Fig. 16.5. The grey and white stripes indicate the time course of flower opening in the direction from upper left to lower right.

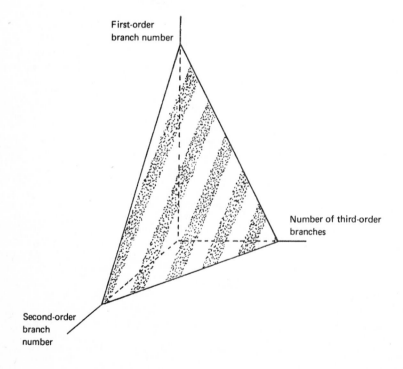

Acropetal flowering sequences on polytelic inflorescences can be easily explained by the statement that flowering occurs in the same order as that in which the successive lateral apices are produced from the base up, on the main axis, and on each paracladium. The apices of these branches remain vegetative throughout the life of the plant. The only morphogenetic mechanism that is necessary to produce such an inflorescence is that of continuing vegetative growth and the production of lateral apices on the main axis and the paracladia. These apices will then flower in the same sequence as they were laid down.

In the case of Compositae, however, this kind of mechanism would not be the correct one. Although they have polytelic inflorescences in the sense that each capitulum has a continually growing apex, the arrangement of capitula on the branching system is monotelic – that is, each branch bears a terminal capitulum. The laterally-placed terminal capitula flower in a basipetal sequence along each branch after the terminal capitulum of the main axis has flowered.

There are two basic possibilities for producing such flowering sequences. The first one relies on a mechanism similar to that previously described for acropetal polytelic inflorescences, namely the induction of an apex to flower would depend on the internal state of that apex, which in turn would only depend on its descent. It is possible to obtain basipetal flowering sequences on the basis of such assumptions (Beunder, 1981). Essentially, one would need built-in time-delays in each of the lateral apices as they are produced from base to apex, and these delays get shorter and shorter as we go higher up on each branch. In this way the main apex can flower first and then the lateral apices flower in basipetal order. But such a position-dependent time-delay mechanism has the disadvantage that the size of the inflorescence has to be fixed in advance. This restriction is not realistic, however, since in most species inflorescences of various sizes can be found, all exhibiting the same basipetal patterns.

Since it does not seem possible to construct a mechanism which depends only on descent relationships among the apices, we have to consider mechanisms based on interactions among the apices and other parts of an extensive branching structure. We have considered two such interactive mechanisms.

In the first (which we shall call Model 1), we assume one morphogenetically active substance to be responsible for the interactions among the inflorescence parts. This substance is to be produced in the basal part of the plant and is transported upwards throughout the branching system.

When it reaches a vegetatively growing apex, it transforms it to a flowering apex. As we have mentioned, these assumptions correspond to the currently accepted hypotheses regarding the production, distribution and activity of the 'florigen' or flowering hormone.

Thus we start with a vegetatively growing and branching shoot system on which the induction of an apex to flowering takes place whenever the active substance arrives there. In order to calculate the flowering patterns obtained by such mechanisms we need to specify the speeds at which such morphogenetic signals travel in various parts and in various directions on the shoot system, and the starting time for the production of the signal with respect to the developmental stage of the vegetatively growing branch system. The rates at which the vegetative apices are growing and producing lateral apices, and possible delays in their growth, must be specified as well.

Various inflorescence forms can be generated in the following way. As we have said before, in general inflorescences exhibit acrotonic, basitonic or mesotonic branching development (vigour) and show acropetal, basipetal or divergent orders of flowering. These attributes must be specified separately for each order of branching, since it is possible to find cases in which the flowering order of first-order branches is acropetal while that of the second-order branches is basipetal. Similarly the vigour of branching may be distributed differently on different orders of branches.

Let us consider now basitonic development on first-order branches. There are three distinct possibilities by which this kind of development can be accounted for on the basis of a flowering hormone travelling from base to top. These possibilities are sketched in three diagrams in Fig. 16.7.

In each diagram time is shown on the horizontal axis and first-order branch number on the vertical axis. The leftmost rising lines in the diagrams (labelled 'gr') indicate the onset of vegetative growth of each of the first-order branches. The middle lines (labelled 'fl') indicate the progress of the flowering hormone up the main axis. The rightmost lines (labelled 's') show the times at which the flowering hormone reaches the apices of the first-order branches transforming them to flowering apices. The horizontal lines symbolize the first-order branches bearing varying numbers of nodes. The lengths of these lines and the corresponding numbers of nodes indicate the branching vigour of first-order branches; this is higher near the base and decreases towards the top, i.e., it is a basitonic pattern, in all three cases. The 'gr' and 'fl' lines meet at the top of the main axis, defining the time at which its terminal flower structure is induced and the 's' lines converge also to this point. For the sake of simplicity straight lines are assumed for the time functions of these processes; in reality more complicated curves could occur.

The flowering sequence is acropetal in the first diagram, basipetal in the third, and flowering occurs simultaneously on all the branches in the centre diagram. These different time sequences result from the rates at which the flowering hormone travels in the main axis and in the first-order branches, from the growth rates of both the main axis and the first-order branches, and from the period which elapses between the start of vegetative growth and the start of production of the flowering hormone at the base of the plant. Formulae have been derived for the slopes of the flowering sequences as functions of the above-named parameters (Janssen & Lindenmayer, in preparation).

The inflorescence development in *Mycelis muralis* exhibits, as we have seen, a basipetal flowering sequence and basitonic branching on the first-order branches. But it cannot be directly comparable to the third case in Fig. 16.7. The reason for this is evident from Fig. 16.8. In this figure we plotted the observed flowering sequence from Fig. 16.4 as a curve labelled 's'. Further, we computed a curve 'v' by subtracting a vegetative growth period from each point on 's' proportional to the number of nodes on each first-order branch as shown in Fig. 16.3. The line 'v' indicates, therefore, the assumed time-course of vegetative growth initiation in each of

Fig. 16.7. Three hypothetical mechanisms by which basitonic branch distribution can be achieved. The lines labelled 'gr' connect points indicating the start of vegetative growth of first-order apices, the lines labelled 'fl' contain the points at which florigen reaches the first-order nodes, and the lines labelled 's' show the times at which the first-order apices are induced to flower. The horizontal lines indicate the lengths of first-order branches and the bars on them represent the second-order branches.

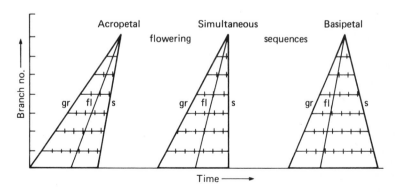

the first-order apices. This line is almost vertical in the upper portions of the main axis but in lower portions curves sharply to the right, indicating that vegetative growth is initiated approximately simultaneously in the upper part of the plant but progresses basipetally in the lower part. This time-course of vegetative growth initiation agrees well with our observations.

But the first-order apices have been laid down in an acropetal time sequence, as shown by the line 'gr' in Fig. 16.8. The flowering stimulus is transported upwards according to line 'fl', its intersection with line 'gr' defining the time and place of flowering at the terminal apex of the main axis. There is now a variable time delay between lines 'gr' and 'v' which has to be accounted for. (In the diagrams in Fig. 16.7 we have assumed that either no such delay occurs, or if it does then there is a constant delay at each node between laying down a branch apex and the initiation of its vegetative growth.) For this variable delay at each node we had to assume an additional control mechanism. The most likely one appeared to be a progressive release from apical dominance, spreading basipetally from the

Fig. 16.8. Curve 's' shows the time values of the observed flowering sequence of first-order apices of *Mycelis muralis* (Fig. 16.4 replotted). Curve 'v' indicates the time values at which first-order apices start their vegetative growth. The points of this curve were calculated by converting the data of Fig. 16.3 into time periods (0.5 days per node produced plus 0.5 days delay), and then subtracting these periods from the corresponding points on curve 's'. The line 'gr' indicates the time course in which the first-order apices were laid down, and the line 'fl' the time course of florigen propagation in the main axis.

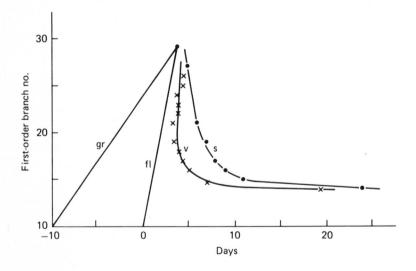

top of the main axis. In other words, we assumed that soon after the terminal apex of the main axis has been transformed into flowering apex, the production of auxin by this apex stops and thereby the apical dominance of this apex is gradually lifted along the main axis. Release from apical dominance allows vegetative outgrowth of the lateral apices, followed by their eventual transformation to the flowering condition by the florigen.

These assumptions by themselves provide for the correct basipetal flowering sequence but not for the basitonic branching development. Since the flowering hormone is already present along the entire length of the main axis, we cannot use its arrival for initiation of terminal flowers on the first-order branches in such a way that longer and longer first-order branches arise in the downward direction. In order to achieve this we assumed additionally that the longer a given first-order apex remained under auxin repression the longer it needed to become transformed to a flowering apex. This additional regulatory mechanism sufficed to provide the desired developmental pattern (Model 2).

We have discussed so far the possible control mechanisms for combining basipetal flowering with basitonic branch development. We could also provide similar mechanisms, based on regulation by a flowering hormone and by apical dominance, which would result in acrotonic branch distribution accompanied by acropetal and basipetal flowering. Mesotonic branching and divergent flowering can also be determined by similar mechanisms. These latter types of inflorescences are found in *Aster novae-angliae* and have been studied by Frijters (1978a) who formulated regulatory mechanisms for their development (see Figure 9 in his paper.).

Our discussion of the various flowering patterns has so far been restricted to the apices of first-order branches. One has to carry out similar investigations concerning each of the higher order branches as well. In cases where each of the branching orders exhibits the same patterns, as in *Mycelis muralis*, the modelling is relatively simple, although problems arise even in this case with the application of certain of the mechanisms to higher and higher orders of branches. Different patterns on different branching orders, or in different parts of the plant, are even more difficult.

Mathematical and computer models

Mathematical models have been constructed, and computer simulations have been carried out by us, on the basis of the above assumptions with respect to inflorescences of various species in the Compositae. The mathematical models are expressed mainly in terms of L-systems. These are discrete functions producing growing numbers of subunits in branching or other spatial structures (Lindenmayer, 1968, 1971, 1975, 1982; Herman & Rozenberg, 1975; Rozenberg & Salomaa, 1980). The computer simulations were carried out with the help of the CELIA program, a computing framework for the above kinds of developmental functions. The construction and results of such simulations have been described in detail previously (Baker & Herman, 1972; Herman & Rozenberg, 1975; Lindenmayer, 1974; Frijters, 1976, 1978a,b; Frijters & Lindenmayer, 1974).

We consider two detailed computer simulations which were carried out on the basis of the physiological mechanisms discussed in the previous section (Janssen & Lindenmayer, in preparation).

(1) First we present a simulation based on Model 1 introduced in the previous section. In this model, vegetative growth of apices follows simply the order of their initiation. The basipetal flowering order is achieved in this case by assuming a flower-inducing signal (e.g. florigen) propagating upwards on the main axis and in each of the branches with rates such that it reaches the top branch apices before it reaches the lower ones. The main axis apex is to be induced first.

We have to specify the following set of parameters for a simulation: The plastochron interval is assumed to be twice as long in the first and second order branches as on the main axis. The lengths of internodes are assumed to have the relationship 2:1:0.5 for the main axis and the first- and second-order branches, respectively. There are no growth delays at the inception of any of the branches. The flowering stimulus originates at the base of the stem at the seventh iteration step (the length of an iteration step is equal to the plastochron of the main axis), and it propagates acropetally with rates of 6.0, 1.0 and 0.375 length units per iteration step in the main axis and in the first- and second-order branches, respectively. Third-order apices cannot be transformed into flowers.

In Fig. 16.9, the structures obtained after 9, 12, 15, 18 and 20 iterations are shown. We see that flowering takes place first at the tip of the main axis and on some of the first-order branches near the top. Flowering of the apices on the first-order branches progresses rapidly downwards; it also does so on the second-order branches. By the 18th iteration only one second-order apex has failed to flower at the very base of the inflorescence and another is in the open flower state. More detailed flowering sequences can be seen by printing-out the structures at more iteration steps and also by using shorter iteration steps in the simulation. The choice of simulation parameters can also be more finely adjusted.

As far as the basitonic branching pattern of the inflorescence is concerned, we see on the last structure (after 20 iterations), that the number of second-order branches on first-order branches increases downwards as follows: 0, 1, 1, 2, 3, 3, 4, 4, 5, 6. This basitonic configuration agrees with

Fig. 16.9. Simulation 1. Basitonic inflorescence with basipetal order of flowering along the main axis and the first-order branches. Flowering signal is introduced at the seventh iteration step at the base and propagates towards the apices of the branches at rates given in the text. Vegetative growth of branch apices follows their acropetal order of initiation. Circles denote flowering capitula and stars withered ones.

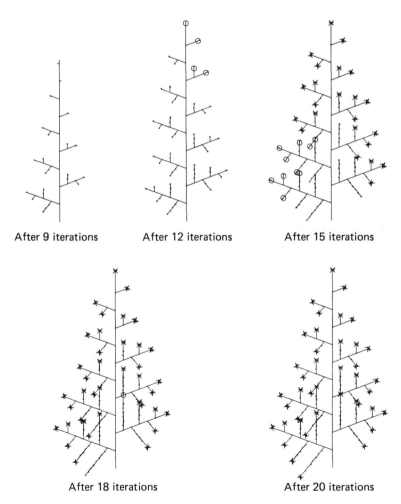

After 9 iterations   After 12 iterations   After 15 iterations

After 18 iterations   After 20 iterations

the one we have seen on *Mycelis muralis*. The basipetal flowering sequence on first- and second-order branches agrees also with that found in
these plants.

There are three aspects in which this simulation model deviates from
the pattern of *Mycelis muralis*. First, the vegetative out-growth of first-
and higher-order branches proceeds basipetally on the plants, and not
acropetally as produced by the model. Secondly the model does not provide for a 'zone of inhibition' below the flowering portion of the plant.
According to the model, all first- and second-order branches flower, while
on the plant the lower nodes do not. It would present no serious problem
to change the simulation accordingly: one would simply need an additional delay built into the program.

A third point to be noted is that such an acropetally-transported flowering signal cannot ensure basipetal flowering sequences on higher-order
branches. In the simulation shown above, proper basipetal flowering was
still produced along first-order branches, but this would not be the case
for second- or higher-order branches. The reason for this lies in the fact
that the differences between the signal propagation rates (6.0, 1.0 and
0.375) have to become narrower as one goes to higher-order branches and
the mechanism eventually becomes ineffective to ensure basipetal flowering on these.

(2) We now come to a model that simulates the development of the
Mycelis muralis inflorescence reasonably well. In this model all the parameters are the same as in the previous model as far as the plastochrons
and the lengths of internodes are concerned. Again there is an acropetal
propagation of a flowering signal which starts from the base at iteration
step 7 and travels with the rates 6.0, 1.0 and 0.375 along the main axis
and the first- and second-order branches respectively. In addition, we
assume that through an apical dominance effect (auxin production) the
vegetative apices suppress the growth of lateral apices to some distance
below them. This blocking effect gradually disappears after a vegetative
apex is transformed into a flowering apex. Finally, it is also assumed that
the length of time for which a lateral apex has been blocked leaves a trace
in that apex, so that its transformation to a flowering apex can take place
only after a delay proportional to the length of time it has been blocked.
These assumptions suffice to give the developmental sequence shown in
Fig. 16.10.

We see that the main axis is still growing vegetatively at the 9th iteration step (not shown in Fig. 16.10), and that the flowering signal has
arrived at its apex by the 12th step. All the lateral buds are blocked at this

stage due to inhibition by the vegetative apex above. At the 15th iteration step the two top-most branches are released from inhibition; since they have been blocked only for short times, and because the flowering signal has arrived there already, they are both induced to flower. Thus, the first

Fig. 16.10. Simulation 2. Basitonic inflorescence with basipetal order of flowering on the main axis and on all higher-order branches. The flowering signal moves into the inflorescence from below. The growth of lateral buds is inhibited by any vegetative apex above them. After transformation of the apex from vegetative to flowering the inhibitory effect ceases. Each lateral bud is blocked from transformation to flowering for a period proportionate to the length of time it has been inhibited. Squares stand for repressed vegetative apices, and triangles for growing ones. Dots indicate apices transformed to flowering, circles of various sizes indicate capitula in different stages of development (the largest ones the fully open stage), and asterisks indicate withered capitula.

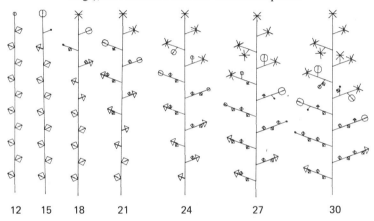

12   15   18   21   24   27   30

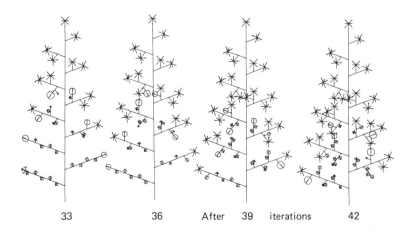

33   36   After   39   iterations   42

flower opens on the top branch at the 18th step. After this, the lateral buds are sequentially released from inhibition; they grow out and produce second-order branches until they too are induced to flower. In order to be induced they must become unblocked and the flowering signal must have arrived at their position. By the 36th iteration all of the first-order apices have flowered, and so have several of the second-order ones. Third-order apices are flowering by the 42nd iteration. As a whole, a developmental sequence and distribution of capitula is achieved which comes close to the observed one.

This mechanism involving apical dominance also agrees with evidence from experimental work on related species (e.g., *Senecio* and *Solidago* (cf. Jauffret (1970)). By cutting off the top portion of an inflorescence at an early stage, we can induce more profuse and faster flowering below. This effect agrees with the idea of a lifting of an inhibitory influence from above, although it would not rule out a stimulatory influence ascending from below.

Discussion

The outline presented here of some of the ideas used in the search for adequate developmental mechanisms of inflorescence patterning has to be, by necessity, very short. Our basic technique was system-theoretical. First we tried to delimit the class of acceptable mechanisms and to group them into a few recognizable sub-classes. Then we attempted to select the minimal ones from among these. Finding minimal sets of assumptions is not a trivial task in this case; it was only after considerable mathematical work and computer modelling that the systems presented here were established with some degree of confidence as both sufficient and minimal.

Identifying the number and nature of control factors was the main purpose of this work. Our early ideas had to do with attempting to control such a process either without any signals being transmitted through the branched structure or to do so with a single signal, the flowering hormone, propagating from base to apex. As we have seen, both of these approaches are inadequate. The first because it would require that the size of the inflorescence be specified beforehand, and the second because the rates of signal propagation in higher-order branches would have to become increasingly lower and could not ensure the basipetal flowering on all orders of branches. Then the case for two signals was considered, with at least one flowering signal propagating from above. In such a system one would need an additional way of determining when the apex should start producing this signal. Thus, besides the growth induction mechanism, provided by the natural order of branch initiation or by a signal, two flower-regulating signals would be needed. A system with a similar number of controls, but with an inhibitory rather than stimulatory mechanism for growth induction, finally satisfied all the requirements and had the advantage over the previous one mentioned in that it could also account for experimental findings concerning apical dominance in inflorescences.

We may ask to what extent the mechanisms proposed here compare with other morphogenetic models. First, we should emphasize that we are dealing with rapidly-growing structures in which specially-differentiated parts (the capitula) have to be positioned and their development has to be timed. Most of the other mechanisms have been proposed for animal development where the size of the field in which morphogenetic processes are taking place remains constant or increases only slowly. Diffusion–reaction mechanisms, such as those proposed by Turing (1952) and more recently by Meinhardt & Gierer (1974), produce regular and repeatable patterns only in fields of constant size. They are based on the generation of standing waves of chemical constituent concentrations in homogeneous media and are not applicable to rapidly growing and branching structures such as those we have been considering. Our models are related to the gradient models, first introduced by Child (1941) and then used by Wolpert (1969) as a basis for 'positional information' as an explanation of regeneration phenonema in *Hydra* and higher animals. According to this model, the establishment of a gradient in a non-growing field leads to the differentiation of certain structures, this mechanism being size-invariant in the sense that the same pattern is generated irrespective of the size of the field. In such systems only the placing of the differentiated structures is important and not the time-span in which such a process takes place.

In the systems we have been discussing, morphogenetic processes and differentiation steps are taking place in rapidly-growing structures rather than in non-growing fields of various sizes. In these developmental systems there has to be an intricate relationship between the timing and the positioning of events. For instance, the vegetatively growing apex of a branch is transformed to a flowering apex when the flowering signal reaches it – an event that takes place at a certain distance from the base of the branch and at a certain time depending on the growth rate of the branch and the propagation rate of the flowering signal. The apices of the first-order and higher-order branches can be transformed to the flowering condition in acropetal, basipetal or divergent sequences. It is important to note here that we are not characterizing the inflorescences according to the *places* where the flowering structures appear but according to the *temporal-spatial* sequences in which they appear. In both the models that we presented, these sequences were determined by dynamically changing concentration values rather than by stationary gradients. Similarly, the acrotonic, basitonic or mesotonic distribution patterns of branches and flowering structures have to be produced by timing as well as by positioning mechanisms, and not by either timing or positioning mechanisms alone. It is quite clear that in plant development dynamic processes which take place in growing fields must play a more essential role than standing waves or stationary gradients in fields of constant size.

The mathematics of non-steady-state processes in growing fields is obviously more difficult than that of the stationary systems. While mathematical formulae can be obtained for gradient fields of constant size, and certain solutions are available for diffusion–reaction processes (Murray, 1978), the behaviour of gradients or diffusion–reaction systems in growing fields cannot be treated analytically. In addition, for extensively branching structures, such as the inflorescences we have described, a separate differential equation would have to be written for at least each individual branch. Since the number of branches increases in the course of inflorescence development, this would imply that we would need to deal with increasingly larger sets of differential equations that could certainly not be handled analytically. There remains the possibility of finding numerical solutions, or to approach the problem through discrete mathematical formalisms (which, rather than exclude the former approach, helps it). A basic discrete theory for such developmental problems is available in the theory of interactive L-systems. This has provided some general results (cf. Herman & Rozenberg, 1975) which can be used to guide the numerical calculations (computer simulations) necessary for the detailed models we have presented.

References

Baker, R. & Herman, G. T. (1972). Simulation of organisms using a developmental model. Parts I and II. *International Journal of Biomedical Computing*, **3**, 201–15, 251–67.

Beunder, H. (1981). A model for closed inflorescences with basipetal or acropetal flowering. *M. S. Thesis, Theoretical Biology Group, University of Utrecht*.

Child, C. M. (1941). *Patterns and Problems of Development*. Chicago: University of Chicago Press.

Erickson, R. O. & Michelini, F. J. (1957). The plastochron index. *American Journal of Botany*, **44**, 297–305.

Favard, A. (1970). Mise en évidence, chez le *Drosera intermedia* Hayne, d'une inhibition exercée par le bourgeon inflorescentiel terminal sur des méristèmes axillaires de la tige sous-jacente pourvue d'une pousse de renfort. *Comptes rendus hebdomadaires des Séances de l'Academie des Sciences Paris, Série D*, **270**, 1685–8.

Frijters, D. (1976). An automata-theoretical model of the vegetative and flowering development of *Hieracium murorum* L. *Biological Cybernetics*, **24**, 1–13.

– (1978a). Mechanisms of developmental integration of *Aster novae-angliae* L. and *Hieracium murorum* L. *Annals of Botany*, **42**, 561–75.

– (1978b). Principles of simulation of inflorescence development. *Annals of Botany*, **42**, 549–60.

Frijters, D. & Lindenmayer, A. (1974). A model for the growth and flowering of *Aster novae-angliae* on the basis of table (1,0) L-systems. In *L Systems*, ed. G. Rozenberg & A. Salomaa, *Lecture Notes in Computer Science*, volume **15**, pp. 24–52, Berlin: Springer-Verlag.

– (1976). Developmental descriptions of branching patterns with paracladial relationships. In *Automata, Languages, Development*, ed. A. Lindenmayer & G. Rozenberg, pp. 57–73, Amsterdam: North-Holland Publishing Co.

Herman, G. T. & Rozenberg, G., with a contribution by Lindenmayer, A. (1975). *Developmental Systems and Languages*, Amsterdam: North-Holland Publishing Co.

Janssen, J. (1983). Models for inflorescence development in *Mycelis muralis*. *M. S. Thesis, Theoretical Biology Group, University of Utrecht*.

Jauffret, F. (1970). Observation et expérimentation sur quelques composées à floraison descendante. *Bulletin de la Société Botanique de France, Mémoires*, **117**, 259–69.

Lindenmayer, A. (1968). Mathematical models for cellular interactions in development. Parts I and II. *Journal of Theoretical Biology*, **18**, 280–99, 300–15.

– (1971). Developmental systems without cellular interactions, their languages and grammars. *Journal of Theoretical Biology*, **30**, 455–84.

– (1974). Adding continuous components to L-systems. In *L Systems*, ed. G. Rozenberg & A. Salomaa, *Lecture Notes in Computer Science*, volume **15**, pp. 53–68, Berlin: Springer-Verlag.

– (1975). Developmental algorithms for multicellular organisms: a survey of L-systems. *Journal of Theoretical Biology*, **54**, 3–22.

– (1977). Paracladial relationships in leaves. *Berichte der deutschen botanischen Gesellschaft*, **90**, 287–301.

– (1982). Developmental algorithms: lineage versus interactive control mechanisms. In *Developmental Order: Its Origin and Regulation*, ed. S. Subtelny & P. B. Green, pp. 219–45, New York: Alan R. Liss.

Meinhardt, H. & Gierer, A. (1974). Applications of a theory of biological pattern formation based on lateral inhibition. *Journal of Cell Science*, **15**, 321–46.

Murray, J. D. (1978). Biological and chemical oscillatory phenomena and their mathematical models. *Bulletin of the Institute of Mathematics and its Applications*, **14**, 162–9.

Rozenberg, G. & Salomaa, A. (1980). *The Mathematical Theory of L Systems*. New York & London: Academic Press.

Schüepp, O. (1942). Beschreibung von Blütenständen auf Grund des zeitlichen Verlauf der Anlage, des Wachstums und des Aufblühens. *Berichte der schweizerischen botanischen Gesellschaft*, **52**, 273–316.

Sell, Y. (1967). Inhibition de floraison de nature corrélative chez *Veronica longifolia* L. *Comptes rendus hebdomadaires des séances de l'Academie des Sciences, Paris, Série D,* **264,** 821–4.

– (1980). Action de l'apex caulinaire et maturation florale dans le cadre de la floraison descendante. *Flora,* **169,** 15–22.

Troll, W. (1964). *Die Infloreszenzen,* Part 1. Jena: Fischer.

Tucker, D. J. & Mansfield, T. A. (1973). Apical dominance in *Xanthium strumarium.* A discussion in relation to current hypotheses of correlative inhibition. *Journal of Experimental Botany,* **24,** 731–40.

Turing, A. M. (1952). The chemical basis of morphogenesis. *Philosophical Transactions of the Royal Society of London,* B, **237,** 37–72.

Vardar, Y. & Kaldewey, H. (1972). Auxin transport and apical dominance in *Pisum sativum.* In *Hormonal Regulation in Plant Growth and Development,* ed. H. Kaldewey & Y. Vardar, pp. 401–11, Weinheim: Verlag Chemie.

Weberling, F. (1965). Typology of inflorescences. *Journal of the Linnean Society (Botany),* **59,** 215–21.

Wolpert, L. (1969). Positional information and the spatial pattern of cellular differentiation. *Journal of Theoretical Biology,* **25,** 1–47.

Index

Bold type indicates bibliographic references.